PRINCIPLES OF
*Microbial
Diversity*

# PRINCIPLES OF
# *Microbial Diversity*

## JAMES W. BROWN
Department of Biological Sciences
North Carolina State University
Raleigh, North Carolina

ASM
PRESS
Washington, DC

**Library of Congress Cataloging-in-Publication Data**

Brown, James W., 1958– author.
    Principles of microbial diversity / James W. Brown, Department of Biological Sciences, North Carolina State University, Raleigh, North Carolina.
        pages cm
    Includes index.
    ISBN 978-1-55581-442-7 (pbk.) -- ISBN 978-1-55581-851-7 (e-book)  1.  Microbial diversity.
2.  Microbial ecology.  I.  Title.
    QR73.B76 2014
    579—dc23
2014000523

eISBN: 978-1-55581-851-7
doi:10.1128/9781555818517

10   9   8   7   6   5   4   3   2   1
All rights reserved
Printed in the United States of America

Address editorial correspondence to ASM Press, 1752 N St. NW, Washington, DC 20036-2904, USA
Send orders to ASM Press, P.O. Box 605, Herndon, VA 20172, USA
Phone: (800) 546-2416 or (703) 661-1593; Fax: (703) 661-1501
E-mail: books@asmusa.org
Online: http://www.asmscience.org

*Cover and interior design:* Susan Brown Schmidler
*Illustrations:* Lineworks, Inc.
*Cover image:* Dark-field image of *Globigerina bulloides*, an abundant and widely ranging planktonic foraminifer. The shell length is ~300 μm. (Image courtesy of Howard Spero, Department of Geology, University of California, Davis.)

*This book is dedicated to the memory of Elizabeth Haas.*
*You are missed by all who knew you.*

# Contents

# Preface

ALTHOUGH IT HAS BEEN RECOMMENDED that undergraduate curricula for microbiology majors require a core course on microbial diversity, microbiology programs most often lack such a course. One reason for this lack is that, unlike the other recommended core microbiology courses, there has been no appropriate textbook on microbial diversity for students at the undergraduate level. *Principles of Microbial Diversity* is intended to fill this gap.

This textbook is intended primarily for junior and senior undergraduate students who are majoring in microbiology or a related field. Students should already have studied a general microbiology course and should have familiarity with genetics and either biochemistry or microbial physiology. The perspective in this book is phylogenetic and organismal, from the Carl Woese school (in contrast to the approach of most general microbiology textbooks) (1). This textbook arose from an existing senior-level lecture/lab course on microbial diversity and so has been in use with success already.

The book comprises four main sections. The first section is introductory, laying out the scope of the text, defining the perspective, and providing a historical context. This is followed by a practical guide to molecular phylogenetic analysis, focusing on how to create and interpret phylogenetic trees, and an overview of "the Tree of Life." The second section is a tour through each of the major familiar phylogenetic groups of *Bacteria* and *Archaea* (microbial eukaryotes and viruses are also covered briefly), discussing the general properties of the organisms in each group, describing some representatives in more detail, and concluding with one or two specific topics on the unique properties of these organisms.

The third section of the book is conceptually and experimentally defined (based on primary literature), beginning with identification of unknown and potentially uncultivable organisms and leading to molecular surveys of populations, linking processes with specific organisms. This sequence leads to the final section, brief discussions of various aspects of microbial genomics and origins.

The most straightforward approach for covering the two large middle sections of the textbook in class is to start with the survey of phylogenetic groups and follow this with the concept/literature chapters. An alternative approach, which I have used with great success, is to intertwine them. In my experience, each lecture begins with the discussion of a particular microbial phylum (a portion of a chapter in section two), with some discussion of general topics raised about these organisms, leading into one of the papers from sections three and four of the textbook (or a more recent paper chosen by the instructor) that highlights organisms in the group discussed in that lecture. For example, a chapter might start out with a discussion of the *Chlamydiae*, describing the members of the group, their phenotype, pathogenicity, and life cycle, and be followed by a discussion of reductive evolution in parasites. It would then shift gears to an introduction to genomics, exemplified by the paper describing the *Protochlamydia amoebophila* genome and what it teaches us about the origin of obligate pathogens. The order of topics, as would be taught in the course, would be defined by the conceptual thread (section four of the text), building in complexity.

1. **Woese CR.** 2006. A new biology for a new century. *Microbiol Mol Biol Rev* **68:**173–186. doi:10.1128/MMBR.68.2.173-186.2004

# *Acknowledgments*

As the sole listed author of this text, I would be negligent if I did not make it absolutely clear that it is the result of a community effort on many levels. The folks listed below all deserve the lion's share of the credit for this work; any errors and shortcoming I claim only for myself.

This book was initiated over the course of a couple of years by the persistent encouragement of Greg Payne at the ASM Press. Once started, Ken April, Production Manager, and John Bell, Senior Production Editor, at the ASM Press made this book happen. Special thanks are also owed to the book's interior and cover designer, Susan Brown Schmidler; Dianna Logan and Peggy Rupp at Dedicated Book Services, Clarinda, Iowa, who assembled this high-quality book from a collection of text files and images; Lindsay Williams, the diligent ASM Press Editorial and Rights Coordinator, who shepherded permissions; and the art renderer, Tom Webster of Lineworks, Inc., who created professional illustrations from what were, in some cases, little more than vague sketches.

This text is based on a course I was hired (in part) to develop and teach in the Department of Microbiology at North Carolina State University. The success of this course is owed to those who recognized its importance before my arrival and encouraged and fostered its development afterwards—especially Leo Parks, Hosni Hassan, and Gerry Luginbuhl, but also the entire faculty of the department.

This book, and the phylogenetic perspective on which it is based, owes everything to Carl Woese, the intellectual father of modern microbiology. The course on which this text is based has its origin not just in Carl's work generally but also very specifically in his fabulously important review article from 1987

(Woese CR. 1987. Bacterial evolution. *Microbiol Rev* **51**:221–271). The importance and utility of the phylogenetic perspective have no better advocate than my postdoctoral mentor, Norm Pace, for whom no amount of thanks can suffice for his mentorship over the years.

Enormous credit goes to those who captured the images of organisms used in this text. A picture is worth at least a thousand words. Photo credits are given with the images, but special thanks are warranted to a few who provided numerous images well beyond anything for which I had the right to ask: Michael Thomm and Reinhard Rachel, John Fuerst and Margaret Lindsay, and D. J. Patterson. A special thanks also goes to Howard Spero for allowing us to use his spectacular image of *G. bulloides* on the cover of this text.

This book also owes its existence to another James W. Brown, my father, for his patient yet persistent encouragement, and to my mother, Phyllis Brown, who nurtured my scientific interests from the earliest possible age. Finally, and most importantly, I am forever grateful for the encouragement and patience of my wife, colleague, and collaborator, Melanie Lee-Brown.

# *About the Author*

From the beginning, Jim Brown had a keen interest in nature, including anything slow or unwary enough to be captured or observed in the woods, rivers, beach, or ocean that was always nearby. A single lecture on microbial diversity in a General Microbiology class while Jim was an undergraduate at Ball State University, and the announcement in that class of the discovery of an entirely new kind of living thing (the "archaebacteria"), sparked his lasting interest in microbiology. That led to undergraduate research examining *Beggiatoa* in a southern Indiana sulfur spring. He later earned his M.S. in Microbiology

at Miami University and joined the MCD Biology Ph.D. program at The Ohio State University, where he worked on the molecular biology of methanogenic archaea with Professor John Reeve. He then moved to Indiana University for a postdoc in Professor Norm Pace's lab, working on the comparative analysis of ribonuclease P RNA in *Bacteria*. Afterwards, Jim joined the Department of Microbiology at North Carolina State University (NCSU) and continued to work on RNase P in *Archaea* and the comparative analysis of RNA. Jim developed and teaches senior-level undergraduate lecture and lab courses in microbial diversity, which are the genesis of this textbook. Jim was awarded the NCSU and Alumni Outstanding Teacher awards in 2005 and the Alumni Association Distinguished Undergraduate Professor award in 2014. He has been a member of the ASM since Graduate School and is a long-time officer of the North Carolina branch of the ASM.

In the well-known children's story by Dr. Seuss, Horton the elephant discovers that a tiny speck of dust contains an entire world of creatures much too small for him to see.

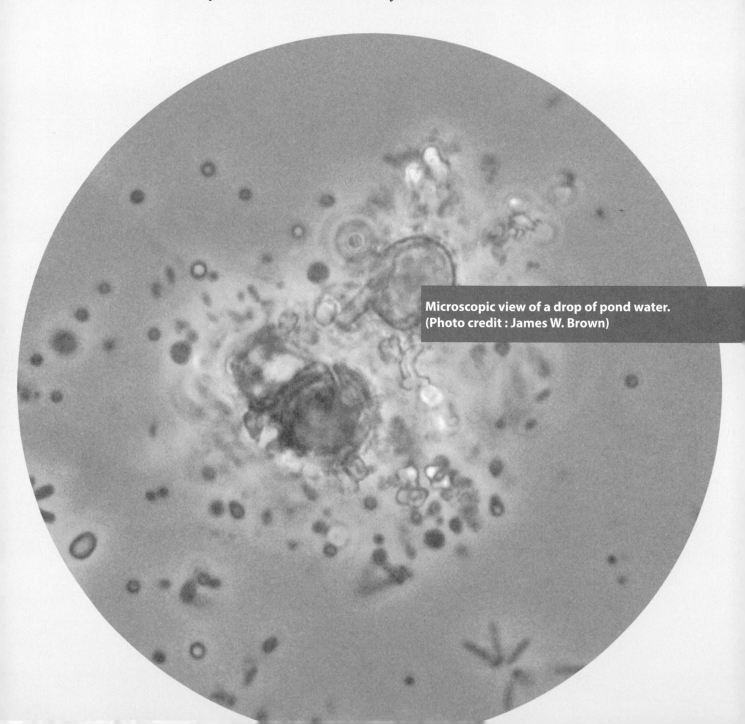

**Microscopic view of a drop of pond water. (Photo credit : James W. Brown)**

# SECTION 1  *Introduction to Microbial Diversity*

It turns out that every speck of dust, every drop of water, every grain of soil, and every part of every plant and animal around us contain their own worlds of microbial inhabitants (facing illustration).

A very tiny fraction of these creatures can do us harm, causing misery, disease, and death, and these few creatures have given the microbial world a bad reputation. But the vast majority of microbes benefit us in essential ways that we fail to recognize. They created, and sustain, the world we live in. The famous paleontologist and evolutionary biologist Stephen J. Gould once wrote, "On any possible, reasonable, or fair criterion, bacteria are—and always have been—the dominant forms of life on Earth."

Welcome to *Principles of Microbial Diversity*. In this book we explore, a little bit, the enormous range of biological diversity in the microbial world.

In the first section of the book, we establish a point of view from which to examine microbial diversity—call this the "phylogenetic perspective." After some background material, this is primarily a problem-solving section, in which we learn how to construct and interpret evolutionary trees from DNA sequences. We finish up with a look at the universal "tree of life" constructed using this process.

In the second section, we climb around in this "tree of life," looking at some examples of microbes on the major branches of the tree—sort of a stroll through the microbial zoo. We extract some conceptual lessons from each group, but this section of the course is mostly about establishing a base of knowledge about the microbial world.

In the third and fourth sections, we learn about how microbiologists are beginning to explore the universe of microbial diversity "in the wild." We do this directly from published research papers, starting with how new organisms are identified (usually without being cultivated) and progressing in steps to broad surveys of entire microbial communities and attempts to get a handle on how specific kinds of organisms contribute to the ecosystem. This is the conceptual and synthetic portion of the book.

In the end, I hope you will have gained an appreciation for the "big picture" of the microbial world, an understanding of the power of the phylogenetic perspective, and a realization that the exploration of this world is just beginning, how this is being done, and the questions that drive this exploration.

We live immersed in an infinite sea of the infinitesimal. Let's have a look.

# 1

# What Is Microbial Diversity?

## Facets of microbial diversity

What is diversity? How *exactly* are organisms either similar to or different from each other? This seems like an easy question in the macroscopic world, but what about microbes?

### Morphological diversity

Microbes are often divided by shape into rods, cocci, and spirals. Although these are the most common cell shapes, bacterial and archaeal cells also come in a wide range of other shapes: filaments (branched or unbranched), irregular, pleomorphic (different shapes under different conditions or even in the same culture), star-shaped, stalked, and many, many others. *Haloquadratum* is a flat, square organism, just like a bathroom tile (Fig. 1.1).

Individual cells of whatever shape can be found in a variety of multicellular arrangements, from simple pairs and tetrads to multicellular filaments, sheets, rosettes, and true multicellular organisms. Many species form highly structured multispecies mats that resemble the tissues of animals and plants that carry out complex biochemical transformations (Fig. 1.2).

**Figure 1.1** The tile-shaped halophilic archaeon *Haloquadratum walsbyi*. (Source: Wikimedia Commons.) doi:10.1128/9781555818517.ch1.f1.1

**Figure 1.2** Section of a stratified microbial mat from Guerrero Negro, Baja California. (Copyright 2007, American Society for Microbiology. Photo by John R. Spear and Norman R. Pace.) doi:10.1128/9781555818517.ch1.f1.2

Most bacteria and archaea measure 1 to 5 μm, but they range from 0.1 μm in thickness to over a millimeter. At the low end, it is hard to understand how everything that is needed for life could fit into the cell. At the high end, they can be easily seen without a microscope (Fig. 1.3).

## Structural diversity

Many bacteria have "typical" gram-positive (single membrane, thick cell wall) or gram-negative (double membrane, thin cell wall) cell envelopes. However, there is wide variation even within these two major types. Many gram-positive bacteria have an outer membrane, made of mycolic acids rather than glycerol-phosphate esters. Many gram-negative bacteria lack the lipopolysaccharide layer. Many archaea and bacteria (both gram positive and gram negative) have an orderly protein coat, the S-layer (Fig. 1.4). In bacteria, cell walls are composed of peptidoglycan, but there is a surprising range of chemical variations within this type of material. Archaea do not have peptidoglycan cell walls, although some archaeal cell walls contain a related material, pseudomurein.

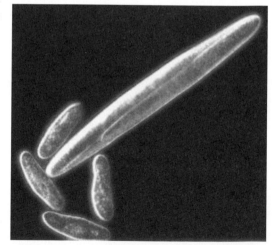

**Figure 1.3** The bacterium *Epulopiscium fishelsoni* (ca. 500 μm long) and four cells of the protist *Paramecium* (ca. 100 μm long). (Courtesy of Esther Angert.) doi:10.1128/9781555818517.ch1.f1.3

Microbes have a wide range of external structures: flagella, pili, fibrils, holdfasts, stalks, buds, capsules, sheaths, and so on. They also have a wide variety of internal structures such as spores, daughter cells, thylakoids, mesosomes, and the nucleoid. In reality, microbial cells are just as structurally organized, and diverse, as are eukaryotic cells.

## Metabolic diversity

Macroscopic eukaryotes are not metabolically diverse; they are either chemoheterotrophic (e.g., animals) or photoautotrophic (e.g., plants). Bacteria and archaea have a much broader range of energy and carbon sources, which can be generally divided into four broad types, chemoheterotrophs, chemoautotrophs, photoheterotrophs, and photoautotrophs.

Chemoheterotrophs obtain both carbon and energy from organic compounds. Some organisms can use a wide range of organic compounds and can either oxidize or ferment them. Others can use only a very narrow range of organic compounds and process them in a specific way. Saprophytes and pathogenic microbes are examples of this group.

Chemoautotrophs obtain cell carbon by fixing $CO_2$. Energy is obtained from inorganic chemical reactions such as the oxidation or reduction of sulfur or nitrogen compounds, iron, hydrogen, etc. These organisms do not need organic compounds for either energy or cell carbon. Sulfur-oxidizing bacteria and methane-producing archaea are examples of this group.

Photoheterotrophs obtain cell carbon from organic compounds, but energy is harvested from light. Halophilic archaea and most purple photosynthetic bacteria are examples of this group.

Photoautotrophs (photosynthetic) obtain cell carbon by fixing $CO_2$. Energy is obtained from light. These organisms do not need organic compounds for either energy or cell carbon. Most cyanobacteria and some purple photosynthetic bacteria are examples of this group.

## Ecological diversity

Microbes live in an amazing range of habitats, from laboratory distilled-water carboys, through freshwater and marine environments, to saturated brines like the Great Salt Lake or the Dead Sea. They grow at temperatures of −5°C to over 118°C; *Pyrodictium* cultures are sometimes incubated in autoclaves! Organisms are known to grow at pH 0 (0.5 M sulfuric acid) and at pH 11 (Drano). Very often, these extremes are combined: *Acidianus* grows in 0.1 M sulfuric acid at 80°C! Some bacteria live in the water droplets that make up the clouds, and others live in deep-underground aquifers or deep-sea sediments. Many microbes live in intimate symbiosis with other creatures, in complex communities, or as permanent intracellular "guests."

In fact, if you are on or around Earth and find liquid water, there is almost certainly something living in it (Fig. 1.5).

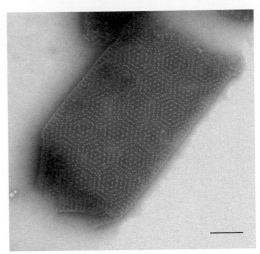

**Figure 1.4** A negative-stain electron micrograph of the S-layer of *Pyrobaculum aerophilum*. Scale bar, 200 nm. (Courtesy of Reinhard Rachel.) doi:10.1128/9781555818517.ch1.f1.4

**Figure 1.5** Moose Pool, Yellowstone National Park, pH ~2, 80°C. doi:10.1128/9781555818517.ch1.f1.5

## Behavioral diversity

It may seem odd to consider the behavior of microscopic organisms, but they do have behavior. Motility and taxis are one form of behavior, both of which come in a variety of forms, from the phototactic *Chlorobium* bacteria that use gas vacuoles and symbiosis with motile bacteria to adjust their place in the water column (Fig. 1.6) to the chemotactic *Rhizobium* bacteria that sense and swim (via flagella) toward chemical signals sent by receptive plant roots. Magnetotactic bacteria have a built-in magnetic compass that allows them to use Earth's magnetic field for orientation.

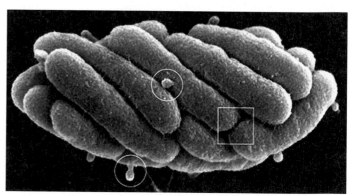

**Figure 1.6** *Chlorobium* symbiotic consortium. (Reprinted from Wanner G, Vogl K, Overmann J, *J Bacteriol* **190:**3721–3730, 2008, with permission.) doi:10.1128/9781555818517.ch9.f9.10

All organisms have developmental cycles; at the very least they can switch between active-growth (i.e., log phase) to resting or slow-growth (i.e., stationary phase) stages. Other developmental cycles include sporulation; the production of swarmer cells, cysts, or akinetes; and even terminal differentiation and

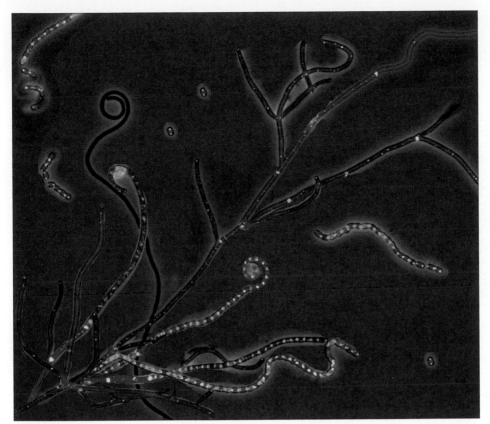

**Figure 1.7** Overlay of phase-contrast and red and green fluorescent images of sporulating *Streptomyces coelicolor* grown on SFM agar plates for 3 days. Red fluorescence results from the DNA stain 7-AAD; chromosomes are stained unevenly because the image was made by using live cells. Green is the fluorescence of SsfA-GFP. SsfA, a protein that is upregulated during sporulation and localizes to sporulation septa, has been fused to the green fluorescent protein GFP. Images were taken with a Zeiss fluorescence microscope and further artistically rendered with Adobe Photoshop. (Courtesy of Nora Ausmees.) doi:10.1128/9781555818517.ch1.f1.7

development into distinct germ and somatic cell types, such as heterocysts in filaments of cyanobacteria, "slugs" in myxobacteria, and the very complex life cycles of *Streptomyces* species (Fig. 1.7).

Microbes also respond to their environments metabolically, by expressing the genes needed to compete for the resources available at the time. An example of this would be converting metabolism from oxidative to fermentative when oxygen is exhausted in a culture or from glucose to galactose use when the glucose is used up in a mixed-sugar medium.

In addition, microbes act communally. Organisms communicate by sending and receiving chemical signals or by direct contact. For example, *Myxococcus* (Fig. 1.8) swarming begins with a chemical signal propagated through the community, which brings the cells into proximity. Direct contact between cells then directs aggregation and formation of fruiting bodies. Microbes also form specific symbioses with other microbes or with macroscopic creatures. Complex communities of microbes associate into "mats" that process and recycle resources throughout the community.

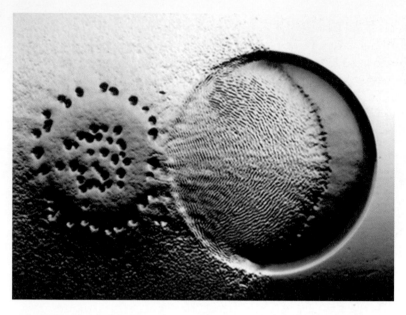

**Figure 1.8** A swarm of *Myxococcus xanthus* (left) invading a colony of *E. coli* (right). (Reprinted from Berleman JE, Scott J, Chumley T, Kirby JR, *Proc Natl Acad Sci USA* **105**:17127–17132, 2008. Copyright 2008 National Academy of Sciences, USA.) doi:10.1128/9781555818517.ch1.f1.8

## Evolutionary diversity

Underlying all of these different aspects of diversity is genetic diversity, perhaps more specifically viewed as evolutionary diversity. Microbes are far more evolutionarily diverse than are macroscopic creatures; the macroscopic world is just the tip of the iceberg of life. Even most plants and animals are microscopic! So microbial diversity is actually the same as biological diversity, with just a few of the more ponderous organisms overlooked.

Evolutionary diversity is usually expressed in terms of trees: branched graphs that trace the genealogies of organisms (Fig. 1.9). When these trees are based on genetic diversity (gene sequences), they can be both quantitative and objective.

This is the perspective on diversity we are using in this book.

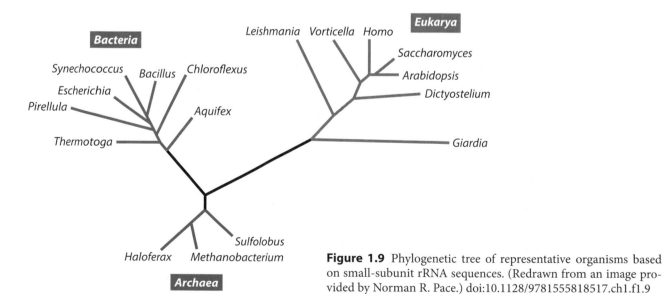

**Figure 1.9** Phylogenetic tree of representative organisms based on small-subunit rRNA sequences. (Redrawn from an image provided by Norman R. Pace.) doi:10.1128/9781555818517.ch1.f1.9

## The fundamental similarity of all living things

Before spending the rest of this book describing diversity, i.e., how organisms differ from one another, it is useful to remember and keep always in mind that *all cells are fundamentally the same in almost every way*. One way to view this is by walking through the flow of information in the cell—the "Central Dogma" (Fig. 1.10)—and point out the fundamental similarities common to all living things.

### DNA

All cells encode information in the form of DNA. The DNA in all cells is composed of the same four bases (G, A, T, and C) and the same sugar (2′-deoxy-D-ribose), assembled with the same chemical structure and stereochemistry. Information in DNA is stored by using a universal three-letter code (e.g., AAA encodes lysine in all cells), and the DNA replication process is handled the same in all organisms (the replication fork complexes are all very much alike). The function of DNA is carried out via transcription into RNA, using RNA polymerases that are all essentially alike.

### RNA

RNA is used primarily to direct protein synthesis based on information in DNA. RNA in all cells has the same structure, the same four bases (G, A, U, and C) and sugar (D-ribose), and the same stereochemistry. Furthermore, all cells have the same types of RNAs, e.g., ribosomal RNA (rRNA), transfer RNA (tRNA), messenger RNA (mRNA), and so forth. These RNAs are very much alike in sequence and structure in all cells; for example, the rRNAs in all organisms share a conserved core of sequence and are very similar in secondary structure (Fig. 1.11).

### Protein

Polypeptides (proteins) direct most of the cell's catalysis and structure. Proteins in all cells use the same 20 (or so) amino acids in the same stereochemical conformations, synthesized in the same way, and they use the same posttranslation modifications. Most of the reactions catalyzed by these proteins are the same (see below), and the enzymes that carry them out are similar in amino acid sequence, cofactors, three-dimensional structure, and mechanism of action.

### Function

With few exceptions, all cells use the same metabolic pathways: the tricarboxylic acid cycle, glycolysis, amino acid biosynthesis, purine and pyrimidine

**Figure 1.10** The Central Dogma: flow of information from archive (DNA) to function (protein function). doi:10.1128/9781555818517.ch1.f1.10

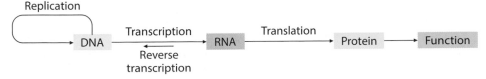

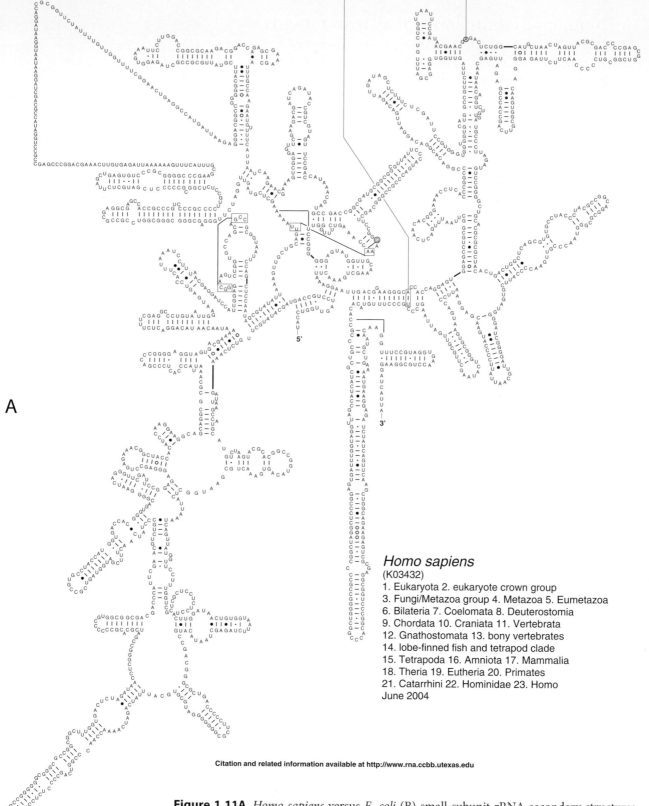

**Figure 1.11A** *Homo sapiens* versus *E. coli* (B) small-subunit rRNA secondary structures. (Courtesy of Robin Gutell.) doi:10.1128/9781555818517.ch1.f1.11A

*Homo sapiens*
(K03432)
1. Eukaryota 2. eukaryote crown group
3. Fungi/Metazoa group 4. Metazoa 5. Eumetazoa
6. Bilateria 7. Coelomata 8. Deuterostomia
9. Chordata 10. Craniata 11. Vertebrata
12. Gnathostomata 13. bony vertebrates
14. lobe-finned fish and tetrapod clade
15. Tetrapoda 16. Amniota 17. Mammalia
18. Theria 19. Eutheria 20. Primates
21. Catarrhini 22. Hominidae 23. Homo
June 2004

Citation and related information available at http://www.rna.ccbb.utexas.edu

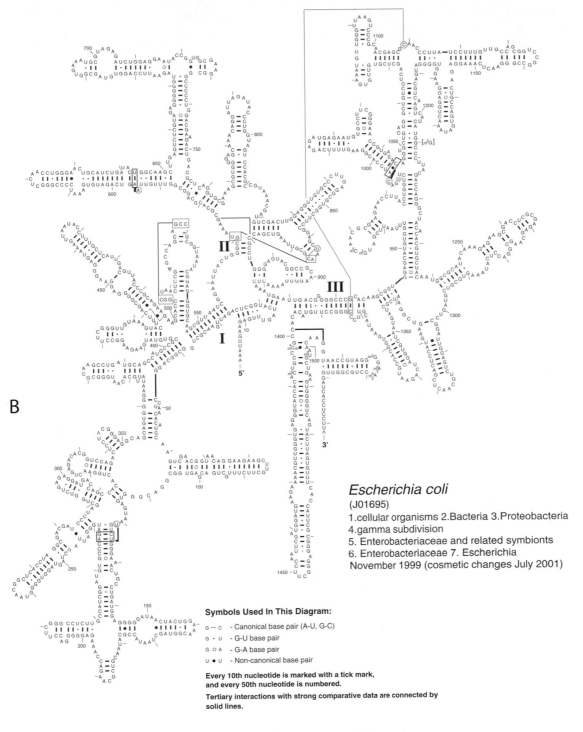

B

*Escherichia coli*
(J01695)
1.cellular organisms 2.Bacteria 3.Proteobacteria
4.gamma subdivision
5. Enterobacteriaceae and related symbionts
6. Enterobacteriaceae 7. Escherichia
November 1999 (cosmetic changes July 2001)

**Symbols Used In This Diagram:**

G — C   - Canonical base pair (A-U, G-C)
G • U   - G-U base pair
G ○ A   - G-A base pair
U • U   - Non-canonical base pair

**Every 10th nucleotide is marked with a tick mark,
and every 50th nucleotide is numbered.**

**Tertiary interactions with strong comparative data are connected by
solid lines.**

Citation and related information available at http://www.rna.icmb.utexas.edu

**Figure 1.11B**   doi:10.1128/9781555818517.ch1.f1.11B

biosynthesis, lipid metabolism, electron transport, and adenosine 5'-triphosphate (ATP) synthesis via the proton gradient. There is perhaps more variation in metabolism than in the previous parts of the Central Dogma, but these are generally minor variations, such as the use of glycolysis versus the Entner-Doudoroff pathway.

## Cells

In addition, all organisms are built of one or more cells, which are bound by a lipoprotein membrane that strictly controls what goes into and what comes out of the cell. This generally defines and separates inside and outside.

## Significance of the similarity between organisms

So, all organisms are mostly the same. What does this mean? Primarily, it means that all organisms share a common ancestry. In other words, all known organisms can trace their history back to a single origin of life. This might not have been the case; other lineages, if they ever existed, seem to be extinct (or perhaps undiscovered or unrecognized?).

The fact that all organisms are very much alike also means that the last common ancestor of all known living things was a complex organism or population of organisms. Most of biochemical evolution predates the last common ancestor. The last common ancestor had all of the biochemistry that is now universal, which means nearly everything. Biochemical evolution occurred very early in the emergence of life. The diversity in extant life (known modern life) is in peripheral biochemistry—just the details!

## Questions for thought

1. What, exactly, does "diversity" mean to you?

2. How are organisms different from one another, and how would you measure objectively (or judge subjectively) how different two or more organisms are? Can you name two animals that are about as different from one another as, say, *Escherichia coli* and *Proteus vulgaris*?

3. If you had isolated a new microbe, what properties would you examine to determine what kind of organism it is? Would this tell you what it is related to evolutionarily?

4. What metabolic pathways can you think of that are unique to specific groups of organisms?

# 2

# Context and Historical Baggage

If, as described in chapter 1, evolutionary diversity underlies the various aspects of "diversity," it is important to review what we mean by evolutionary diversity. This is especially so because even some biologists have a relatively poor understanding of modern evolutionary theory and it is important to at least dig up and expose outdated but widely held notions about how organisms are related.

## The evolution of evolutionary thought

### *The chain of being*

Well before people discovered that species changed over time, both living and nonliving things were organized (in Western culture, including Aristotle) into the "chain of being" (Fig. 2.1).

All species and substances were placed onto individual slices of a vertical scale, ascending from "inferior" to "superior." Much thought and energy were put into deciding exactly how to order the major categories of materials, living things, species, races, social classes, and even individuals onto this linear arrangement. (It will come as no surprise to the cynical that the ethnic group placed at the top of the human region of the chain generally matched that of whoever was organizing the chain.) This was an impossible task, with lots of problematic and contradictory issues. In retrospect, it is obvious that the reason for this was that there *is* no linear relationship between things in nature.

As an aside, this linear arrangement of substances survives, in general form, in the periodic table, in which the elements are arranged in order of increasing

doi:10.1128/9781555818517.ch2

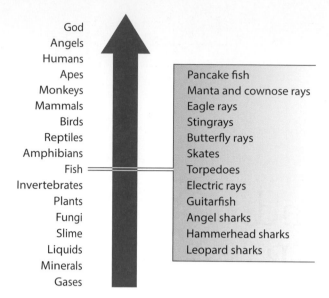

**Figure 2.1** The "chain of being," with some representative organisms shown. To the right is a closer view of one thin slice of this "chain." doi:10.1128/9781555818517.ch2.f2.1

atomic number. The chain is broken into segments, and the segments are stacked in such a way that elements with similar properties form columns. The reason this works is because elements *are* structured in a linear series based on atomic number.

## The evolutionary ladder

The earliest forms of evolutionary thought had each species (up to but not including humans) moving up the "evolutionary ladder," a climbable form of the chain of being. Evolution, in this view, was linear and progressive; species evolved from one rung to the next as an effect of a progressive force (Fig. 2.2).

Although the notion of an evolutionary ladder is pre-Darwinian and hopelessly incorrect, we are discussing it here because this scheme is firmly embedded in modern biological unconscious thought. You even hear the term "evolutionary ladder" used fairly often by research scientists who actually know better (at least, one hopes they do). The terms "higher" and "lower" eukaryotes, and "missing link," which are in common use, are relics of this view.

## Early phylogenetic trees

By the time of Charles Darwin, it was clear that the evolutionary ladder was not a reasonable view of evolution. Darwin describes a much better view, which has proven to be essentially correct, in which species originate by divergence, as shown in Fig. 2.3 (this diagram is the only illustration in *On the Origin of Species*).

It turns out that species actually evolve by diversification, not by progression. Eukaryotes did not evolve from bacteria, animals did not evolve from ciliates, plants did not evolve from fungi, and humans did not evolve from chimpanzees. Each of these pairs of modern organisms

**Figure 2.2** The "evolutionary ladder," an evolutionary transformation of the chain of being. (Redrawn from Prothero DR, Buell CD, *Evolution: What the Fossils Say and Why It Matters*. Copyright 2007 Columbia University Press. Reprinted with permission of the publisher.) doi:10.1128/9781555818517.ch2.f2.2

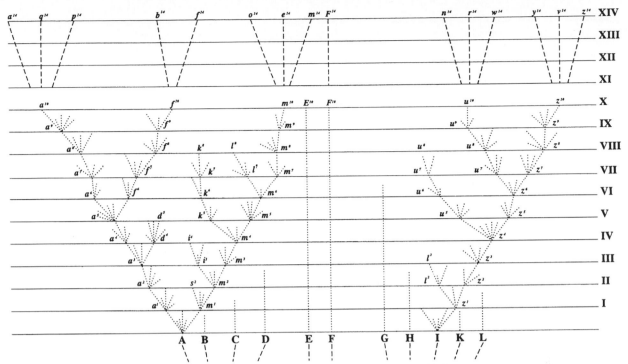

**Figure 2.3** Evolution by diversification. In this diagram, species A and I at the beginning (bottom) split many times and diverge constantly. Most of these divergences do not go anywhere (they become extinct), but some do make it, at least for a while, resulting in this case in species A splitting into three separate surviving species and species I splitting into two surviving species at time X. Species A and I no longer exist at time X. Note that most of the original species (B through H, K, and L) are in stasis, remaining unchanged through the span of time shown here (or at least the part of it during which they survive). (Reprinted from Charles Darwin, *On the Origin of Species*. John Murray, London, 1859.) doi:10.1128/9781555818517.ch2.f2.3

shares a common ancestor, from which each diverged. This is a fundamental aspect of evolution that is poorly understood even by many biologists.

One of the best developed of the early divergent evolutionary trees was that of Ernst Haeckel. In this tree, there are three major, equivalent divisions of life—plants, animals, and protists (Fig. 2.4). This tree is a huge improvement over the evolutionary ladder. It is a tree, species are not ranked, and modern species are not considered to be the ancestors of other modern organisms. Plants and animals are not thought of as having evolved from modern prokaryotes (monerans) but are separate groups.

The Whittaker five-kingdom tree being taught in various forms in most classrooms today is a refinement of this tree (Fig. 2.5). In some ways, however, this five-kingdom tree is actually a step backward toward the evolutionary ladder. In most versions of this scheme (such as the original tree by Whittaker, above), eukaryotes are shown to be descended from within the bacteria (not true), and in many representations eukaryotic algae are shown as descendants of cyanobacteria (not true), fungi are shown as descendants of filamentous gram-positive bacteria (not true), and protists are shown as descendants of

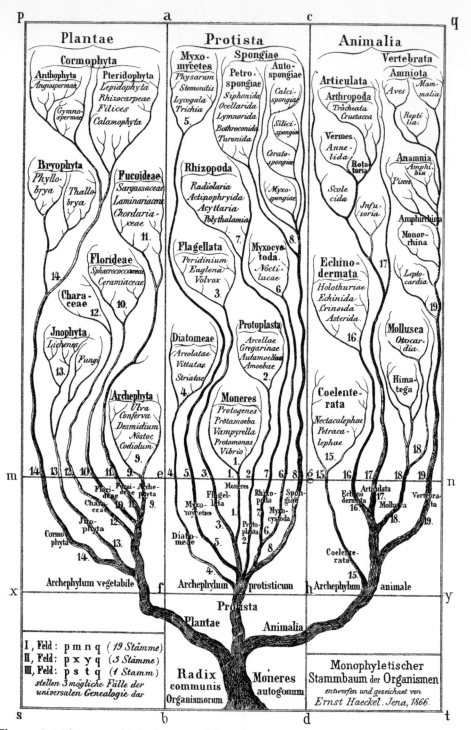

**Figure 2.4** The tree of life. (Reprinted from Ernst Haeckel, *Generelle Morphologie der Organismen*. G. Reimer, Berlin, Germany, 1866.) doi:10.1128/9781555818517.ch2.f2.4

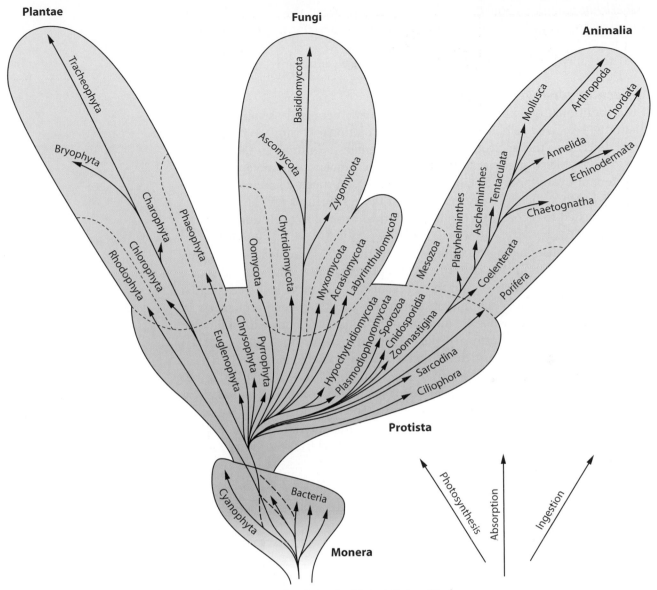

**Figure 2.5** The five-kingdom tree taught in most schools in the United States. (Redrawn from Whittaker RH, *Science* **163**:150–160, 1969. Used with permission from AAAS.) doi:10.1128/9781555818517.ch2.f2.5

wall-less gram-positive bacteria (also not true). Also notice the implied vertical axis: either superiority (sometimes expressed as "complexity") or time (usually labeled "time of origin"). But if this axis is complexity, what exactly is being measured? Eukaryotes are sometimes morphologically complex, but what about parasites that have simplified; why are these organisms not drawn as downward-pointing arrows? What about metabolic complexity, which would place animals close to the bottom? Is morphological complexity the only factor being considered? Why? If the vertical axis is time of origin, why are the recent emergences of bacterial families, genera, and species not considered? The genus

*Escherichia* emerged about 100 million years ago, about the same time as the primates; why is *Escherichia* (along with all other bacteria), a modern organism, shown as a relic of the past?

The reality is that the five-kingdom tree is entirely qualitative and subjective.

## Molecular phylogenetic trees

If the traditional five-kingdom tree is problematic because it is subjective and qualitative, we need an objective measure of evolutionary history. With the ability to determine the nucleotide sequences of genes beginning in the 1970s, it became possible to use variation in these gene sequences as molecular chronometers of evolutionary distance. We cover this in much detail in chapters 3 through 6, but suffice it to say for now that these sequences provide the information needed to reconstruct evolutionary trees both objectively and quantitatively (Fig. 2.6).

This example of a molecular phylogenetic tree is an unrooted dendrogram. The length of the branches quantitatively represents the evolutionary distance separating gene sequences within these organisms. This particular tree is based on the analysis of small-subunit ribosomal RNA (rRNA) gene sequences. In this tree, the tips of branches are modern organisms. Each node within the tree represents a common ancestor. The last common ancestor (the root) is marked with a star. The way this was determined is described in chapter 7.

Notice that there is no explicit or implied ranking of above (superior) or below (inferior) in the tree. Evolutionary distance (divergence) is measured along the lengths of the branches connecting species. There are no axes in this graph.

One of the most exciting early outcomes of this method was the discovery of a new type of organism: the *Archaea* (archaebacteria). Previously it was

**Figure 2.6** Phylogenetic tree of representative organisms based on small-subunit rRNA sequences. (Redrawn from a figure provided by Norman R. Pace.) doi:10.1128/9781555818517.ch2.f2.6

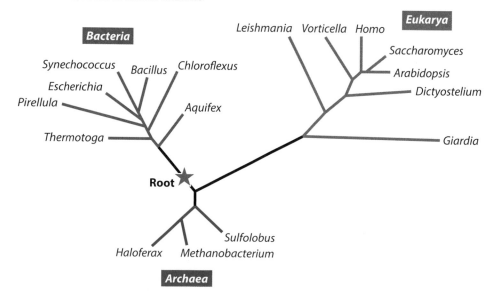

thought that all living things were members of either the *Bacteria* (eubacteria) or the *Eukarya* (eukaryotes). Archaeal species had previously been scattered haphazardly among whatever bacteria they superficially resembled. Indeed, in terms of superficial phenotype, the *Archaea* are generally similar to the *Bacteria*, but biochemically they are just as similar to the *Eukarya*, and in evolutionary terms they form a distinct group that is probably more closely related to the *Eukarya* than to the *Bacteria*. The *Archaea* as a group have changed less since their common ancestry than either the *Bacteria* or *Eukarya* (they are primitive), and so they more closely resemble our common ancestry.

Multicellular eukaryotes, the plants (e.g., *Arabidopsis*), animals (e.g., *Homo*), and fungi (e.g., *Saccharomyces*), are a very small portion of evolutionary diversity in this tree: just the tip of one or two branches of the *Eukarya*, not three-fifths of evolutionary diversity as the five-kingdom scheme has it. Notice that *Eukarya* is as ancient a group as is either *Bacteria* or *Archaea* and that it did not evolve from either of these other groups. Bacteria are not primitive ancestors of "higher organisms."

The tree also offers confirmation of the endosymbiont theory for the origin of mitochondria and chloroplasts. These organelles have their own DNA and genes, including small-subunit rRNA genes, and so they can be analyzed separately from the nucleus (*Eukarya*) by molecular phylogenetic analysis. Mitochondria turn out to be members of the proteobacteria (exemplified by *Escherichia* in this tree), and chloroplasts are members of the cyanobacteria (*Synechococcus* in this tree).

## Taxonomy and phylogeny

A taxonomy is a classification scheme for species (or any other collection of objects, for that matter). There are three related components of a biological taxonomy:

1. *Grouping*: The organization of organisms into groups based on similarity
2. *Naming*: The labeling of organisms and groups of organisms with names
3. *Identifying*: The identification of organisms when they are found

Taxonomies are artificial constructions, methods created and used by humans to organize species. Any self-consistent taxonomy is valid, whether it reflects the natural relationships of the organisms or not.

For example, wildflower field guides organize species by features that are readily observed in the field. The first division might be by flower color, a trivial feature of the plants in evolutionary terms but perfectly reasonable for taxonomy. There is no implication that plants with flowers of the same color are related phylogenetically or that plants with flowers of different colors are not related. The field guide is designed for grouping, naming, and identifying species, and so are useful taxonomies.

A phylogeny is the evolutionary pathway relating species. Think of a phylogeny as a large-scale genealogy of species. Phylogenies represent the actual natural relationships between organisms. They are most commonly displayed graphically in the form of phylogenetic trees.

Unfortunately, there are generally no natural delineations between groups in a phylogenetic tree. However, taxonomists can start with a phylogenetic tree and try to divide it into reasonable groups based on the branches of the tree and the phenotypes of the organisms. In doing so, they attempt to devise a taxonomy that reflects, or at least is consistent with, the phylogenetic relationships between the species.

The classical taxonomies of plants and animals are fairly good representations of the phylogenetic relationships between their members because the intricate morphology of these organisms reveals their ancestry. This is not true for microbes. There was no way to determine evolutionary relationships between *Bacteria* and *Archaea* or even between protists, algae, and fungi until the development of molecular phylogenetics, in which phylogenetic relationships are inferred on the basis of gene sequences.

## *Why is an understanding of phylogeny important?*

Thinking about organisms from a phylogenetic perspective is critically important for two reasons:

First, so that you can predict the properties of organisms based on the properties of their relatives. Think about how much insight you can get into a person by getting to know his or her family—this is why meeting the family of someone you're dating is so important! Understanding an organism's relationships to other species is the key to understanding its properties.

Second, so that you can prevent inappropriate comparisons based on nonexistent relationships. For example, *Euglena* was used for years as a unicellular model system to study photosynthesis in green plants. However, it turns out that *Euglena* is related not to plants but to trypanosomes. Its chlorophyll is not like those of plant chloroplasts and was acquired independently. *Chlamydomonas* is a better system because it is phylogenetically a unicellular green plant.

## The false eukaryote-prokaryote dichotomy

The classic major division of living things is between eukaryotes and prokaryotes. Most textbooks have a table (Table 2.1) indicating how different these kinds of living things are.

The problem with the prokaryote-eukaryote dichotomy is that it is exclusionary; it is a lot like the vertebrate-invertebrate dichotomy of animals. The terms "prokaryote" and "invertebrate" describe what an organism is *not*, but they do not describe what it *is*. "Prokaryote" simply means "not a eukaryote"! Therefore, the term "prokaryote" as a label for a group of organisms is scientifically invalid. The notions of "prokaryotes" arose over the years for no real reason other than default; it became an assumption that all noneukaryotes were of a single kind, because the existing tools could not distinguish different "kinds" among them, but as we see, this is not the case, any more than it is the case that all "invertebrates" are of a kind. There are in fact two fundamentally different kinds of "prokaryotes," *Bacteria* and *Archaea*, which are at least as different from each other as either is from eukaryotes. Banish the term "prokaryote" from your

**Table 2.1** Sample "eukaryotes versus prokaryotes" table common to biology and even microbiology textbooks

| Eukaryotes | "Prokaryotes" |
|---|---|
| Large (20–100 μm) | Small (1–5 μm) |
| Contain nucleus | No nucleus |
| Contain many large linear chromosomes | Contain one small circular DNA chromosome |
| Contain organelles | No organelles |
| Diploid | Haploid |
| Cell cycle includes mitosis | No mitosis |
| Reproduce sexually or by budding | Reproduce by binary fission |
| Cells contain a cytoskeleton | No internal skeleton |
| Ingestive or photosynthetic | Absorptive |
| Multicellular | Unicellular |
| Complex life cycles and cellular differentiation | Simple division cycle, no differentiation |
| mRNAs are polyadenylated | No polyadenylation |
| Genes transcribed separately | Genes transcribed together in operons |
| Genes contain introns | No introns |
| DNA packaged with histones | DNA not packaged |

vocabulary, along with "evolutionary ladder" and "higher eukaryote," except to use (in quotation marks) in reference to a quaint, outdated anachronism. File it in your mind alongside the term "phlogiston."

In addition, many of the stark contrasts between bacteria ("prokaryotes" in the case of this table) and eukaryotes come from falsely assuming that plants and animals are typical eukaryotes and that *E. coli* and *Bacillus subtilis* are typical prokaryotes, and from an active striving to identify differences, no matter how trivial, in order to make eukaryotes "higher" and prokaryotes "lower." But something that is true for *E. coli* and *B. subtilis* is not necessarily true for other bacteria, and something that is true for plants and animals is not necessarily true for other eukaryotes. Bacteria are not primitive—they are modern organisms, the result of over 3.6 billion years of evolution, just like eukaryotes. As a matter of fact, *Eukarya*, *Bacteria*, and *Archaea* are not at all as different as these tables suggest. All of the apparent differences listed above are bogus in one way or another: at best overgeneralizations and at worse simply false.

## Questions for thought

**1.** Can you think of a positive trait (i.e., something that is present or has a positive characteristic) that is typical of bacteria but not eukaryotes?

**2.** Can you think of a positive trait that is characteristic of protists but not higher eukaryotes? What is a "higher" eukaryote? Why? Would you still think this if you were an egocentric trypanosome?

**3.** What is the difference between a multicellular organism and just a lump of cells of the same species?

**4.** For how many of the groups shown in the three-kingdom molecular phylogenetic tree can you name species?

**5.** In the molecular phylogenetic tree shown above, how would you add a novel organism whose ancestors diverged from other living things *before* the last common ancestor of the *Bacteria*, *Archaea*, and *Eukarya*?

**6.** Where is the last common ancestor of all known bacteria?

**7.** Which modern organism(s) is closest in evolutionary distance in this tree to the last common ancestor?

# 3

# Phylogenetic Information

Molecular phylogenetic analysis is the use of macromolecular structure (usually nucleotide or amino acid sequences) to reconstruct the phylogenetic relationships between organisms. The extent of difference between homologous DNA, RNA, or protein sequences in different organisms is used as a measure of how much these organisms have diverged from one another in evolutionary history.

The typical scenario where a phylogenetic analysis is needed is the characterization of a novel organism: for example, determining the phylogenetic placement (phylotype) of a novel organism, in order to make predictions about its unknown properties. This might be a clinical isolate of a potential pathogen, an organism that carries out useful biochemistry, an organism that seems to be abundant in an interesting environment, or anything else of interest.

The process of molecular phylogenetic analysis can be divided into four critical parts, each of which, of course, also has various subparts:

1. Decide which organisms and sequences to use in the analysis
2. Obtain the required sequence experimentally or from databases
3. Assemble these sequences in a multiple-sequence alignment
4. Use this alignment to generate phylogenetic trees

In this chapter, we walk through the first three steps of this process.

## Deciding which organisms and sequences to use in the analysis

The sequences of genes, RNAs, or proteins contain two very different kinds of information: structural/functional information and historical information. Think of it this way: any particular amino acid in a specific protein is what it is (say, for example, an alanine) in part because it facilitates the formation of the correct structure and function of the protein. But usually there are a number of alternatives that might function just as well. The reason it is what it is, and not any of these alternatives, is that it was inherited from a successful ancestor. *This* is historical information. Comparisons among an aligned collection of homologous sequences can be used to sort out both the structure of the functional molecule (especially for RNAs) and their historical relationships: a phylogenetic tree.

Phylogenetic trees are usually generated by using alignments of single genes, RNAs, or proteins, but no such sequence is either ideal or universally useful for the generation of informative phylogenetic trees. This being said, some sequences do carry more phylogenetic information than others; these sequences can be called "molecular clocks."

### *Features required of a good molecular clock*

#### Clock-like behavior

The sequences of genes, RNAs, and proteins change over time. If this change is entirely random (within the constraint of the structure and function of the molecule; i.e., by genetic drift), the amount of divergence between any particular sequence in two organisms should be a measure of how long ago these organisms diverged from their common ancestor. If this is true, these sequences can be said to exhibit clock-like behavior (Fig. 3.1).

Clock-like behavior depends mostly on functional constancy of the sequence; a change in the function (or functional properties) leads to large, selected (and therefore nonrandom) sequence change, i.e., adaptation. Clock-like behavior also depends on the sequence being long enough to provide statistically significant information and being made up of a large number of independently evolving "bits" so that random changes in one part of the sequence

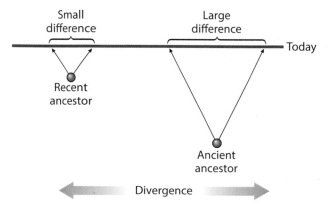

**Figure 3.1** Clock-like behavior. The extent of sequence divergence between a pair of specific sequences should be a measure of how long ago they separated. doi:10.1128/9781555818517.ch3.f3.1

do not influence changes in other parts of the sequence. The sequence must also have an appropriate amount of sequence variation; too little variation does not provide enough difference to be statistically meaningful, whereas too much makes alignment difficult or impossible and decreases the reliability of the treeing algorithm (see chapters 4 and 5 on evolutionary models). Nonfunctional sequences (e.g., some introns) usually change too fast for analysis except of the very closest of relatives.

## Phylogenetic range

In order to be useful for a phylogenetic analysis, a sequence must be present and identifiable in all of the organisms to be analyzed and must exhibit clock-like behavior within this range. Watch out for gene families, because each member of the family is probably specialized for a slightly different function and it is often difficult to identify the correct ortholog or confirm that it really does have the same function.

## Absence of horizontal transfer

Absence of horizontal transfer means that the gene must be acquired only by inheritance from parent to offspring, not by transfer from one organism to another except by descent. Examples of frequently horizontally transferred genes are those encoding antibiotic resistance, but any gene has the potential to be transferred horizontally. You can still generate a tree with sequences that have been horizontally transferred, but if the sequence is otherwise a good molecular clock, the resulting perfectly valid tree will reflect the phylogenetic relationships between the sequences but not the organisms that carry these sequences.

## Availability of sequence information

It is of great pragmatic importance to choose a sequence, whenever possible, for which a great deal of the sequence data required is already available and annotated and perhaps already aligned. If you are interested in the phylogenetic placement of organism X, it is better if you do not have to obtain or identify the sequence data yourself for a large number of organisms to which it might (or might not) be related.

## *The standard: small-subunit ribosomal RNA*

In most cases, the best molecular clock for phylogenetic analysis is the small-subunit ribosomal RNA (SSU rRNA) (Fig. 3.2). This sequence is always the best starting point; only after you know where your organism resides in an SSU rRNA phylogenetic tree can you decide what other sequences might provide additional information (see chapter 6 for alternatives).

The SSU rRNA is so often the best sequence of choice for the following reasons.

- It is present in all living cells.
- It has the same function in all cells.
- It comprises 1,500 to 2,000 residues—large enough to be statistically useful but not too large to be onerous to sequence.

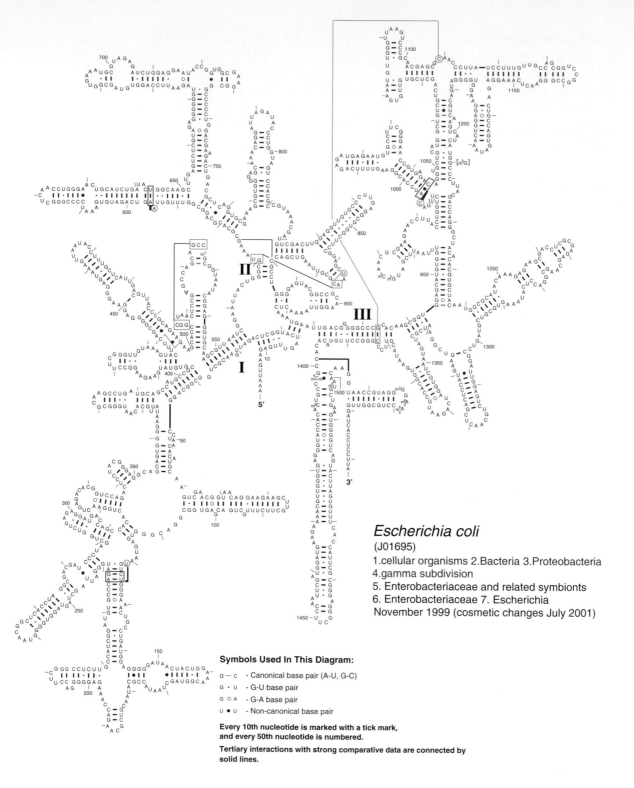

Citation and related information available at http://www.rna.icmb.utexas.edu

**Figure 3.2** The *Escherichia coli* SSU rRNA secondary structure. (Courtesy of Robin Gutell. Adapted from Cannone JJ, Subramanian S, Schnare MN, Collett JR, D'Souza LM, Du Y, Feng B, Lin N, Madabusi LV, Müller KM, Pande N, Shang Z, Yu N, Gutell RR, *BMC Bioinformatics* 3:2, 2002. doi:10.1186/1471-2105-3-2) doi:10.1128/9781555818517.ch1.f1.11B

- It is made up of ca. 50 independently evolving helices and ca. 500 independently evolving base pairs.
- It is conserved highly enough in sequence and structure to be easily and accurately aligned.
- It contains both rapidly and slowly evolving regions—the rapidly evolving regions are useful for determining close relationships, whereas the slowly evolving regions are useful for determining distant relationships.
- Horizontal transfer of rRNA genes is exceedingly rare (most genes of the central information-processing pathways of the cell are also resistant to horizontal transfer).
- Huge data sets of sequences, alignments, and analysis tools are available.

### Deciding which organisms to include

Usually, deciding which organisms to include is part of the treeing process rather than something done in advance. As is explained in chapter 4, most often you start out by generating a tree with representatives from a wide range of organisms scattered around the tree in order to identify what kind of organism it is in very general terms, and then you replace most of these disparate representatives with representatives that you now know are likely to be closely related. The resulting tree gives you more specific information about the group to which your organism belongs, which can be used again to choose even closer relatives, and so on until you are satisfied with the representation of the tree. For example, if you have a new organism to identify, an initial tree containing one or two representatives from each bacterial phylum might show you that your organism is a member of the *Firmicutes*. With this information in hand, a second tree populated by representatives of each order and family of *Firmicutes* might show you that you might have a member of the family *Veillonellaceae*. From there, you could populate a final tree with most of the species in this family.

## Obtaining the required sequence data

Sequences for a phylogenetic analysis usually come from two sources: electronic databases and your own experimental results. Ideally, all of the sequence data needed can be obtained from databases, or at least all of the data needed except that of the specific organism(s) of interest. However, you might find that some other sequences you need for comparison are unavailable; if this is the case, you may need to obtain them yourself experimentally.

### Databases

For most sequences commonly used for phylogenetic analysis, there are specialized databases of pre-aligned sequences. There are several databases of SSU rRNAs and their alignments: the Ribosomal Database Project is a prime example and is specialized for phylogenetic analysis. As of this writing, the Ribosomal Database Project contains 1.3 million aligned SSU rRNA sequences and a suite of software tools to access these data, including a taxonomic browser that can be used to collect any desired aligned sequences for further analysis. Databases of this type are usually the best starting point from which to collect an initial data set.

Very often, however, there is additional useful information that is not yet available in these specialized databases. BLAST searches of general sequences databases (e.g., GenBank), most often through the National Center for Biotechnology Information website, often identify additional useful sequences.

## Obtaining sequences experimentally

The commonly used method to obtain DNA for sequence analysis is the polymerase chain reaction (PCR). PCR amplifies genes exponentially—a single

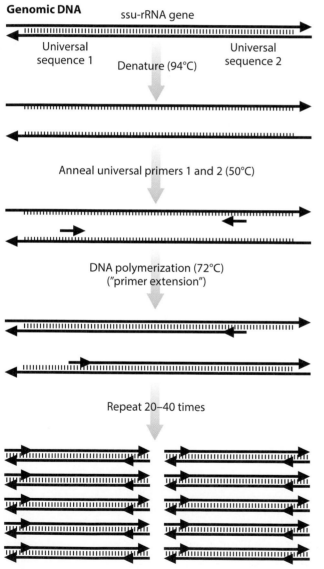

**Figure 3.3** The polymerase chain reaction (PCR). doi:10.1128/9781555818517.ch3.f3.3

molecule of a gene, embedded in the rest of the genomic DNA, is specifically amplified to up to a million molecules in just a couple of hours. In a PCR, three steps (denaturation, primer annealing, and DNA polymerization) are cycled over and over, each time doubling the amount of the specific DNA fragment (Fig. 3.3).

The PCR product DNA is then sequenced (i.e., its nucleotide sequence is determined), often using the same oligonucleotide primers that were used in the PCR. Sequencing involves denaturing the DNA, annealing an oligonucleotide primer, and extending from this primer with DNA polymerase in the presence of deoxynucleoside triphosphates (dNTPs) and small amounts of chain terminator dideoxynucleotides (analogs of dNTPs from which DNA polymerase cannot continue extending) (Fig. 3.4). Usually this process is carried out by a

**Figure 3.4** Chain termination sequencing. doi:10.1128/9781555818517.ch3.f3.4

Add dATP, dGTP, dCTP, dTTP, DNA polymerase, and buffer, plus 1:1,000 of each fluorescently labeled ddNTP

**ddATP in red**    **ddGTP in blue**    **ddCTP in green**    ddTTP in yellow

Elongated and terminated product

For example, if the primer/template looks like this:

5′ ——▶ 3′
————TCCGATGCGGTCTTGAA . . . . . .

the products will look like this:

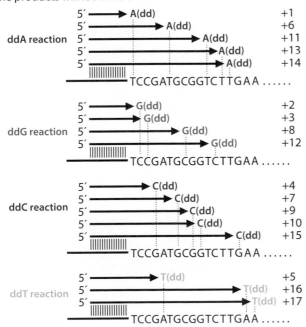

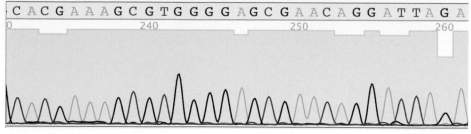

C A C G A A A G C G T G G G G A G C G A A C A G G A T T A G A

**Figure 3.5** Example sequence data from a DNA sequencing reaction.
doi:10.1128/9781555818517.ch3.f3.5

commercial service rather than in a research lab. A fluorometer at the bottom of the gel or end of the capillary detects the termination dyes as they run past. The connected computer collects this data and reads the sequence from the pattern of peaks (Fig. 3.5).

Each reaction typically yields 500 to 1,000 bases of reliable sequence data, so it is usually necessary to use several primers spaced along the length of the molecule to get the complete sequence of an rRNA gene. It is also usually expected that both strands of the DNA will be sequenced for confirmation.

## Assembling sequences in a multiple-sequence alignment

The raw material used by a phylogenetic tree-generating program is an alignment (Fig. 3.6). A sequence alignment is a two-dimensional matrix of multiple sequences. Each sequence is in a line (row) of the matrix. Each position (column) in an alignment contains homologous (corresponding) residues of each sequence. Gaps (usually shown as dashes) are added where needed to maintain the alignment; these gaps represent the absence of bases in the sequence that are present in some other sequence(s) in the alignment.

### *Alignment based on conserved sequence*

Most alignments are generated by using computer programs that align sequences from algorithms (e.g., CLUSTAL) that attempt to maximize the similarity (measured in a variety of ways) of all of the sequences. Where the sequences in an alignment are very similar, this approach can generate very good alignments. This is especially true for protein-encoding sequences, with 20 possible amino acids and good scoring matrices to count how similar or different any two amino acids are from each other. This is less true for DNA or RNA sequences, with only four possible bases and where similarity between pairs of bases is less meaningful in the context of the encoded macromolecule.

Very often, however, RNA alignments are either created by hand or at least adjusted manually. Sequences must be fairly similar in sequence and length to be readily aligned by eye or even by computer alignment programs. However, most of the length of SSU rRNAs is highly conserved and can with experience be manually aligned without much trouble.

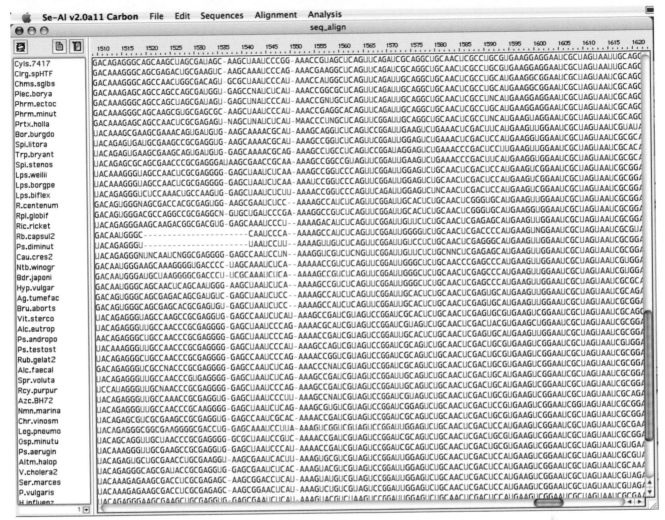

**Figure 3.6** A small window into an alignment of SSU rRNA sequences.
doi:10.1128/9781555818517.ch3.f3.6

Some of the tricks to aligning sequences by hand are the following.

- Sequences are often aligned sequentially; start by aligning the two most similar sequences, then add sequences to the alignment one at a time, starting with the sequences most similar to those already aligned and finishing with the most distantly related sequences. Likewise, if you are adding a single sequence to an existing alignment, start by identifying the most similar sequence in the alignment and use that sequence as a guide.

- Alternatively, you can identify conserved blocks of sequence in all of the sequences and align these. You have now broken the alignment problem into smaller, easier chunks. Add gaps as needed to align the space between prealigned chunks according to the criteria below.

- Start by finding patches of very similar sequences and align these, then work out in both directions from these, adding gaps sparingly when

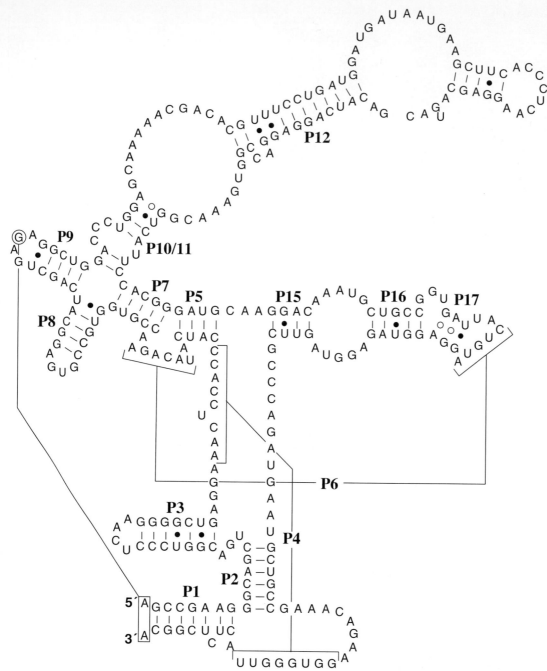

**Figure 3.7A** Comparison of two RNase P RNAs with very different sequences and very similar secondary structures. RNase P RNAs are the catalytic subunits, associated with one or more accessory proteins, that remove the 5′ leaders from tRNA and other RNA precursors. (Adapted from Harris JK, Haas ES, Williams D, Frank DN, Brown JW, *RNA* 7:220–232, 2001, with permission.) doi:10.1128/9781555818517.ch3.f3.7A

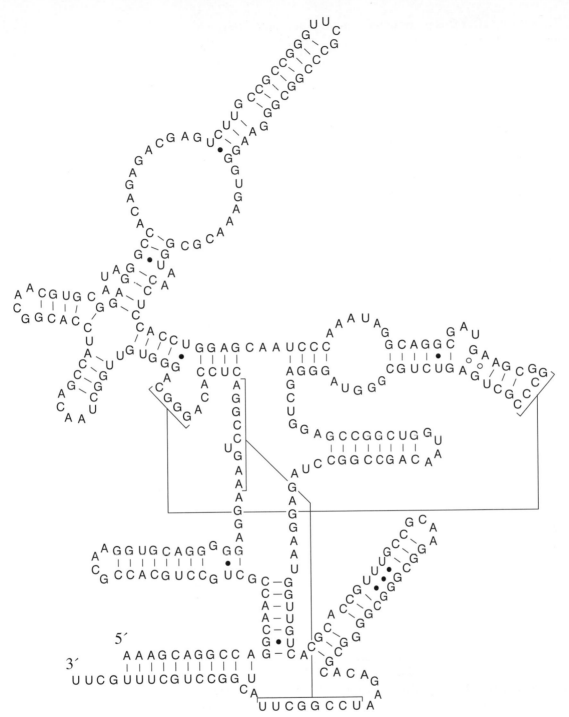

**Figure 3.7B**  doi:10.1128/9781555818517.ch3.f3.7B

needed. Everything after this is about rearranging (and potentially adding or removing) these gaps.

- Where there are sequence differences, slide the gaps around to keep purines (G, A) aligned with purines, and pyrimidines (C, U/T) aligned with pyrimidines.
- Try also to keep differences together in variable sequence positions, and align gaps together in columns wherever possible. A single gap of two positions is a lot better than two separate gaps of one position each.
- Try to keep what look like conserved positions (columns) conserved, and all things being equal, put differences into positions already known to be variable.

### Alignment based on conserved structure

In the case of RNAs, however, advanced alignment algorithms (e.g., infeRNAl) can use the secondary structures of the RNAs to align sequences. The ability to use well-defined secondary structures to identify homologous residues (i.e., to align sequences) is one of the key advantages of RNA over protein for phylogenetic analysis. In other words, you can use the secondary structure of the RNA to identify homologous parts of the RNA, rather than relying only on sequence similarity (Fig. 3.7).

**Figure 3.8** An RNA alignment based on secondary structure. If residue *n* (e.g., 24, highlighted) of any sequence pairs to residue *m* (e.g., 29, also highlighted), then so should the corresponding homologous residues in all sequences. This is an RNA alignment based on secondary structure: stem-loop P3 of RNase P RNA. In this example, the first six rows are not sequences, they are annotations. The first three are just a reference numbering; in this case, the *Methanothermobacter thermautotrophicus* (Mthermo) sequence is the reference sequence. The row marked "helices" indicates the secondary structure: the 5′ strand of P3 followed by the loop and then the 3′ strand. Each base pair in this stem-loop is indicated by matching right- and left-facing parentheses in the following row and is labeled alphabetically (for human readability) in the subsequent row. doi:10.1128/9781555818517.ch3.f3.8

```
hundred  000.0000000................0000..................0..00.00000000
tens     111.1122222................2222..................2..33.33333333
ones     567.8901234................5678..................9..01.23456789
helices  ....<------|---P3-5'------->..L3..<----------P3-3'-|--------->....
pairing  ----((((((|(((((((((----(((((------)))))----)))))))))|--))-))))----
pairing  ----ABCDEF|GHIJKLMN----OPQRS------SRQPO-----NMLKJIHG|--FE-DCBA----
Ssolfat  UAA-CGGGG-|--------------CAAA-----------------------|---C-CCUGAGGA
Sacidoc  UUA-CGGGA-|--------------AUA------------------------|---U-CCUGAGGA
Msedula  CCA-CGG---|--------------GAAA----------------------|-----CUGGGGA
Apernix  CCA-CGGCCC|CC----------AGCCA-------------------GGG|--GG-GCUGAGGA
Pfurios  UGC-CGGGC-|-----------UUUAU-----------------------|---G-CCCGAGGA
Tlittor  CCU-CGGGU-|-----------AUUUG-----------------------|---A-CCCGAGGA
Mthermo  UGA-CGGUCC|------------UCAA----------------------G-|-GG-GCUGAGGA
MthMarb  UGA-CGGCCC|A------------UUUU---------------------U-|-GG-GCUGAGGA
Mformic  UAC-CGGUUU|CUAUAGAU---------UUAAU----------GUCUGUAG|UUAA-ACUGAGGA
Tvolcan  UGA-CGCC--|-----------------GUAA-----------------|-----GGUGAGGA
Mbarker  UGA-CGGGC-|-----------------UUCG---------------GG|-UCUGAGGA
Hcutiru  UGCCCGUGCC|-----------------GUGA--------------GG|-CAUGAGGA
Hvolcan  UCC-CGUGCC|G-----------------AGA----------------CG|-GG-CAUGAGGA
Hmorrhu  CAC-CGCGGC|GUACC---GACAGGCAC-ACAC-GUGCCAGCG----GGUAC|--GCACGCGAGGA
Ngregor  UGC-CGCGGG|CGUC--------------GUGC-------------GACG|--CG-CGCGAGGA
```

This works because in general it does not matter (usually) to the RNA what the bases in the helices are; what matters is that opposing bases are complementary so that they can form the helix. As a result, the secondary structure of an RNA is much more highly conserved than its sequence, because coevolution of bases that form base pairs maintains the secondary structure as the sequence changes. Variation in the length of the RNA is usually in hairpin lengthening or shortening. Therefore, it is usually possible to keep track of homologous parts of RNA structures even if the sequences are quite different.

In this type of alignment, the secondary structures of all of the RNAs are directly encoded in the alignment (Fig. 3.8). If residue *n* (e.g., 24 in Fig. 3.8) of any sequence pairs to residue *m* (e.g., 29), then so should the corresponding homologous residues in all sequences (Fig. 3.9).

**Figure 3.9** RNase P RNA helix "P3" in a variety of *Archaea*. The base pairs corresponding to the highlighted bases in the sequence alignment in Fig. 3.8 are highlighted. P3 is present in all archaeal (and bacterial) RNase P RNAs, but both the sequence and structure of this helix are highly variable. doi:10.1128/9781555818517.ch3.f3.9

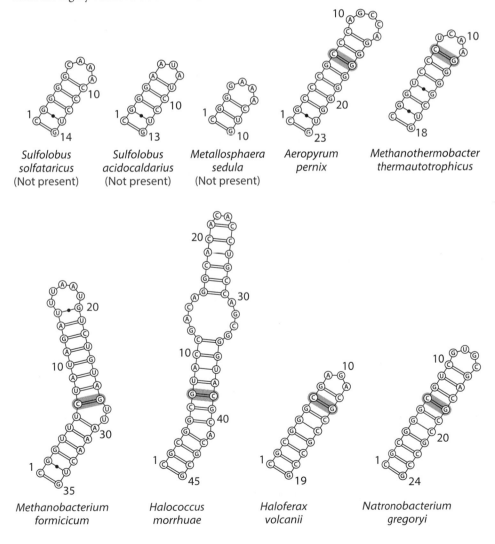

Given this type of alignment, a computer can readily compute any of the RNAs as secondary structures. Inversely, given a preexisting alignment and an RNA sequence with the same secondary structure, a computer algorithm can add this sequence correctly to the alignment. This is what infeRNAl does; it takes a sequence and tries to fold it into the correct secondary structure. If it can do so, it then threads this sequence into the alignment based on this structure.

## PROBLEMS

1. Align the following two sequences:

> Sequence A = GGCCUUCCGGCCACA and Sequence B = GCCCUUCCGGGCGCA

Now add the following sequence to this alignment:

> Sequence C = GUCCUUUGGACGC

Now add the following sequence to this alignment:

> Sequence D = GACCUUUCGGUCAC

2. Align the following sequences:

> Sequence A: GGAGCAGUCCGUGGAUC
> Sequence B: UAGGAGCAGCCGUGGAUC
> Sequence C: GGAGCAGGCCGCGGUACC

3. Align the following sequences:

> Sequence A: CUCGAGUUAACCCGGCACCCG
> Sequence B: GCUCGGGUUAACACGGACCCG
> Sequence C: UCGAGCCAACUCGGACCCG

4. Align the following sequences (note that these are in Fasta format, commonly used for the electronic transfer of sequence data):

```
>tRNA-A
GGGCUCAUAGCUCAGCGGUAGAGUGCCUCCUUUGCAAGGAGGAUGCCCUGGGUUCGAAUCCCAGUGAGUCCA
>tRNA-B
GGGCUCAUCGCUCAGCGGUAGAGUGCCUCCCUUGCAAGGAGGAUGCCCUGGGUUCGAAUCCCAGUGAGUCCA
>tRNA-C
GGGCUCGUAGCUCAGCGGGAGAGCGCCGCCUUUGCGAGGCGGAGGCCGCGGGUUCAAAUCCCGCCGAGUCCA
>tRNA-D
GGGCUCGUAGCUCAGCGGGAGAGCGCCGCCUUCGCGAGGCGGAGGCCGCGGGUUCAAAUCCCGCCGAGUCCA
>tRNA-E
GGGCCGGUAGCUCAGUCUGGUAGAGCGUCGCCUUGGCAUGGCGAAGGCCGGGGUUCAAAUCCCCACCGGU
```

**5.** Draw the secondary structures of the sequences in this alignment:

```
      A A              ( ( ( ( - - - ) ) ) ) - -
   G     A     Seq V   G G G G G A U A C U U C U A
     C - G     Seq W   A U G C U U C G G C A U U A
     U - A     Seq X   G U U U U U U - A A G C U A
     G - C     Seq Y   G G C - C U U G - G C C - A
     G - C U A  Seq Z  U U U U U U U U A A A A A A
```

**6.** Create an alignment of the following RNA structures:

```
                                  A C            U U
   U C           U C C         U      G       C      G
  U     G       G      A      G - C       G - C               U
  C - G         G - C         C - G       C - G       U     U
  U - G         C - G             A       U - A       U - A
  C - G         A - U         A - U       C - G       U - A
  C - G U      A G - C U A   A G - C U A  C - G U     C - G U A
```

**7.** Add the following Seq V RNA structure to the preexisting alignment:

```
  U C
 U     G     U               ( ( ( ( - - - ) ) ) ) - - ( ( ( - - - - ) ) ) - - -
 C - G     U     U    Seq W  C U U C G A G A G A A G - G C U C U U C G G A G U A U
 U - G     U - A      Seq X  C C C - U G C - - G G G U G C G C U U C - G C G U A -
 C - G     U - A      Seq Y  U C C - C U U C - G G G - A C U C C U U G G G G U G C
 C - G U A C - G U A  Seq Z  G A A G G A G A U U U U U G G G G U U U A C C C U A G
```

## Questions for thought

**1.** What are some DNA sequences that would not be useful for phylogenetic analysis? Why?

**2.** What are some other sequences that would be useful for phylogenetic analysis, and in what situations would they be useful?

**3.** How did people get large amounts of a specific DNA for sequencing before PCR was invented?

**4.** In an episode of *The X-Files* (an old TV show), FBI Agent Dana Scully sequences some extraterrestrial DNA and finds "missing bands" in the sequences that she interprets to correspond to bases that are unique to aliens (not found in Earthling DNA). Why is this not technically reasonable?

**5.** Given the variation of sequences in the context of the same secondary structure, how do scientists solve these secondary structures by comparative sequence analysis?

**6.** Mutations occur one at a time. How, then, could the base pairs in a helix change without disrupting the structure of the RNA? Does this explain (at least in part) why base-pair changes that keep the purines and pyrimidines in the same positions (transitions) are more common than those that switch them (transversions)?

**7.** Although RNA three-dimensional structures are scarce, there are hundreds of protein three-dimensional structures, determined by X-ray crystallography. Can you imagine a way to use these structures, analogous to the use of RNA secondary structures, to align protein sequences more meaningfully?

# 4

# Constructing a Phylogenetic Tree

In chapter 3, we covered the first three steps of a phylogenetic analysis, leaving the final step toward which the others build. The steps in a phylogenetic analysis are as follows:

1. Decide which gene and species to analyze (small-subunit ribosomal RNA [SSU rRNA])
2. Determine the gene sequences (polymerase chain reaction [PCR] and DNA sequencing, database "mining")
3. Identify homologous residues (sequence alignment)
4. Perform the phylogenetic analysis

The most common type of phylogenetic analysis is tree construction. A tree is nothing more than a graph representing the similarity relationships between the sequences in an alignment. This is why we'll be going through this process in such detail, to show that tree construction is not rocket science but involves straightforward mathematical transformations of sequence data.

There are several methods for building trees. In this chapter, we cover the neighbor-joining method in some detail as an example, because it is conceptually straightforward and commonly used. In the next chapter, we briefly cover some other approaches.

doi:10.1128/9781555818517.ch4

# Tree construction: the neighbor-joining method

Tree construction starts with an alignment. Neighbor joining is a distance matrix method, meaning that the alignment is first reduced to a table of evolutionary distances, a distance matrix. The distance matrix cannot be generated directly from the alignment, however, because actual evolutionary distance cannot be directly measured. Instead, the alignment is reduced to a table of observed (measurable) similarity, the similarity matrix. The distance matrix is calculated from the similarity matrix, and then the tree is generated from the distance matrix.

## *Generating a similarity matrix*

The similarity matrix is just a table of fractional similarities, for example, in this alignment of six sequences with 20 positions.

```
Seq A   G C C A U G C C G A C G A U U G G U C C
Seq B   G C C A U G C C A A C G A U U A G U C C
Seq C   G C C A U G C U A G C G G U U G G U U C
Seq D   G C C A U G C U A A C G A C U G G U U C
Seq E   G G C A C G C U A A U G A U U G G U C C
Seq F   G C G G C G C U A A C G A U U G A U C C
```

Just count the fraction of identical bases in every pair of sequences in the alignment.

```
Seq A   G C C A U G C C G A C G A U U G G U C C
        | | | | | | | | X | | | | | | | X | | |   → 18/20 = 0.90 similarity
Seq B   G C C A U G C C A A C G A U U A G U C C
```

The similarity values for all pairs of sequences are calculated in the same way and assembled into a table:

```
        Similarity matrix:
          A      B      C      D      E      F
A        --     --     --     --     --     --
B       0.90    --     --     --     --     --
C       0.75   0.75    --     --     --     --
D       0.80   0.80   0.85    --     --     --
E       0.75   0.75   0.70   0.75    --     --
F       0.70   0.70   0.65   0.70   0.75    --
```

In this example, sequences A and B are 0.90 (90%) similar, A and C are 0.75 similar, B and C are 0.75 similar, and so forth. Note that values on the diagonal (A:A, B:B, . . .) do not need to be calculated; they are always 1. Likewise, there is no reason to calculate both above and below the diagonal; the value for X:Y is the same as that for Y:X, so the second calculation would be redundant.

## Converting a similarity matrix into an evolutionary distance matrix

Next is the estimation of evolutionary distances from their sequence similarity. You might think that the distance would just be 1 − similarity (i.e., "difference"), and you would be right except that the number of differences you count between any two sequences misses some of the changes that probably have occurred between them. More than one evolutionary change at a single position (e.g., A to G to U, or A to G in one sequence and the same A to U in another) counts as only one difference between the two sequences, and in the case of reversion or convergence it counts as no change at all (e.g., A to G to A, or A to G in one organism and the same A to G in another). As a result, the observed similarity between two sequences underestimates the evolutionary distance that separates them.

One common way to estimate evolutionary distances from similarity is the Jukes and Cantor method, which uses the following equation:

$$\text{Evolutionary distance} = -3/4 \ln[1 - 4(1 - \text{similarity})/3]$$

As shown graphically in Fig. 4.1, similarity and distance are very closely related initially (e.g., 0.90 similarity ≈ 0.10 distance) but level off to 0.25 similarity, where evolutionary distance is infinite. This makes sense; for two sequences that are very similar, the probable frequency of more than one change at a single site is low, requiring only a small correction, whereas two sequences that have changed beyond all recognition (infinite evolutionary distance) are still approximately 25% similar just because there are only four bases and so approximately one of the four will match entirely by chance.

**Figure 4.1** The Jukes and Cantor equation plotted as observed sequence similarity (from the similarity matrix) versus estimated evolutionary distance.
doi:10.1128/9781555818517.ch4.f4.1

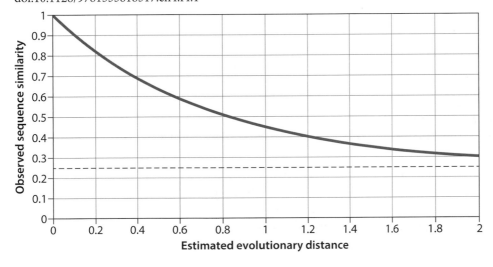

To convert a similarity matrix to a distance matrix, just convert each value in the similarity matrix to evolutionary distance using either the graph or the equation. In our example:

```
    Similarity matrix:                ---> Distance matrix:
      A     B     C     D     E     F      A     B     C     D     E     F
A   ----  ----  ----  ----  ----  ----   ----  ----  ----  ----  ----  ----
B   0.90  ----  ----  ----  ----  ----   0.11  ----  ----  ----  ----  ----
C   0.70  0.70  ----  ----  ----  ----   0.30  0.30  ----  ----  ----  ----
D   0.75  0.75  0.85  ----  ----  ----   0.23  0.23  0.17  ----  ----  ----
E   0.40  0.40  0.40  0.45  ----  ----   0.30  0.30  0.38  0.30  ----  ----
F   0.45  0.45  0.45  0.50  0.75  ----   0.38  0.38  0.47  0.38  0.30  ----
```

## Generating a tree from a distance matrix

In the neighbor-joining method, the structure of the tree is determined first and then the branch lengths are fit to this skeleton.

### Solving the tree structure

The tree starts out with a single internal node and a branch out to each sequence: an *n*-pointed star, where *n* is the number of sequences in the alignment. The pair of sequences with the smallest evolutionary distance separating them is joined onto a single branch (i.e., the neighbors are joined, hence the name of the method), and then the process is repeated after merging these two sequences in the distance matrix by averaging their distances from every other sequence in the matrix.

Using our distance matrix, the tree starts out like this (remember that we are sorting out the structure of the tree, not yet the branch lengths).

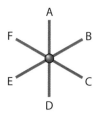

The closest neighbors in the distance matrix are A and B (0.11 evolutionary distance), so these branches are joined:

The distances to A and B from each of the other sequences are then averaged to reduce the distance matrix:

Starting matrix:                    ---> Reduced matrix:

|   | A    | B    | C    | D    | E    | F    |
|---|------|------|------|------|------|------|
| A | ---- | ---- | ---- | ---- | ---- | ---- |
| B | 0.11 | ---- | ---- | ---- | ---- | ---- |
| C | 0.30 | 0.30 | ---- | ---- | ---- | ---- |
| D | 0.23 | 0.23 | 0.17 | ---- | ---- | ---- |
| E | 0.30 | 0.30 | 0.38 | 0.30 | ---- | ---- |
| F | 0.38 | 0.38 | 0.47 | 0.38 | 0.30 | ---- |

|     | A/B  | C    | D    | E    | F    |
|-----|------|------|------|------|------|
| A/B | ---- | ---- | ---- | ---- | ---- |
| C   | 0.30 | ---- | ---- | ---- | ---- |
| D   | 0.23 | 0.17 | ---- | ---- | ---- |
| E   | 0.30 | 0.38 | 0.30 | ---- | ---- |
| F   | 0.38 | 0.47 | 0.38 | 0.30 | ---- |

In this case, the averages are trivial; the average of 0.30 and 0.30 is, of course, 0.30, and so forth. This is not usually the case. Then the process is repeated. The closest neighbors in the reduced matrix are D with C (0.17):

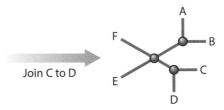

Join C to D

Once again, the distance matrix is reduced by averaging. But be sure to average from the original distance matrix, not the previously reduced matrix:

Starting matrix:                    ---> Reduced matrix:

|   | A    | B    | C    | D    | E    | F    |
|---|------|------|------|------|------|------|
| A | ---- | ---- | ---- | ---- | ---- | ---- |
| B | 0.11 | ---- | ---- | ---- | ---- | ---- |
| C | 0.30 | 0.30 | ---- | ---- | ---- | ---- |
| D | 0.23 | 0.23 | 0.17 | ---- | ---- | ---- |
| E | 0.30 | 0.30 | 0.38 | 0.30 | ---- | ---- |
| F | 0.38 | 0.38 | 0.47 | 0.38 | 0.30 | ---- |

|     | A/B   | C/D   | E    | F    |
|-----|-------|-------|------|------|
| A/B | ----  | ----  | ---- | ---- |
| C/D | 0.265 | ----  | ---- | ---- |
| E   | 0.30  | 0.34  | ---- | ---- |
| F   | 0.38  | 0.425 | 0.30 | ---- |

Note that the value listed for A/B with C/D (0.265) is the average of four values in the original matrix: A to C (0.30), A to D (0.23), B to C (0.30), and B to D (0.23). If averaged AB/C (0.30) and AB/D (0.23) were averaged from the reduced matrix, the same number would be obtained in this case, but usually this is a more complex average and the numbers do not come out the same.

Once again, the smallest number in the matrix represents the nearest neighbors, in this case A/B with C/D (0.265), so these two branches are joined:

Join AB to CD

Each node on this tree has only three branches connecting to it; all of the nodes are completely resolved. This means that the structure of the tree has been determined. If there were more sequences, it would be necessary to reduce the matrix (joining A/B with C/D) and repeat the process until all of the nodes were resolved. The internal nodes have been arbitrarily labeled *w*, *x*, *y*, and *z* for reference when sorting out branch lengths (below).

## Determining branch length

The next step is to determine the lengths of the branches on this tree. Basically, this is done by going through each node and finding where along the branch it is by figuring out the average difference in length along each of two branches. By choosing various sets of three sequences in a tree, the branch lengths can be sorted out just like a puzzle.

### Branches *w*/A and *w*/B (*w* is the common node between A and B)

In our example, the distance between A and B is 0.11, and so the lengths of the two branches connecting them (A/*w* and *w*/B) must add up to 0.11. But where along this branch is the node (*w*)? If you look at the distance from any other sequence (C, D, or F) to A and to B, it is always the same. For example, C/A is 0.30 and so is C/B. This means that node *w* must be midway between A and B; each branch, then, has a length of 0.055. For example, with C used as reference:

### Branches *x*/C and *x*/D

C and D are also simple neighbors, so we can easily solve these two connecting branches as well. The distance between C and D is 0.17. However, the distance to either C or D from elsewhere in the tree differs; this means that the node connecting C and D is *not* equidistant between them. In fact, this difference in distance between C and D varies a bit depending on which reference we use. These differences occur because we can only estimate evolutionary distance. As a result, we use the average (although most computer algorithms would use a least-squares average):

A/C = 0.30

A/D = 0.23     so A/C − A/D = 0.30 − 0.23 = 0.07

Therefore, 0.07 is the difference in branch length between A/C and A/D

B/C = 0.30

B/D = 0.23      B/C − B/D = 0.30 − 0.23 = 0.07

E/C = 0.38

E/D = 0.30      E/C − E/D = 0.08

F/C − F/D = 0.47 − 0.38 = 0.09

Average (0.07, 0.07, 0.08, 0.09) = 0.08

This value, 0.08, is the average amount by which $x$/C is longer than $x$/D. Because the total length of C/D (think of this as C/$x$/D) is 0.170 and $x$/C is 0.08 longer than $x$/D, then:

$x$/D = (0.170 − 0.08)/2 = 0.045

$x$/C is the same 0.045 plus the 0.08 by which we already determined it was longer = 0.125:

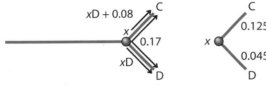

x/C is 0.08 greater than x/D

### Branches z/E and z/F

The distance between E and F can be calculated in the same way:

A/E − A/F = 0.30 − 0.38 = −0.08

B/E − B/F = 0.30 − 0.38 = −0.08

C/E − C/F = 0.38 − 0.47 = −0.09

D/E − D/F = 0.30 − 0.38 = −0.08

Average = −0.08

The length of E/$z$/F (E/F) is 0.30, and the average difference between anywhere else in the tree and E or F is −0.08. The negative sign means that the branch to F is longer than the branch to E. So $z$/E = (0.30 − 0.08)/2 = 0.11, and $z$/F − 0.11 + 0.08 = 0.1.

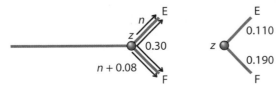

z/F is 0.08 greater than z/E

### Internal branches *w/x*, *x/y*, and *x/z*

The tree so far is:

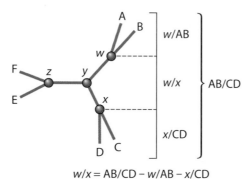

To solve the lengths of internal branches, the same process is used for collections of branches and then the average length outside the internal nodes is subtracted. To determine the lengths of *w/y*, *x/y*, and *z/y*, we need to determine the total lengths of *w/x*, *x/z*, and *z/w* and then figure out where along these to place *y*.

To solve for the length of *w/x*, the average distance between AB and CD in the reduced matrix (0.265) is used, and then the average *w*/A and *w*/B distance (AB/2 = 0.11/2 = 0.055) and the average *x*/C and *x*/D distance (CD/2 = 0.17/2 = 0.085) are subtracted to leave the *w/x* distance: $0.265 - 0.055 - 0.085 = 0.125$.

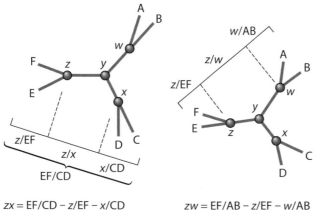

*w/x = AB/CD − w/AB − x/CD*

To solve the length of *x/z*, the average CD/EF distance is calculated (this was not done in the reduced matrices, but it is the same process; it is 0.38) and then *x*/CD (CD/2 = 0.085) and *x*/EF (EF/2 = 0.30/2 = 0.150) are subtracted to give *x/z* = 0.145. Likewise, *z/w* = EF/AB (0.34) − *w*/AB (0.055) − *z*/EF (0.15) = 0.135.

*zx = EF/CD − z/EF − x/CD*          *zw = EF/AB − z/EF − w/AB*

Thus, $w/x = 0.125$, $x/z = 0.145$, and $z/w = 0.135$.

To determine where on any of these line segments node $y$ is, we need to pick any of the line segments and calculate the difference to each end of this segment from the other branch. We do this just like we did for solving the lengths of the branches between A/B, C/D, or E/F. For example, for branch $w/x$:

$z/w - z/x = 0.135 - 0.145 = -0.010$ ($z/w$ is shorter than $z/x$ by 0.010)

$y/w = (w/x - 0.10)/2 = (0.125 - 0.010)/2 = 0.058$

$y/x = y/w + 0.010 = 0.068$

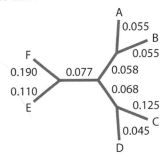

Now we know all of the line segment lengths but one, $y/z$. We can get this by subtracting all of the known lengths of $w/y$ from $w/z$, or all the known lengths of $x/y$ from $x/z$; either way we get the same answer:

$y/z = w/z - w/y = 0.135 - 0.058 = 0.077$

$y/z = x/z - y/x = 0.045 - 0.068 = 0.077$

So, our final tree looks like this:

or like this with the branches drawn to scale:

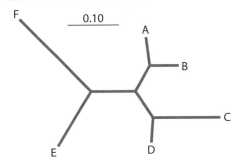

Notice that in the final tree, the evolutionary distance between any two sequences is approximately equal (any difference is due to the need to average branch lengths) to the sum of the lengths of the line segments joining those two sequences; in other words, the tree is additive. This should be true on any quantitative phylogenetic tree.

The neighbor-joining method is very fast and so can be used on trees with much larger numbers of sequences than can other methods, which are discussed in chapter 5.

## Rooting a tree with an outgroup

Whenever possible, an outgroup sequence should be included in the analysis; an outgroup is a sequence that is known to be outside of the group you are interested in treeing. Only by including an outgroup can the root of a tree be located. For example, when building trees from mammalian sequences, the sequence from a reptile might be included as an outgroup. Outgroups provide the root to the rest of the tree; although no tree generated by these methods has a real root, if other information shows that one (or a set) of the sequences is unrelated to the rest, wherever that branch connects to the rest of the tree defines the root (common ancestor) of that portion of the tree. In our example tree above, sequences A to D might be mammalian whereas E and F might be reptilian. If the tree included only mammalian sequences, it would be impossible to know where the root is, but the inclusion of an outgroup provides that information.

## Molecular phylogenetic trees depict the relationships between gene sequences

The final step in the tree-building process is the leap of faith that the tree depicting the relationships between the sequence similarities in an alignment also depicts the evolutionary history of the organisms from which the sequences were derived. You can go wrong for many reasons at this step.

The most common reason is the choice of sequence for use in the analysis, the first step of the whole process. What would happen if, for example, you made a tree of mammals by using globin genes, but used alpha globin sequences from some species and beta globin genes from others? What might happen if, in an analysis of plants, you used nuclear SSU rRNA sequences from some plants, chloroplast SSU rRNAs from others, and mitochondrial sequences from still others? What if, in an analysis of bacterial species, you used a gene like *amp*, encoding penicillin resistance, which clearly moves readily from species to species? The trees you would build with these sequences might be perfectly valid, accurately representing the relationships between the genes used, but they would not represent the relationships between the organisms.

So, if a tree just fundamentally does not look right, check the alignment first and then the quality of the sequences, but then think about the sequences used and how a tree of these sequences might not reflect the relationships between the organisms.

## PROBLEMS

1. Construct a similarity matrix from this alignment:

```
Seq A   G A U C U U U G G A U C
Seq B   A A U C U C U G G A U U
Seq C   C A U C U U U U G A U G
Seq D   A A U C U U U G G A U U
Seq E   C A U C U C U G A A U G
```

2. Given the Jukes and Cantor conversion graph in Fig. 4.1, convert the similarity matrix from question 1 to a distance matrix.

3. Given this distance matrix, draw a tree relating these sequences and label the lengths of the branches of the tree:

|       | Seq A | Seq B | Seq C | Seq D |
|-------|-------|-------|-------|-------|
| Seq A | –     | –     | –     | –     |
| Seq B | 0.2   | –     | –     | –     |
| Seq C | 0.6   | 0.6   | –     | –     |
| Seq D | 0.9   | 0.9   | 0.9   | –     |

4. Given this distance matrix, draw the structure of a tree relating the sequences using the neighbor-joining method, and label the lengths of the branches:

|       | Seq A | Seq B | Seq C | Seq D | Seq E | Seq F |
|-------|-------|-------|-------|-------|-------|-------|
| Seq A | –     | –     | –     | –     | –     | –     |
| Seq B | 0.2   | –     | –     | –     | –     | –     |
| Seq C | 0.7   | 0.7   | –     | –     | –     | –     |
| Seq D | 0.6   | 0.6   | 0.5   | –     | –     | –     |
| Seq E | 0.7   | 0.7   | 0.8   | 0.7   | –     | –     |
| Seq F | 0.6   | 0.6   | 0.7   | 0.6   | 0.3   | –     |

5. Generate a tree from this alignment using the neighbor-joining method, with approximate branch lengths.

| | |
|---|---|
| *Thiobacillus ferrooxidans* | GAAUUCCCGGGAG-GGGCCAGGCGACCCCCGAAUUCCCGG |
| *Escherichia coli* | GAAUUCCCGGAAGCAGACCAGACAGUCGCCGAAUUCCCGG |
| *Serratia marcescens* | GAAUUCCCGGAAGUAGACCAGACAGUCACCGAAUUCCCGG |
| *Chromatium vinosum* | GAAUUCCCGGGAG-GGGCCAGACAGUCCCUGAAUUCCCG- |

## How to read a phylogenetic tree

A phylogenetic tree is a representation of the evolutionary/genealogical relationships among a collection of organisms (or molecular sequences). There are many different ways to draw these trees, but they share a common set of features: terminal nodes, branches connected by internal nodes, perhaps a root, and a scale (Fig. 4.2).

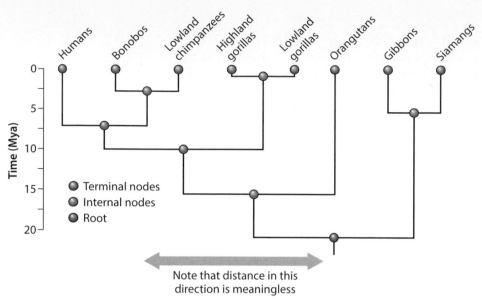

**Figure 4.2** An example phylogenetic tree of the relationships between great apes.
doi:10.1128/9781555818517.ch4.f4.2

## Scale

The scale typically is either time or evolutionary distance. Trees with a time scale are based on some form of physical data, such as a fossil record, that provide dating information. If the scale is time, all modern organisms should obviously be shown at the same part of the scale (the present). More often, the scale is evolutionary distance, i.e., some measure of change in the organisms (or molecules). Because the extent of divergence is usually different in various parts of the tree, this is depicted by the lengths of the branches and a scale bar.

## Terminal nodes

Terminal nodes are the ends of the branches of the evolutionary tree; typically these are the modern organisms (or molecules) being compared, but in some cases they are the ends of evolutionary branches that became extinct or provided museum samples.

## Internal nodes

Internal nodes represent the last common ancestors of all of the organisms (or molecules) bound by this node.

## Root

The root is the "base" of the tree—the last common ancestor of all of the organisms (or molecules) in the tree. It is not always possible to identify the root of a tree; typically, this requires either physical data (e.g., a fossil record) or data about organisms outside the part of the tree shown, i.e., an outgroup.

## Branches

Branches are the connections between nodes in the tree. They represent the evolutionary pathways between common ancestors (internal nodes) and modern organisms (terminal nodes). The lengths of these branches are defined by the scale; each branch represents a certain length of historical time if time is the scale used in the tree, or a certain amount of evolutionary change if evolutionary distance is the scale used in the tree.

## Tree representations

There are a variety of ways of drawing the same tree. Figure 4.3 shows two trees of the same organisms. The scales (time) in these two trees are horizontal rather than vertical, and the branches are simple diagonal lines connecting nodes, but the information in these trees is the same as in Fig. 4.2. These trees are phenograms; the scale is read from an axis, in this case the horizontal axis. Also notice that the order of the terminal nodes is irrelevant: only the topology of the tree and the lengths of the connections count. The positions of the branches and nodes can be switched around at will as long as the nodes and their connections are not broken and remain true to the scale.

Figure 4.4 shows trees in which the scale, in this case phylogenetic distance (the extent of divergence of some sequence), is measured along the lengths of the branches. Notice that the lengths of the branches are uneven, because the rates of evolutionary change in these sequences are not constant. The tree on the right is rootless; no root is shown. In order to root a tree (as in the tree to the left), data from the fossil record or other physical information is needed, or in a molecularly based tree, an outgroup must be included in the tree to place the root. For example, in this tree of apes, the tree could be rooted with data from an Old World monkey.

These trees are dendrograms; the scale is measured along the branch lengths rather than horizontally or vertically (Fig. 4.5). The evolutionary distance between any two organisms is the total of the lengths of all of the branches that

**Figure 4.3** Two different representations of the same phenogram of phylogenetic relationships among great apes. doi:10.1128/9781555818517.ch4.f4.3

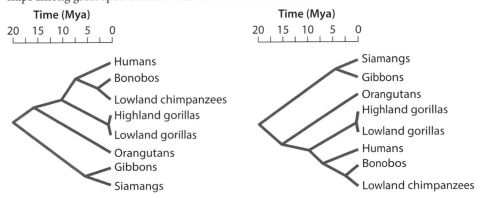

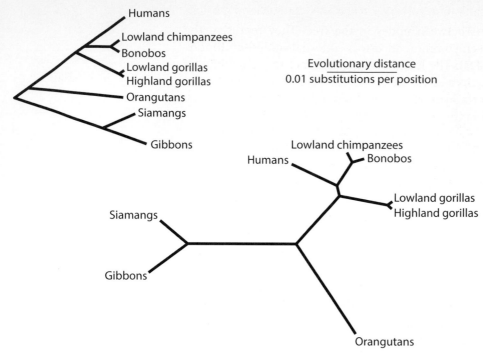

**Figure 4.4** Two different dendrogram representations of the same phylogenetic relationships. doi:10.1128/9781555818517.ch4.f4.4

**Figure 4.5** Measuring phylogenetic distances in dendrograms.
doi:10.1128/9781555818517.ch4.f4.5

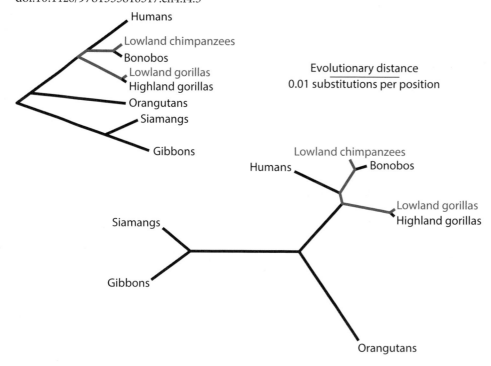

connect them. For example, the evolutionary distance between lowland gorillas and lowland chimps would be about 0.0143.

A phenogram can also be used with an evolutionary distance scale. In this case, remember that the scale (evolutionary distance) is measured only in horizontal (or vertical) distance (Fig. 4.6).

Because evolutionary rates are not constant, some organisms have changed more than others since their common ancestor. In the example above, the sequences of humans have changed more than those of lowland chimps since their last common ancestor. Lowland chimps, then, are primitive relative to humans with respect to this sequence. Humans are more highly derived than chimps, again with respect to this sequence. If the traits of an organism overall are more similar to the ancestor than in the other members of that group, that organism is thought of as a primitive organism. This is very useful information, but it can be dangerous—in most cases, the traits of an organism are not evolving at similar or constant rates, and so an organism might be primitive in most traits but highly derived in others. Sharks, for example, are primitive fish with respect to many traits (e.g., cartilaginous skeletons and placoid scales) but are highly derived with respect to others (immunologically, and their electrosensory system). The danger is a tendency to confuse generally primitive organisms with ancestors. For example, chimps are morphologically more primitive than humans, but chimps are not ancestors of humans; the common ancestor of humans and chimps was not a chimp. Chimps are modern organisms! They just have more morphological similarity to the common ancestor of humans and chimps than do humans.

**Figure 4.6** Measuring phylogenetic distances in phenograms.
doi:10.1128/9781555818517.ch4.f4.6

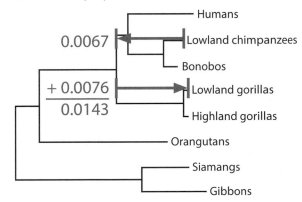

## PROBLEMS

**1.** Answer the following questions on this tree:

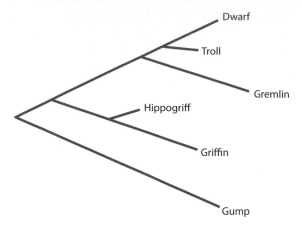

**a.** Which are the two most closely related species?
**b.** Which is the most primitive of these species?
**c.** Which of these is probably the outgroup?
**d.** Which is most distantly related to the Gremlin other than the outgroup species?
**e.** Circle the last common ancestor of the Hippogriff and the Gremlin.
**f.** Redraw this tree to scale as a phenogram.

**2.** Answer the following questions about this tree:

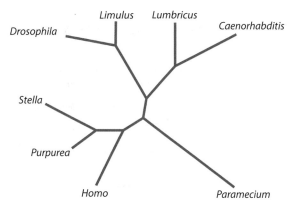

**a.** Which are the two most closely related species?
**b.** Which of these is probably the outgroup?
**c.** Based on the outgroup from question b, circle the last common ancestor of the remaining creatures.
**d.** Now circle the root of the tree (again, assuming the same outgroup).
**e.** Circle the two main groups in this tree (other than the outgroup).
**f.** Which is most distantly related to the *Drosophila* other than the outgroup species?
**g.** Which species in this tree is most closely related to *Homo*?
**h.** Redraw this tree as a phenogram.

3. Answer the following questions about this tree:

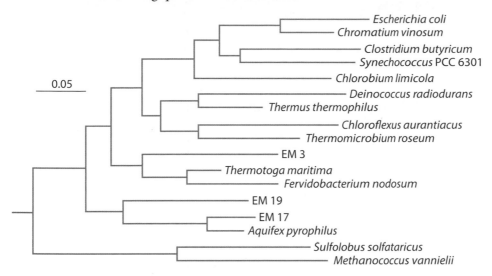

a. Which organism is the closest relative of *Chromatium vinosum*?
b. Which organism(s) is the outgroup?
c. Circle the root of the tree based on this outgroup?
d. What is the evolutionary distance between *Thermus thermophilus* and *Thermomicrobium roseum*?
e. What is the evolutionary distance between *Chloroflexus aurantiacus* and *Methanococcus vannielii*?
f. Which organism(s) is/are most closely related to *Chloroflexus aurantiacus* and *Thermomicrobium roseum*?
g. EM3 is an unknown organism. What is its closest relative on this tree?
h. Excluding the outgroup, which is the most primitive organism (or at least sequence) in this tree?
i. Excluding the outgroup, which is the most highly evolved organism (or sequence) in this tree?
j. Circle the last common ancestor of *Chromatium vinosum* and *Deinococcus radiodurans*.
k. Redraw this tree as a dendrogram.

## Example analysis

In about 1990, an organism designated ES-2 was isolated from a deep-sea hydrothermal vent sample. ES-2 grows heterotrophically at 65°C. A lipid analysis of the isolate was unusual, showing that it contained a number of apparently branched lipids as well as fatty acids. Electron microscopy and standard microbiological tests were not helpful in identifying the organism.

Phylogenetic analysis of the organism was performed essentially as above. DNA was isolated from cells, and the SSU rRNA was amplified by PCR using primers near both the 5′ and 3′ ends of the gene. The amplified DNA was cloned and sequenced (Fig. 4.7), and the secondary structure of the encoded RNA was determined for use in the alignment process (Fig. 4.8).

```
LOCUS        Eub.tmarin      893 bp       RNA              12-FEB-1993
DEFINITION   Eubacterium thermomarinus.
 ACCESSION   diversity:1834:811629716
   AUTHORS   Pledger,R.J., Brown,J.W., Hedrick,D.R., Pace,N.R., White,D.C.
REFERENCE    1  TITLE Preliminary characterization of a rod-shaped thermophilic,
               auth line 2:and Baross,J.A.
               title line 2:anaerobic eubacterium isolated from a submarine
               title line 3:hydrothermal vent environment
BASE COUNT      345 a     335 c     465 g     263 t    1438 others
ORIGIN
        1  GCCUAACACA UGCAAGUCGA GCGGGUAGGU GCUUUAAUGA ACCUUCGGGG
       51  GAUUUAAAGU ACUGAAAGCG GCGGACGGGU GAGUAACGCG UGGGGAACCU
      101  GCCCUAUGCA GGGGGAUAGC CUCGGGAAAC CGGGAUUAAU GGCCCAUAAC
      151  ACUUAUGGAC CGCAUGGUGC AUAAGUCAAA GCGUUUAGCG GCAUAGGAUG
      201  GCCCCGCGUC CCAUUAGCUA GUUGGUGAGG UAACGGCUCA CCAAGGCGAC
      251  GAUGGGUAGC CGGCCUGAGA GGGUGGACGG CCACACUGGG CCUGAGACAC
      301  GGCCCAGACU CCUACGGGAG GCAGCAGUGG GGAAUAUUGC ACAAUGGGCG
      351  AAGCCUGAUG CAGCGACGCC GCGUGAGCGA AGAAGGCCUU CGGGUCGUAA
      401  AGCUCUGUCC UAAGGGAAGA AAGAUGUCGG UACCUUAGGA GGAAGCCCCG
      451  GCUAACUACG UGCCAGCAGC CGCGGUAAUA CGUAGGGGGC GAGCGUUGUC
      501  CGGAAUCACU GGGCGUAAAG GGAGCGUAGG CGGCCCUGCA AGUCAGGUGU
      551  GAAAGGCAUC GGCUCAACCG AUGUGAGCAC UUGAAACUGU AGGGCUUGAG
      601  UGCAGGAGAG GAGAGCGGAA UUCCUAGUGU AGCGGUGAAA UGCGUAGAUA
      651  UUAGGAGGAA CACCAGUGGC GAAGGCGGCU CUCUGGACUG UAACUGACGC
      701  UGAGGCACGA AAGCGUGGGG AGCGAACAGG AUUAGAUACC CUGGUAGUCC
      751  ACGCCGUAAA CGAUGAGUGC UAGGUGUUGG GGGGUAACUC CUUCAGUGCC
      801  GCAGUUAACA CAUUAAGCAC UCCGCCUGGG GAGUACGGUC GCAAGACUGA
      851  AACUCAAAGG AAUUGACGGG GACCCGCACA AGCAGCGGAG CAUGUGGUUU
      901  AAUUCGAAGC AACGCGAAGA ACCUUACCAA GACUUGACAU CCUCCGGACG
      951  GCUCUGGAGA CAGGGCUUUC CCUUCGGGUA CUGGAGAGGC AGGUGGUGCA
     1001  UGGUUGUCGU CAGCUCGUGU CGUGAGAUGU UGGGUUAAGU CCCGCAACGA
     1051  GCGCAACCCU UGCCUUUAGU UGCCAUCAGG UAAAGCUGGG CACUCUAUGA
     1101  GGACUGCCGG CUAAAAGUCG GAGGAAGGUG GGGAUGACGU CAAAUCAUCA
     1151  UGCCCCUUAU GUCUUGGGCU ACACACGUGC UACAAUGGCC GGUACAGAGG
     1201  GCAGCGAACC CGUAAGGGGG AGCGAAUCCC AAAAAGCCGG UCCCAGUUCG
     1251  GAUUGUGGGC UGCAACUCGC CCACAUGAAG CUGGAGUUGC UAGUAAUCGC
     1301  GAAUCAGAAU GUCGCGGUGA UGCGUUCCCG GGUCUUGUAC ACACCGCCCG
     1351  UCACUCCAUG AGAGUCGGCA ACACCCGAAG CCAGUGGGCU AACCGAAGGA
     1401  GGCAGCUG
//
```

**Figure 4.7** Sequence of the ES-2 SSU rRNA, in GenBank format.
doi:10.1128/9781555818517.ch4.f4.7

   With the sequence in hand, the next step in the process was to align the
sequence with the SSU rRNA database. This was originally done by hand in a
local database, but these days it can be done automatically at the Ribosomal
Database Project (RDP), using an algorithm called "infeRNAl" (from INFER
RNA ALignment). After this, the next step was to calculate phylogenetic trees;
the RDP uses a weighted neighbor-joining method. First, a "representative" tree
was generated using representative sequences from each major phylogenetic
group of bacteria to see which group ES-2 seems to belong to (Fig. 4.9).

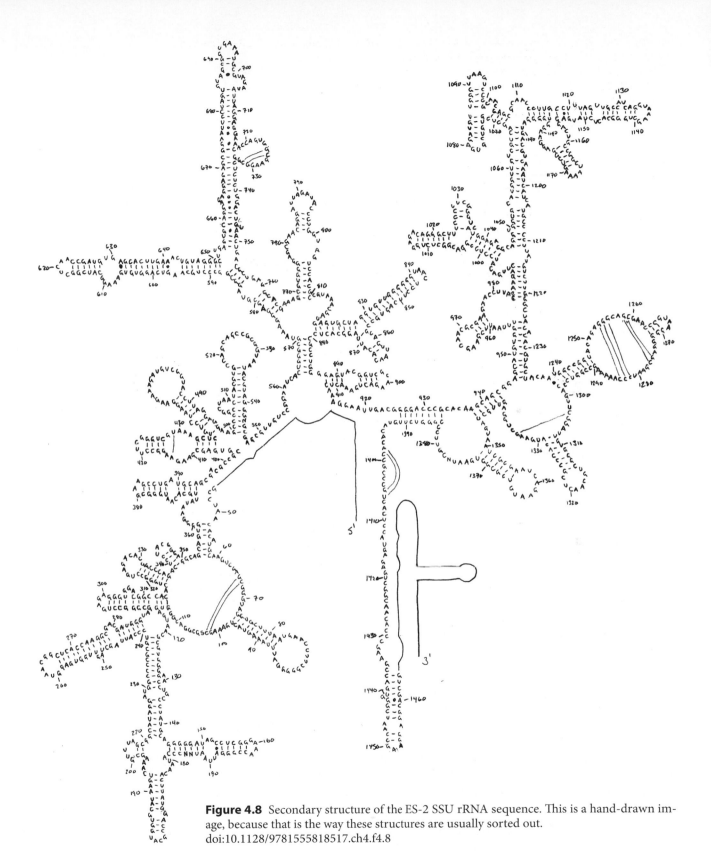

**Figure 4.8** Secondary structure of the ES-2 SSU rRNA sequence. This is a hand-drawn image, because that is the way these structures are usually sorted out.
doi:10.1128/9781555818517.ch4.f4.8

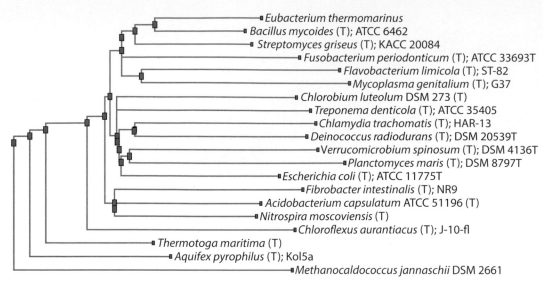

**Figure 4.9** Phylum-scale tree including ES-2, generated using the RDP II website. doi:10.1128/9781555818517.ch4.f4.9

ES-2 is clearly most closely related to the representative of the phylum Firmicutes (represented by *Bacillus mycoides*). So, the next step was to generate another tree by using representative sequences from the Firmicutes (Fig. 4.10). At the time this analysis was performed, only a few such sequences were available. Now there are thousands of Firmicute sequences in the database.

This tree shows that ES-2 is a member of the *Clostridium/Eubacterium* group of the Firmicutes. So a final tree with representatives of this group, and some especially close relatives identified using BLAST was generated (Fig. 4.11).

**Figure 4.10** Tree of representative Firmicutes, including ES-2, generated using the RDP II website. doi:10.1128/9781555818517.ch4.f4.10

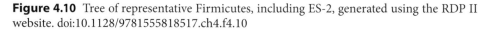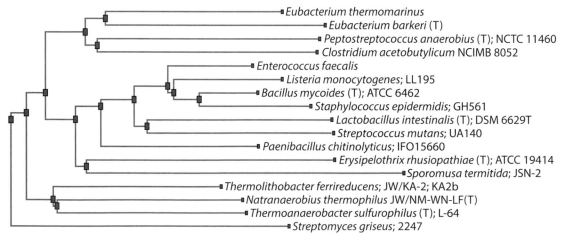

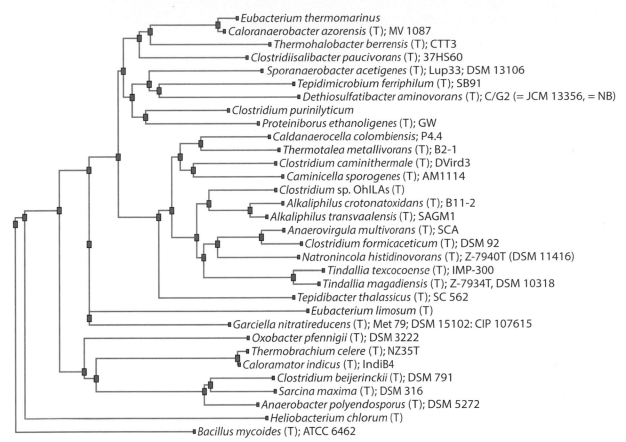

**Figure 4.11** Fine-scale phylogenetic tree of ES-2, generated using the RDP II website. doi:10.1128/9781555818517.ch4.f4.11

## Interpretation

ES-2 is a member of the *Clostridium/Eubacterium* group of the Firmicutes, and is particularly close (probably deserves to be in the same genus or even species) to *Caloranaerobacter azorensis*, a more recent deep-sea vent isolate. At the time of this analysis (but no longer), organisms from this group that produce spores were considered to be in the genus *Clostridium*, while those that did not produce spores were classified as *Eubacterium*. Because of the environment from which it was isolated, the new species was named *Eubacterium thermomarinus*.

## What good was this information?

This information was very useful; we were interested in identifying organisms that were very distinct from those already in hand for analysis of RNase P RNA structure. Given that this was basically a member of the genus *Clostridium*, from which we already had several sequences, this organism became a low priority and was not pursued further. Knowing the phylotype of ES-2 prevented us from spending a fair amount of effort for what would have been trivial gain.

## Questions for thought

**1.** Which organism would you choose for an outgroup for an rRNA tree of mammals? Does it matter which nonmammal you choose? Why? What might you choose as an outgroup for an rRNA tree of Bacteria? What about for a "universal tree" containing sequences of all kinds of organisms?

**2.** What would a tree of some animals look like if constructed from globin genes where some of the sequences were alpha globins and others were beta globins? What if some of these were adult alpha or beta globins and others were juvenile or fetal globins?

**3.** What would a tree (no pun intended) of plants look like if some of the sequences (rRNAs) were accidentally taken from the chloroplast instead of the nucleus? What if all of the sequences were from the chloroplasts?

**4.** On the initial "phylum-scale" representative tree of Bacteria shown above, can you show where we might have hoped ES-2 would be?

**5.** Which properties can you predict for ES-2 based on its phylotype? Which properties can you not predict?

# 5 Tree Construction Complexities

The process of generating phylogenetic trees as described in chapter 4 is straightforward. This is a gross simplification. Phylogenetic analysis is an entire scientific area of study, and the material that has been presented is very highly simplified. In this chapter we touch (just touch) on some of the complexities.

However, before thinking about more refined substitution models, treeing algorithms, or alternative sequences, keep one thing in mind: the most important thing by far that is needed to get a good, robust tree is to start with a good alignment. The simple and fast neighbor-joining method, using the Jukes and Cantor substitution model, usually gives perfectly usable trees if given a good small-subunit ribosomal RNA alignment to work with. Combined with bootstrapping (see below), this method is probably used more than any other for the creation of published trees.

## Substitution models

In chapter 4 we talked about the Jukes and Cantor method to estimate evolutionary distance from sequence similarity. This is a simple method, but there are several other more sophisticated methods. The Jukes and Cantor method and other methods for estimating evolutionary distance amount to an attempt to describe how sequences change. In other words, they are mathematical models of the process of evolution of these sequences, and they are therefore usually called "substitution models." The choice of an appropriate substitution model is critical and often underappreciated.

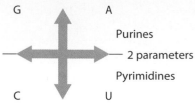

**Figure 5.1** Two-parameter substitution models distinguish between transitions and transversions and score them differently. Each parameter is represented by an arrow. The values of these parameters can be predetermined (typically 1.0 for transversions and 0.5 for transitions) or determined by presifting the alignments to count the observed ratios of differences. doi:10.1128/9781555818517.ch5.f5.1

In the Jukes and Cantor method, any difference in two sequences is scored equivalently; for each position in a pairwise comparison, the bases are either a match or they are not. A commonly used alternative is the Kimura two-parameter model, in which transitions (purine to purine or pyrimidine to pyrimidine) and transversions (purine to pyrimidine or pyrimidine to purine) are scored differently because transitions are much more common than transversions (Fig. 5.1). These scores are based on presifting the alignment to determine the relative frequency of transitions to transversions, and these different types of changes are scored accordingly. It is even possible to have a six-parameter model, in which each type of substitution (G:A, G:C, G:U, A:U, A:C, and U:C) is scored differently (Fig. 5.2).

It is also possible to "weigh" the score of each position (column) in an alignment differently based on how conserved that position is; a difference in a conserved position is then scored as a greater difference than a difference in more variable positions. This requires alignments with many sequences so that variability at each position can be measured reliably, and so very often these are predetermined for the class of RNA being analyzed. The Weighbor algorithm used by the Ribosomal Database Project does this; the name stands for "weighted neighbor joining." Distance matrices from protein alignments usually use a scoring table derived from the observed relative frequency with which any amino acid is substituted by another from a huge collection of aligned protein sequences, e.g., the PAM tables.

There are also different ways in which gaps can be dealt with. In most treeing algorithms, gaps are ignored; these positions are counted as neither a match nor a mismatch. This is not because they are unimportant; in fact, because insertions and deletions are less common than nucleotide substitutions, they are

**Figure 5.2** A six-parameter substitution model scores each possible substitution differently. doi:10.1128/9781555818517.ch5.f5.2

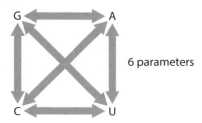

potentially more important than substitutions. However, it is not clear how to deal with gaps for a variety of reasons. The obvious case is where the alignment contains sequence fragments, i.e., partial sequences, instead of full-length sequences (Fig. 5.3). Partial sequences have two kinds of gaps: gaps that represent bases that are not present in that sequence (indels), and gaps that represent regions of sequence outside of the region for which the sequence is available. The algorithm cannot distinguish between these because alignments do not distinguish between different kinds of gaps.

It is also difficult to deal with the fact that adjacent gaps are not independent. A string of gaps probably represents the insertion or deletion of more than one base at the same time, not the one-at-a-time insertion/deletion of individual bases. For example, a five-base string of gaps most likely represents a single insertion/deletion of five nucleotides, not five independent insertions/deletions of single nucleotides. Sophisticated algorithms use a large scoring penalty for a single gap but then only a very small additional penalty for additional adjacent gaps.

In addition, it is not clear how to deal with variation at the 5′ and 3′ ends of the RNA; for example, some RNase P RNAs have the rho-independent

**Figure 5.3** An alignment showing two fundamentally different types of gaps. All of the gaps in the upper half of the sequences are indels; at least one sequence in the database has nucleotides at these positions, but these sequences do not. Some of the sequences in the bottom half of the alignment are partial sequences, i.e., sequence fragments that use gaps wherever there are no sequence data. doi:10.1128/9781555818517.ch5.f5.3

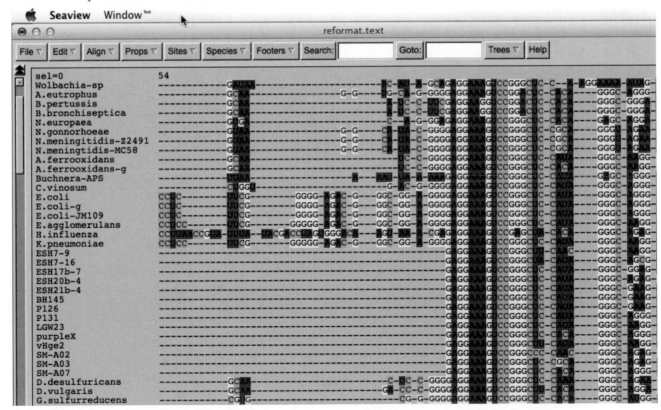

terminator stem-loop at the end of the RNA removed while some do not (and in at least some organisms, the RNA exists in both versions). What to do with all the gaps in aligned RNAs in which this structural element is removed? And because most RNA sequences are determined from their genes, the exact ends of the encoded RNAs are most often not even known.

## The special case of G+C bias

Sometimes even rRNA sequences change adaptively—the bane of phylogenetic analysis. The most common example is the tendency of sequences to differ in G+C content, either because the genome has an unusual G+C content (i.e., there is pressure toward either G+C or A+T richness in the genome) or because the organism is a thermophile and so might prefer G=C over A=U base pairs in its RNAs. This can cause havoc in a tree. One way around this is to do a transversion analysis, which ignores transitions and only scores transversions. The common way to do this is simply to convert all of the A's in the alignment to G's and all U's to C's. Trees are generated from these alignments in the usual fashion. These trees are, of course, based on fewer data since more than half of the phylogenetic information in the alignment has been discarded, but they should be free of G+C bias artifacts.

## Long-branch attraction

One of the things substitution models fight is a treeing artifact called long-branch attraction. Long-branch attraction is the result (primarily) of an underestimation of the evolutionary distance of distantly related sequences. This underestimation results in a tendency for the longest branches in a tree to artificially cluster together; this also results in the artificial clustering of short branches. Figure 5.4 shows a very simple demonstration of how long-branch attraction can result in incorrect trees.

Long-branch attraction happens because of the difference in evolutionary rates in the branches. Therefore, it is always worth worrying about the details of trees containing branches with very different evolutionary rates, i.e., those with branches of very different lengths.

**Figure 5.4** Generation of a "long-branch attraction" artifact in a phylogenetic tree. If the sub-tree to the left is the representation of how these sequences are actually related, imagine what would happen in a neighbor-joining analysis. Sequences A and B are more alike (i.e., they have a smaller evolutionary distance between them) than either is to C, and so they will be erroneously joined, as shown on the right. doi:10.1128/9781555818517.ch5.f5.4

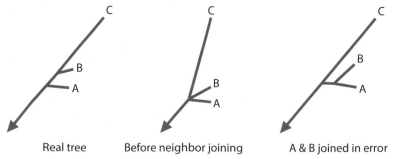

Real tree        Before neighbor joining        A & B joined in error

One of the primary causes of strikingly long branches, by the way, is bad sequence or poor alignment. If the primary sequence data are poor, every mistake in the data will be counted as an evolutionary change by the treeing algorithm. Likewise, poor alignment causes most of the bases in the poorly aligned region to be counted as evolutionary changes, lengthening the branch leading to that sequence. Again, beware of trees with unexpectedly long branches! Poor alignment or bad sequence data, resulting in long branches, can combine with long-branch attraction to make trees meaningless.

## Treeing algorithms

### Fitch-Margoliash: an alternative distance-matrix treeing method

Another useful method for generating trees from distance matrices is that of Fitch and Margoliash, commonly called Fitch. This algorithm starts with two of the sequences, separated by a line equal to the length of the evolutionary distance between them. For example, for this distance matrix:

|   | A | B | C | D | E |
|---|---|---|---|---|---|
|   |   | Evolutionary distance | | | |
| A | – | – | – | – | – |
| B | 0.18 | – | – | – | – |
| C | 0.23 | 0.04 | – | – | – |
| D | 0.29 | 0.35 | 0.29 | – | – |
| E | 0.77 | 0.77 | 0.77 | 0.77 | – |

one might start with sequences A and B (usually in practice, the sequences are chosen randomly):

$$A \underline{\qquad 0.18 \qquad} B$$

Then the next sequence is added to the tree such that the distances between A, B, and C are approximately equal to the evolutionary distances:

A ——— 0.18 / 0.23 ——— B 0.04 / C

Note that the fit is not perfect. If we could determine the evolutionary distances exactly, they would fit the tree exactly, but since we have to estimate these distances, the numbers are fit to the tree as closely as possible using averaging or least-squares best fit.

The next step is to add the next sequence, again readjusting the tree to fit the distances as well as possible:

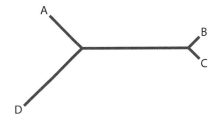

And at last we can add the final sequence and readjust the branch lengths one last time using least-squares:

Neighbor joining and Fitch are both least-squares distance-matrix methods, but a big difference is that neighbor joining separates the determination of the tree structure from solving branch lengths, whereas Fitch solves them together but does so by adding branches (sequences) one at a time.

## Parsimony

Parsimony is actually a collection of related methods based on the premise of Occam's razor. In other words, the tree that requires the smallest number of sequence changes to generate the sequences in the alignment is the most likely tree.

For example:

```
                              +-----GGGAUCCC
             +--GGGACCCC--+
             |                +-----GGGACCCC
--GGGRCCCC--+
             |                +-----GGGGCCCC
             +--GGGGCCCC--+
                              +-----GGGGGCCC
```

No distance matrix is calculated; instead, trees are searched and each ancestral sequence is calculated, allowing for all uncertainties, in a process analogous to Sudoku puzzles. The number of "mutations" required is added up, and the tree with the best score wins. Testing every possible tree is not usually possible (the number of trees grows exponentially with the number of sequences), so a variety of search algorithms are used to examine only the most likely trees. Likewise, there are a variety of ways of counting (scoring) sequence changes.

Parsimony methods are typically slower than distance-matrix methods but very much faster than the maximum-likelihood methods described below. Parsimony uses more of the information in an alignment, since it does not reduce all of the individual sequence differences to a distance matrix, but it seems to work best with relatively closely related sequences and is not usually used for rRNA sequences.

## Maximum likelihood

The maximum-likelihood method turns the tree construction process on its head, starting with a cluster analysis to generate a "guide" tree, from which a very complete substitution model is calculated. The algorithm then goes back and calculates the likelihood of any particular tree by summing the probabilities of all of the possible intermediates required to get to the observed sequences. Rather than try to calculate this for all possible trees, a heuristic search is used starting with the guide tree. Sound complicated? It is, and maximum-likelihood tree construction is by far the most computationally intensive of the methods in common use. However, it is generally also the best, in the sense that the trees are more consistent and robust. The limitation is that fewer and shorter sequences can be analyzed by the maximum-likelihood method because of its computational demands. A tree that might take a few seconds by neighbor joining or a few minutes by parsimony or Fitch can take a few hours or a couple of days by maximum likelihood. This is serious; it means that you cannot usually "play" with trees, testing various changes in the data or treeing parameters and seeing the result immediately.

### Bayesian inference

Bayesian inference is a relatively new approach to tree construction. This approach starts with a random tree structure, random branch lengths, and random substitution parameters for an alignment, and the probability of the tree being generated from the alignment with these parameters is scored. Obviously the initial score is likely to be very poor. Then a random change is made in this tree (branch order, branch length, or substitution parameter) and the result is rescored. Then a choice is made whether to accept the change; this choice is partially random, but the greater the improvement in tree score, the more likely it is to be accepted. If the change is accepted, the process is repeated starting with this new tree; if the change is rejected, the process is repeated starting with the old tree. After many, many cycles of this process, the algorithm settles in to a collection of trees that are nearly optimal. Various tricks are used to keep the algorithm from getting stuck in local-scoring minimum zones.

## Bootstrapping

Unfortunately, it is mathematically impossible to extract confidence intervals ($\chi^2$ or $p$ values) for nodes in a tree generated using the standard treeing methods; in other words, we cannot determine how sure we should be of the placement of each node in a tree (commonly referred to as branching order). This is not true of branch lengths, from which confidence intervals can be readily calculated. The standard method of evaluating tree branching-order reliability is called bootstrapping.

In a bootstrap analysis, the columns of a sequence alignment are randomly sampled to create a jumbled and haphazard alignment of the same length as the original. Typically 100 or 1,000 such alignments are generated from the initial alignment, and trees are generated from each. The reliability of a particular branching arrangement in a tree is judged by the frequency that the branch appears in all of the resulting trees.

The random sampling starts with the input alignment. In the standard tree generation process, the similarity matrix is generated by checking sequences against each other pairwise, tallying the similarity to each position in the

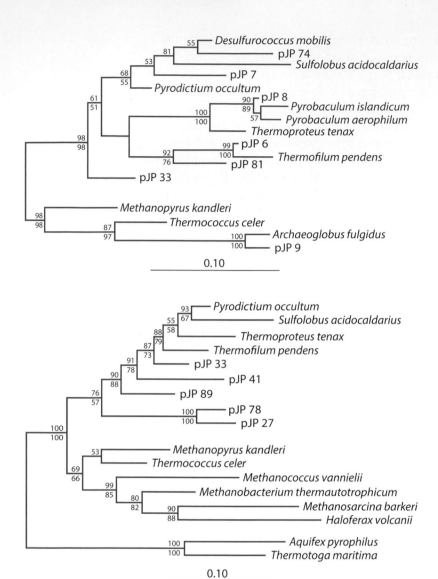

**Figure 5.5** Example of a published phylogenetic tree with bootstrap values included. In this case, this is a maximum-likelihood tree, with bootstrap values from both maximum-likelihood (above the branches) and maximum-parsimony (below the branches) analyses shown. (Reprinted from Barns SM, Fundyga RE, Jeffries MW, Pace NR, *Proc Natl Acad Sci USA* **91:**1609–1613, 1994. Copyright 1994 National Academy of Sciences, USA.) doi:10.1128/9781555818517.ch5.f5.5

alignment. Each position in the alignment is therefore counted once and only once. In a bootstrap sampling, a similarity matrix would be generated by comparing randomly selected positions in the alignment, so that some positions are compared more than once and some are not compared at all. For example, the starting alignment

```
Position   1 2 3 4 5 6 7 8 9
Sequence 1 g g u u c g c c u
Sequence 2 g c u u u g g c u
Sequence 3 g c u u u – g c u
```

would be randomly sampled, ending up with a bootstrapped alignment that might look like this:

```
Position   9 2 5 8 3 9 2 1 9
Sequence 1 u g c c u u g g u
Sequence 2 u c u c u u c g u
Sequence 3 u c u c u u c g u
```

from which a similarity matrix and tree would be generated.

Notice that some positions in the alignment are included multiple times (9 and 2) and some are not included (4, 6, and 7). In realistically large alignments, such randomly sampled alignments yield good trees if the branching arrangements are well supported by the sequence data. Therefore, in a bootstrap analysis, a tree is generated from each of 100 to 1,000 alignments generated by random sampling of the alignment, and the number of these trees that contain each branch of the reference tree (generated in the usual way) is determined. In other words, the presence or absence of every branch in the tree is scored. The percentage of trees from a bootstrapped alignment that contain each branch in the reference tree is used to label the branches (Fig. 5.5). Often, the same type of analysis is performed using more than one method of tree construction, e.g., neighbor joining and maximum likelihood.

The evaluation of bootstrap scores is subjective, but generally branches that show up in at least 50% of trees generated from bootstrapped data sets are considered to be reliable.

## Questions for thought

**1.** Compare the bootstrap values on the tree in the bootstrap section above. Which treeing algorithm seems to have generated the more robust trees, maximum likelihood or maximum parsimony? Are long or short branches more reliably predicted?

**2.** What kind of artifact would you predict from a substitution model that over-estimated, instead of underestimated, the relative evolutionary distance of long branches?

**3.** During bootstrapping of an alignment, the sequences become scrambled (the residues appear out of order). Does this matter? Why or why not? (Think about it this way—would a tree come out the same if the alignment was scrambled, or not?)

**4.** In the substitution models we talked about, we focused on substitutions (obviously). How do you suppose these models deal with gaps? What about under-specified bases (e.g., R, Y, or N), or instances where sequence data are absent (i.e., in the case where only a piece of the sequence is available)?

**5.** In a parsimony analysis, a by-product of the analysis is predicted ancestral sequences. Can you think of a situation in which this might be useful?

# 6 Alternatives to Small-Subunit rRNA Analysis

## SSU rRNA cannot be used to distinguish closely related organisms

Molecular phylogenetic analysis using small-subunit ribosomal RNA (SSU rRNA) sequences is generally the most useful method for determining the phylogenic relationship between organisms and is also one of the most definitive ways to verify the identity of an organism. For example, in some laboratories, whenever a new organism comes from another laboratory, whether from across the hall or across the world, whether from someone well known and trusted or someone asked out of the blue, from the American Type Culture Collection (ATCC) or DSMZ (the German equivalent), the very first thing done is to streak it out for purity and then sequence its SSU rRNA to make sure it is what it should be. The importance of this is usually a lesson learned the hard way.

But although SSU rRNA molecular phylogenetic analysis is usually the most useful method for identifying organisms, and the method that should at least be considered as a starting point, it is not always the best method or the final word.

The main problem with using SSU rRNA to identify organisms is that close relatives cannot be distinguished using rRNA sequences. As a general rule of thumb, SSU rRNA analysis can determine the genus, but not the species (at least not reliably), of an organism. Sometimes phenotypically distinct but closely related species have very similar or even identical rRNA sequences; this occurs in the genus *Bacillus*, for example. On the other hand, the rRNA operons in the same organism often vary to some extent. Usually this variation amounts to just a couple of differences, but its extent and the way it clusters in the populations are usually unexplored. In a very few cases, different rRNA operons in the same

doi:10.1128/9781555818517.ch6

organism may be quite different, either because the gene conversion process has failed to keep them evolving in concert or because the organism is a recent hybrid of two species or has recently acquired rRNA genes by horizontal transfer (yes, it does happen, if only rarely).

Another part of the problem is the larger issue of how species are defined in non-sexually reproducing organisms. As problematic as the "species concept" is in the world of sexually reproducing species (a topic of discussion at length in *On the Origin of Species*), no rational species concept exists for organisms in which reproduction and sexual exchange are not linked, including *Bacteria*, *Archaea*, and most unicellular eukaryotes. The result is that species are typically defined arbitrarily in the microbial world. These arbitrary boundaries are not very closely related to how different these "species" are in terms of rRNA sequence.

Generally speaking, SSU rRNA sequences that are greater than about 97% identical cannot be meaningfully distinguished and their relationships cannot be reliably determined in trees. The ability to distinguish between and determine the relationships of very closely related organisms is critical for medical and food microbiologists; for example, different strains of *Staphylococcus aureus* are very different in their pathogenicity and must be carefully identified. Likewise, in food fermentations using natural microbes, such as in cheese-making, different very closely related strains make the difference between a perfect Stilton and spoiled milk.

So, although SSU rRNA is the standard method for general phylogenetic analysis, other methods are often needed for fine-scale analyses or identifications. In many cases, alternative sequences are useful. In other cases, entirely different approaches are needed. Let's touch (just touch) on a few of these, especially those you are already familiar with or are likely to run into in reference materials.

## Alternative sequences

In cases where SSU rRNA sequences are too alike to give reliable trees, it is common practice to use other molecular sequences in the same way to create phylogenetic trees.

### *Other RNAs*

In most cases, it would be best to use other RNA-encoding genes that retain the advantages of SSU rRNA sequences, with sequences that evolve faster than SSU rRNA. The large-subunit rRNA (LSU rRNA) is sometimes used but has not proven to be as robust or useful as the SSU rRNA. The "other" LSU rRNA, the 5S rRNA, is too short at ca. 120 nt to be useful in most circumstances. Transfer RNAs (tRNAs) are both too short and too highly conserved to be generally useful for phylogenetic analysis. Many other larger RNAs, such as the signal recognition particle RNA and the tmRNA (involved in the release of ribosomes from truncated messenger RNAs [mRNAs]), are not conserved enough in structure to be reliably aligned with the precision required for good phylogenetic analysis.

However, the RNA subunit of ribonuclease P (RNase P, a catalytic RNA involved in tRNA biosynthesis) has been used extensively to examine relationships in the *Bacteria* and *Archaea* (Fig. 6.1 and 6.2). The RNase P RNA in

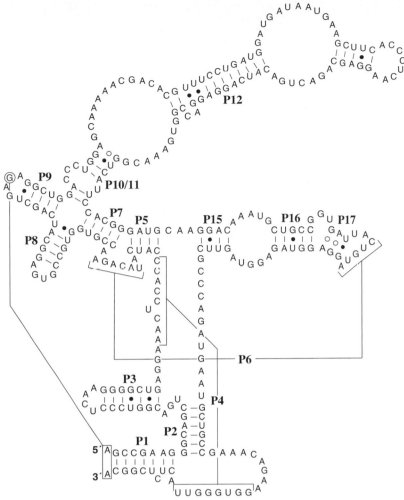

**Figure 6.1** A representative archaeal RNase P RNA, from the methanogen *Methanother-mobacter thermoautotrophicus* (previously *Methanobacterium thermoautotrophicum* strain delta H). doi:10.1128/9781555818517.ch3.f3.7Λ

*Bacteria* evolves about sixfold faster than does the SSU rRNA overall, too fast to be useful for probing the deep branches of the tree. However, RNase P RNA provides resolution among close relatives not available by analysis of SSU rRNA, and is also sometimes used to test relationships in SSU rRNA trees.

It is common, however, that even these RNAs are too highly conserved for molecular phylogenetic analysis to give the resolution required. In this situation, alternative approaches are required.

## rRNA spacer sequence analysis

rRNA spacer sequence analysis is a lot like SSU rRNA analysis, but it takes advantage of the fact that, in most organisms, the SSU and LSU rRNAs are directly adjacent to each other in the genome and in this order, with a relatively small spacer in between (Fig. 6.3). This makes it easy to use the polymerase chain reaction (PCR) to amplify and sequence this highly variable spacer with primers

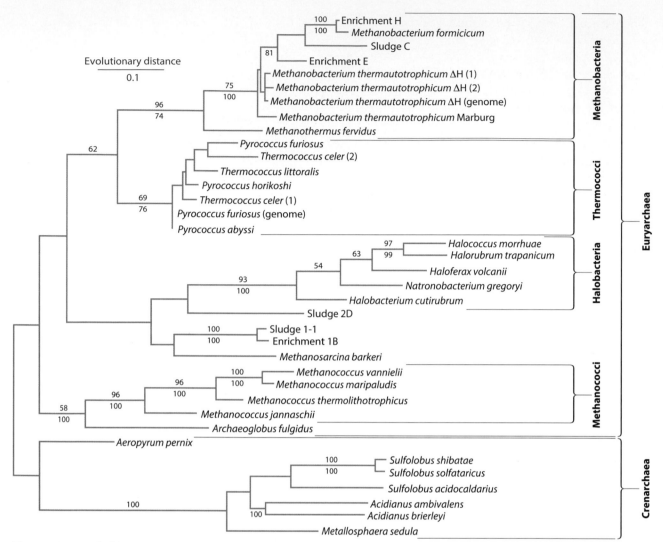

**Figure 6.2** A phylogenetic tree generated from an alignment of archaeal RNase P RNA sequences. (Reprinted from Harris JK, Haas ES, Williams D, Frank DN, Brown JE, *RNA* 7:220–232, 2001, with permission.) doi:10.1128/9781555818517.ch6.f6.2

**Figure 6.3** Location of the "spacer" sequence in rRNA gene clusters. doi:10.1128/9781555818517.ch6.f6.3

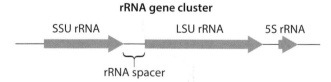

that hybridize to highly conserved sequences at the 3′ end of the SSU rRNA and the 5′ end of the LSU rRNA.

This small-spacer sequence evolves very quickly and can be used to distinguish between different closely related species or even strains of the same species. This sequence is often used to analyze the relationships between animal species, using the spacer sequence from the mitochondrial DNA; mitochondrial genes evolve much faster than nuclear genes.

The main disadvantage of this approach is that there are usually many copies of the rRNA gene cluster and the spacer sequences usually are not the same in the different clusters. For example, *Escherichia coli* has seven rRNA operons, three of which have one spacer sequence and four of which have another sequence. Therefore it is necessary to obtain and compare the corresponding spacers in order for the analysis to work. This is usually easy; these spacers generally contain tRNA genes (in *Bacteria* and *Archaea*), and these can serve to indicate which spacer sequence corresponds to which. In most *Bacteria*, one spacer type contains a single gene for tRNA$^{Glu}$ and the other spacer type contains genes for tRNA$^{Ile}$ and tRNA$^{Val}$. Although a genome might have several of each, all of the spacers within a class generally evolve together, and so it is not necessary to sequence all of the spacers in the genome and identify which rRNA operons specifically correspond.

## Protein sequence analysis

One might imagine that alignments and trees based on the amino acid sequences of proteins would be more informative than those based on DNA or RNA sequences, because there are only 4 nucleic acid bases but 20 standard amino acids (incidentally, there are at least 2 known "nonstandard" amino acids that are built into proteins during translation, selenocysteine and pyrolysine). However, despite the lower information content of nucleic acid sequences, there are fundamental reasons why non-mRNAs are generally better for phylogenetic analysis:

1. Protein sequences are almost always aligned automatically by programs like CLUSTAL that maximize the pairwise similarities between sequences. This is not an advantage. It is objective but is far less accurate than alignment based on structure, which is how RNAs are aligned unless the sequences are very similar. If a new sequence is added, the rest of the alignment will change, unlike an RNA alignment. RNAs but not proteins can be aligned on the basis of structure because RNAs (but not proteins) have a predictable, well-defined secondary structure. Alignments have been made on the basis of solved or predicted three-dimensional structure, but this is a very rare exception.

2. The base pairs and helices of an RNA molecule are independently evolving substructures, a prerequisite for a molecule to exhibit clock-like behavior. Protein structure is much less hierarchical than is RNA structure—a change in any internal amino acid can have significant effects throughout the domain in which it resides. In other words, the amino acid changes in a protein are not as independent as are base (actually base-pair) changes in an RNA molecule.

3. Although there is more information per position in a protein alignment, there are usually more phylogenetically informative positions in an RNA (at least SSU rRNA) sequence. These RNAs are longer (in terms of the number of residues) than most proteins, and more of the residues can be reliably aligned and yet are variable.

4. Families of functionally distinct genes with similar sequences are more of a problem for protein-encoding genes than for RNA-encoding genes. Appropriate homologs of these functional alternatives are problematic to identify, and sequence differences are often the result of non-clock-like evolutionary pressures.

5. Horizontal transfer seems to be more frequent for protein-encoding genes than for RNA-encoding genes. Among protein-encoding genes, those involved in metabolism are more likely to have been horizontally transferred than those of "information processing" (DNA replication, transcription, and translation).

Nevertheless, in some situations, protein-based trees can be informative. Usually the inferred amino acid sequence rather than the DNA sequence of the gene is used for phylogenetic analysis. Highly conserved proteins that have proven to be useful for phylogenetic analysis include RNA polymerase subunits, DNA polymerase subunits, ATPase subunits, glycolytic enzymes, electron transport enzymes, and stress response proteins.

### Catenated alignments

It is also possible to use more than one gene at a time for an analysis. The alignment just contains more than one molecular sequence in each row; i.e., the sequences are catenated. Very often such catenated alignments yield trees that are more reliable than the trees of the individual sequences from which they are made. For example, the alignment might be composed of SSU rRNA, LSU rRNA, and RNase P RNAs all together. Alternatively, the entire set of ribosomal proteins and RNase polymerase subunits could be used.

The extreme version of using catenated alignments is to make alignments composed of *all* of the genes of the organisms to be analyzed, i.e., the entire genome. This requires a lot of data but is possible (just) with modern computing power. The problem is that it uses good molecular clocks together with poor ones, and genes that transfer frequently horizontally together with those that do not, and it is not clear that the resulting trees are generally better than rRNA trees. This is, however, a very promising area of active research.

## Alternatives to sequence-based methods

### DNA:DNA hybridization

The DNA:DNA hybridization method was commonly used in the past and is still often required when new species are defined. The extent to which the genomic DNAs of two species hybridize is a general measure of how much sequence similarity there is between them and therefore of how closely related the organisms are. This method was traditionally used to define bacterial species;

in general, two organisms were considered to belong to the same species if the DNA:DNA hybridization was ca. 70% or greater, or to different species of the same genus if they had measurable hybridization of less than 70%.

## DNA base composition

DNA base composition is also an older technology. Because every G is paired to a C and every A is paired to a T, the ratios of G and C are always the same, and likewise for A and T. Furthermore, because the sum of all four is 100%, there is only one degree of freedom in base composition, and it is usually expressed as %G+C. The %G+C in most organisms' genomes is somewhere between 40 and 65, averaging about 55. Two closely related strains are considered to be members of different species if their DNA base compositions (%G+C) are reliably distinguishable.

There are two commonly used methods for determining the relative amounts of A, G, C, and T in the genome of an organism. DNA can be hydrolyzed to individual nucleosides, which can be separated (typically by two-dimensional thin-layer chromatography) and quantitated. An alternative method is to carefully measure the denaturation midpoint of the DNA; because G≡C pairs are more resistant to denaturation than A=T pairs, melting point can be related more or less directly to the ratio of G≡C/A=T base pairs.

However, DNA base composition is rarely used any more—it turns out that DNA base compositions can change very rapidly and unpredictably during evolution. Nevertheless, differences in %G+C content have traditionally been used to distinguish different species within a genus where the species are phenotypically very similar, and %G+C is one of the standard values listed for species in their formal descriptions.

## Serology

Serology is used primarily to identify very closely related clinical isolates, usually different strains of a single species. This method uses antisera developed from various strains of *Bacteria* to identify the strain to which a new isolate belongs. For example, when *Salmonella* is isolated from a patient, a bank of antisera is used to determine which of the hundreds of serotypes that particular isolate is. A rapid serological test is used to test for virulent strains of *Streptococcus pyogenes* in throat swabs to see if a patient has strep throat (Fig. 6.4). This is an old but still widely used method, since the antisera are easy to make and the assay is very quick, easily automated, and reliable. Perhaps as important, both the virulence and the serological type of pathogens are dependent on variable surface proteins. This means that the serotype of a strain can be a good predictor of virulence. The commonly used enzyme-linked immunosorbent assay (ELISA) is a serological method.

## Lipid profiling

Fatty acid methyl ester (FAME) analysis is also a fairly quick and easy method. A sample of growing culture is extracted with a base to lyse the cells and saponify the fatty acids, and is then treated with a strong acid and methanol to convert these acids to FAMEs. FAMEs are extracted with an organic solvent, and a

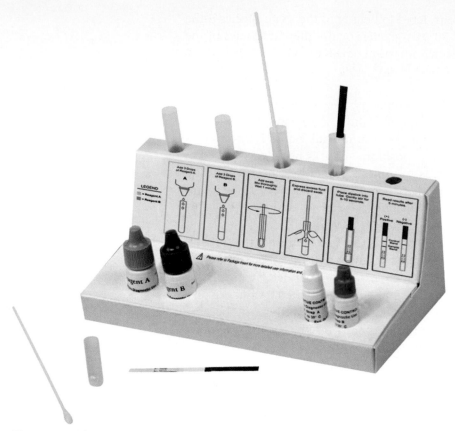

**Figure 6.4** The iScreen Strep A test kit uses an ELISA. (Image supplied by CLIAwaived, Inc.) doi:10.1128/9781555818517.ch6.f6.4

**Figure 6.5** FAME profile of *Mycobacterium szulgai.* (Redrawn from Müller KD, Schmid EN, Kroppenstedt RM, *J Clin Microbiol* **36:**2477–2480, 1998, with permission.) doi:10.1128/9781555818517.ch6.f6.5

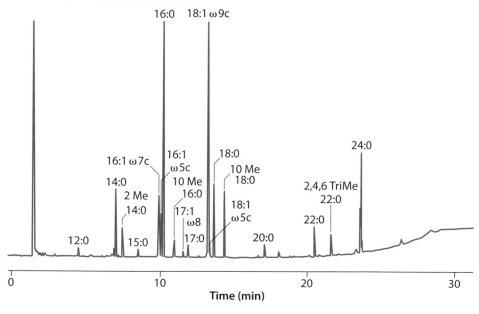

sample is analyzed by gas chromatography (Fig. 6.5). The FAME profile (both the identity of the FAMEs and their relative ratios) is compared to a database of standard profiles for identification; it can also be analyzed using tree-building methods. This method is very fast (just over 24 hours), standardized, and automatable. For medically relevant organisms, the resolution of FAME analysis is better than that of SSU rRNA analysis; species within a genus are routinely distinguishable. However, cultures have to be grown under very strictly controlled conditions, because organisms alter their membrane lipids in response to growth conditions. The analysis is fundamentally limited to the quality and coverage of the standard database; little information is derived from organisms not related to those in the FAME database. As a result, this approach is most commonly used in medical microbiology.

## RFLP methods

Restriction fragment length polymorphism (RFLP) analysis is a widely used group of technologies, commonly used by animal and plant geneticists to determine paternity, to identify the source of forensic tissue samples, etc. When a layman hears about "DNA testing," it is almost certainly some form of RFLP analysis. When this method is used with samples from humans or other animals, it is primarily done to distinguish or identify individuals, but in the microbial world it is more often used to differentiate strains of the same species.

In traditional RFLP, genomic DNA from the organism is digested with restriction enzymes and separated by size on a gel. The gel is then transferred to a membrane and hybridized with a set of oligonucleotide probes complementary to variable regions in the genome. Variation in these sequences results in differences in the presence or absence of these restriction sites or in the lengths of the fragments that hybridize to the probes. If the probes are carefully designed, the RFLP banding patterns are unique genetic fingerprints of the individual or strain. The patterns of bands are compared to those of other strains (or parents, suspects, etc.) to test identity or specific genetic relationships. With primers or probes designed to yield very complex banding patterns, it is possible to generate trees based on these patterns.

A very common RFLP method involves performing PCR to amplify a specific variable region of genomic DNA (this is often the SSU rRNA or some other region of the rRNA operon, such as the intergenic spacer), using restriction enzymes to digest this DNA fragment into several smaller fragments, and separating the fragments by electrophoresis. The number and size of the resulting DNA bands are diagnostic for specific kinds of organisms.

Although not strictly speaking RFLP, related methods are now far more common in which PCR amplification, rather than restriction digestion, is used to generate complex banding patterns (Fig. 6.6). This method often uses many arbitrary or "random" primers together in a complex PCR mix; good primer sets are identified empirically. This method is more formally called amplified fragment length polymorphism (AFLP).

Notice in Fig. 6.6 that the serotype and the AFLP groups do not correspond. This is the rule, not the exception. What we refer to as a strain is directly related to how we distinguish between strains.

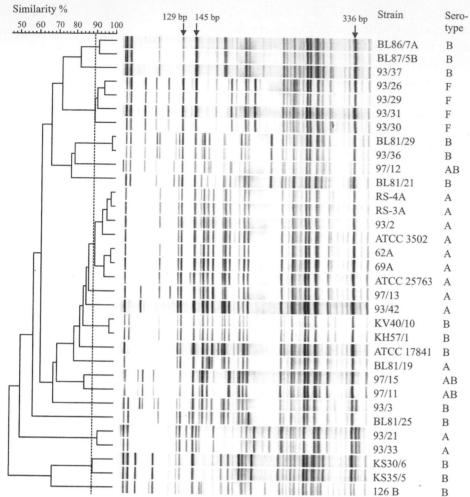

Figure 6.6 AFLP analysis of *Clostridium botulinum* type I strains. A phenogram showing the relationships between banding patterns, generated from a similarity matrix using the same approaches used for generating phylogenetic trees from sequence data, is shown on the left. (Reprinted from Keto-Timonen R, Nevas M, Korkeala H, *Appl Environ Microbiol* **71**:1148–1154, 2005, with permission.) doi:10.1128/9781555818517.ch6.f6.6

Figure 6.7 API 20E strip. This result identifies the culture as *E. coli*. The pattern above decodes into + − + + − − − − + − / − + + − + + + − + − +, which can be looked up for identification. (Image supplied by BioMérieux.) doi:10.1128/9781555818517.ch6.f6.7

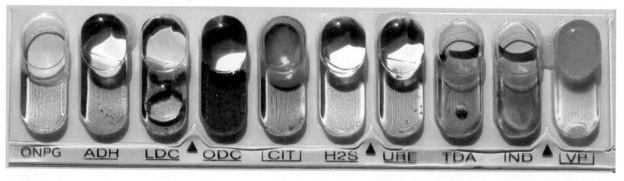

## Phenotype

Phenotypic markers are what most people think of when they want to compare different microorganisms. What carbon sources can it use? Is it motile or nonmotile? Is it aerobic, anaerobic, or facultative? What is its shape and size? What is its optimal growth temperature? Is it gram negative or gram positive?

Although gross phenotypic markers are not always very useful in determining phylogeny, they are still perfectly viable markers for taxonomies. Until relatively recently, phenotype was all microbiologists could go by.

You are probably familiar with these sorts of tests and markers from an "unknown" identified in a beginning microbiology teaching laboratory. In most research or clinical environments, you would not be using giant tubes of broth but instead more rapid systems that let you measure a number of phenotypic properties at the same time. A classic system is the API strip, by BioMérieux in Durham, NC (Fig. 6.7). The API strip is a plastic strip with little wells of dehydrated media, into which a specified dilution of inoculum is added. These grow up and change color (or some other property), and the identification of the organism is read from the pattern of results on the strip.

Because different kinds of tests are required to distinguish organisms in different phylogenetic groups, this approach usually starts with a preliminary set of tests to determine very generally what kind of organism is in hand: enteric bacteria, nonenteric gram-negative rods, staphylococci, streptococci, yeasts, etc. Then the appropriate specific API strip is used for the identification.

For example, the API 20E test strip is used to identify enteric bacteria. Organisms are first tested by the Gram stain (enteric bacteria are gram-negative rods) and cytochrome oxidase test (enteric bacteria are negative). Once the organisms are confirmed to be enteric bacteria, they can be tested on the API 20E strip, which contains tests for (reading in order from left to right on the strip) $o$-nitrophenyl-β-D-galactopyranoside, arginine decarboxylase, lysine decarboxylase, ornithine decarboxylase, citrate utilization, $H_2S$ production, urease, tryptophane deaminase, indole production, Voges-Proskauer test, gelatinase, and fermentation/oxidation tests for glucose, mannitol, inositol, sorbitol, rhamnose, saccharose, melibiose, amygdalin, and arabinose.

Organisms are identified by scoring the strip with the analytical profile index (hence the abbreviation API).

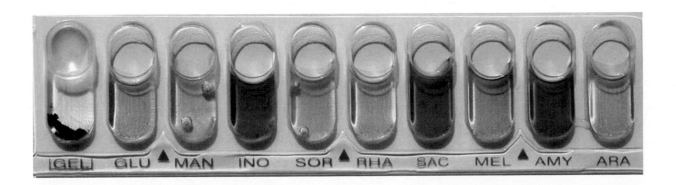

## Questions for thought

**1.** rRNA sequence divergence is a measure of phylogenetic relatedness, i.e., biological diversity. How is this related to what people usually think of as biological diversity?

**2.** What other types of phylogenetic analysis can you think of?

**3.** Other than alignment gaps, what complexities can you think of in trying to create a reasonable evolutionary model for phylogenetic analysis of entire genome sequences?

**4.** Standard microbiological taxonomy relies heavily on antiquated ways to distinguish or classify organisms, such as genomic %G+C and DNA:DNA hybridization. How can you imagine this changing? What are the hurdles to such a change?

**5.** Why do you think RFLP methods are more likely to be used to identify individuals when used for plants and animals but to identify strains when used for microbes?

**6.** Different API strips, with different arrays of tests, are used to identify different kinds of organisms. Why do you suppose this might be necessary? Do you think you could design a more complete method, with many more tests, to identify any kind of organism?

# 7

# The Tree of Life

Most of the time until now, we have been working with sequences and trees in the abstract: sequence A, B, C, and D. But what do you get if you use the small-subunit ribosomal RNA (SSU rRNA) molecular phylogenetic methods we have discussed with representatives from as wide a range of actual organisms as possible? You get the sort of unrooted dendrogram shown in Fig. 7.1.

Remember that the previously common trees, such as the Whittaker five-kingdom tree, were subjective and qualitative. This molecular phylogenetic tree is quantitative and objective, being based on statistical analysis of gene sequences. In this case, of course, the sequence used is the SSU rRNA gene, for reasons already covered (see chapter 3), but other appropriate molecules yield much the same result.

## Major lessons of the "Big Tree of Life"

### *The discovery of the* Archaea

To people's great surprise when these molecular phylogenetic trees first began to come out in 1977, and to the continuing consternation of many people, it turns out that there are three major, distinct evolutionary groups of known living things:

- *Bacteria:* also known as eubacteria
- *Archaea:* also known as archaebacteria
- *Eukarya:* the nuclear/cytoplasmic component of eukaryotes (does not include the mitochondria or chloroplasts [see below])

doi:10.1128/9781555818517.ch7

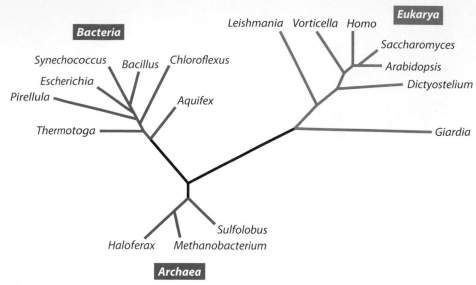

**Figure 7.1** A phylogenetic tree of representative organisms based on SSU rRNA sequences. (Redrawn from artwork supplied by Norman R. Pace.) doi:10.1128/9781555818517.ch7.f7.1

These major groups are sometimes referred to as kingdoms, and thus this tree is sometimes known as the three-kingdom tree. To allow at least some of the classically defined kingdoms to remain unscathed (plants, animals, and fungi, to be specific), these three major phylogenetic groups are more often referred to as domains, a level added to the Linnaean taxonomy higher than kingdoms. And so this tree is also sometimes referred to as the three-domain tree, which is how it is referred to in this book.

The species that turned out to be members of the previously unrecognized group *Archaea* had been scattered haphazardly among the bacterial taxonomy. For example, halobacteria were considered to be odd pseudomonads and *Sulfolobus* was thought to be a weird relative of corynebacteria. Phenotypically, these organisms are grossly similar to *Bacteria*, but it turns out that in many ways (see chapter 15) they are more similar to *Eukarya*, especially in their central information systems (think of the Central Dogma). They are generally primitive, having apparently changed less since their common ancestry than either the *Bacteria* or *Eukarya*, and so more closely resemble our common ancestry than do other known organisms. This makes them an ideal system to study the early evolution of organisms, and especially that of eukaryotes.

## Big eukaryotes represent a small portion of biological diversity

Notice that the so-called "prokaryotes" (*Archaea* plus *Bacteria*) represent two-thirds of the tree, that is, two of the three domains. But it turns out that these two entirely separate groups of "prokaryotes" are very different, and the tree shows that they are, in fact, not really "of a kind." This is why we argued in chapter 2 that the term "prokaryotes" is invalid, meaning nothing more than "not a eukaryote."

Even within the *Eukarya*, multicellular eukaryotes are delegated to a very small portion of evolutionary diversity—just the tips of some branches, not three-fifths of evolutionary diversity that the five-kingdom scheme implies. And let me reiterate that even most fungi, plants, and animals are microscopic, and so "microbial."

It also turns out that *Eukarya* is as ancient a group as are *Bacteria* and *Archaea*, and did not evolve from within either of the other groups. This is counter to the view of the traditional five-kingdom tree, in which eukaryotes are shown to have evolved from bacteria. Interestingly, all of the known branches of eukaryotes are "late" (there is a long bare branch connecting eukaryotes to the other domains). There are no known primitive eukaryotes or early branches in the eukaryotic group, unless you count the *Archaea* (see below). It is not clear how many main eukaryotic groups there might be, as there are huge numbers of visually described protists that have not been analyzed genetically. In fact, the eukaryotes are probably the least well studied (or at least the least well understood) of the domains in terms of molecular phylogeny.

## Gram-positive versus gram-negative is not the major division in Bacteria

The *Bacteria* turns out to be composed of a handful of major groups and an ever-growing list of minor groups. Gram-negative-type cell envelopes, with both an outer membrane and a cytoplasmic membrane, are the norm for most all but two of these groups: the Firmicutes and Actinobacteria (Fig. 7.2). (One potential addition is the phylum Planctomycetes, depending on the homology of its membranes to those of other bacteria [see chapter 13].) The majority of the other familiar (gram-negative) bacteria are members of the Proteobacteria, just a single one of the major branches. Cyanobacteria (blue-green algae) also represent a single bacterial lineage.

**Figure 7.2** Phylogenetic tree of the bacterial phyla, with the two phyla of predominantly gram-positive organisms highlighted. doi:10.1128/9781555818517.ch7.f7.2

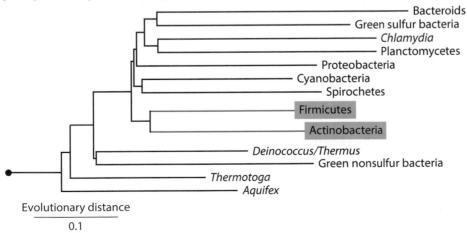

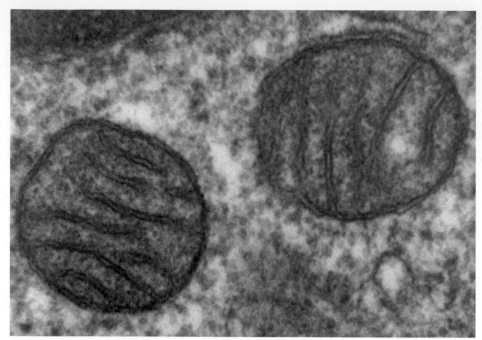

**Figure 7.3** Electron micrograph (thin section) of mammalian lung mitochondria. (Source: Wikimedia Commons, Mitochondria,mammalianlungTEM.jpg.) doi:10.1128/9781555818517.ch7.f7.3

### *Mitochondria and chloroplasts are bacterial endosymbionts*

The three-domain tree provided the definitive confirmation of the endosymbiont theory for the origin of mitochondria and chloroplasts, i.e., that these organelles are permanent internal bacterial symbionts. These organelles contain their own DNA, distinct from that of the nucleus; replicate by fission; and in many other ways look like bacteria living inside of eukaryal hosts. Mitochondrial and chloroplast genomes contain, among other things, genes encoding SSU rRNA, and so they can be analyzed separately from nuclear SSU rRNA sequences. The mitochondria turn out to be proteobacteria, and more specifically the alphaproteobacteria, such as *Wolbachia* and *Rickettsia*, that are commonly symbionts of eukaryotes. Chloroplasts (and plastids generally) are cyanobacteria, and specifically related to prochloral bacteria that, like chloroplasts, contain chlorophyll *a* and *b* (Fig. 7.3).

## Rooting the "Tree of Life"

Where in the "Big Universal Tree of Life" is the last common ancestor? To answer this question, we need to root the tree, but as we saw earlier (chapter 4), in order to root a tree we need an outgroup. How is it possible to have an *outgroup* in a *universal* tree?

The solution to this quandary lies in the fact that the trees generated in a molecular phylogenetic analysis are trees of gene sequences, not organisms. This distinction can be used to create a rooted phylogenetic tree of all organisms by using

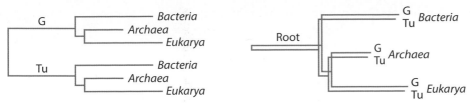

**Figure 7.4** Schematic of how trees of early-duplicated gene alignments can be used to root a "universal" tree. In this example, the gene pair used is EF-Tu and EF-G. doi:10.1128/9781555818517.ch7.f7.4

gene sequences that span a greater evolutionary breadth than do the organisms in the tree. In other words, we can use gene sequences that are outgroups of a tree of other gene sequences that span the entire phylogenetic diversity of organisms.

This is done by using gene sequences from gene family pairs that split before the appearance of the last common ancestor. Good examples of these are adenosine triphosphatase (ATPase) subunits alpha and beta, initiator and elongator tRNA$^{Met}$, EF-Tu and EF-G, or any pair of related aminoacyl-tRNA-synthetases (e.g., Leu and Ile). We then end up with a dendrogram with two main clusters, one cluster for each type of sequence (Fig. 7.4). The subtree from one member of the gene family serves as root for the other.

The root turns out to be very close to the base of all three domains, on the branch dividing the *Bacteria* from the *Archaea* and *Eukarya* (Fig. 7.5). In other words, the *Archaea* and *Eukarya* are specifically related, not the *Bacteria* and *Archaea* (i.e., prokaryotes), as most people thought. This conclusion has been substantiated by analysis of genome sequences.

**Figure 7.5** Rooted "universal" dendrogram showing the specific affiliation of *Archaea* and *Eukarya*. doi:10.1128/9781555818517.ch7.f7.5

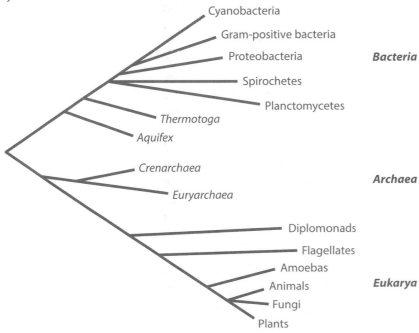

The *Archaea* are probably, from this perspective, an early, primitive branch of the eukaryotic lineage. This is the reason for a lot of the interest in *Archaea* these days; these organisms probably resemble, more than anything else we know of, the ancestor of eukaryotes, and they certainly have much to tell us about where eukaryotes (including humans) came from.

## The caveat of horizontal transfer

The fly in the ointment of molecular phylogenetic analysis, especially in attempts to sort out its deepest parts, is the issue of horizontal transfer. The trees discussed so far are based on SSU rRNA, and they have been substantiated by a number of other highly conserved genes. However, we know that genes have moved from one organism to another across phylogeny; in fact, gene flow at about the species level is probably the most common source of phenotypic variation in bacteria. But horizontal transfer is not limited to near relatives; genes on rare occasions have moved across the farthest reaches of the tree (Fig. 7.6). The most substantial example of this is the movement of bacterial genomes into eukaryotes in the endosymbioses that resulted in the mitochondria and chloroplasts. Most of the genes of these organelles, especially mitochondria, have moved to the nucleus.

**Figure 7.6** Schematic view of the "Tree of Life" incorporating horizontal gene transfer. (Courtesy of Perdita Phillips.) doi:10.1128/9781555818517.ch7.f7.6

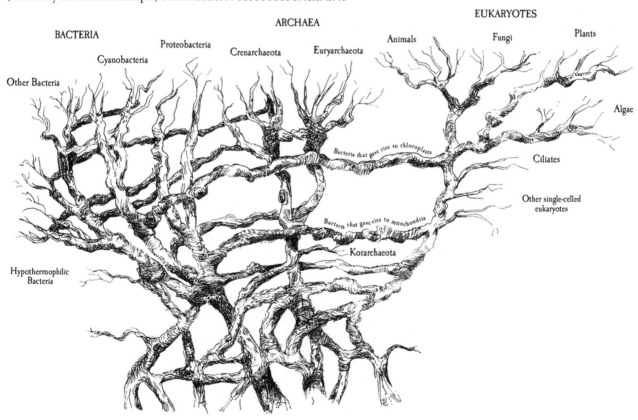

## What impact does horizontal transfer have for the "Big Tree"?

Horizontal transfer is a contentious issue, and a lot of contention is the result of historical baggage. For a long time, horizontal transfer was a ready excuse used, without evidence, to explain any unexpected gene in an organism. As a result, horizontal transfer acquired a bad name; it was seen as thoughtless hand-waving. When SSU rRNA trees began to revolutionize how we thought about microbial evolution, the focus was on understanding general phylogenetic relationships, and the issue of horizontal transfer was set aside as being sufficiently rare to not be significant in the evolution of these organisms. But then genome sequences began to become available, and it became clear that horizontal transfer is a general and significant aspect of microbial evolution; some scientists went so far (as the pendulum swung) as to declare that phylogenetic relationships are meaningless for *Bacteria* and *Archaea*, that every gene in an organism has its own evolutionary history, and that microbes as a whole are just transient subsets of a big "prokaryotic" gene pool. The notion of species within *Bacteria* and *Archaea* is meaningless in this view. This remains a topic of harsh argument and personal attacks.

However, it seems most likely that although horizontal transfer is a major factor in the evolution of microbes, a core of the genome generally reflects the vertical evolution of the organism. This core is predominantly the genes for the "Central Dogma" of the cell; in other words, the information-processing system (Fig. 7.7). In contrast, genes for metabolism (and therefore phenotype) seem to be far more susceptible to horizontal transfer. Since these are the things we see and measure about an organism, they are certainly important.

But if the properties of an organism can move horizontally, does this make phylogeny meaningless? Not in my view. When a gene is acquired from some other source by an organism, it begins the rapid process of adapting to its new genetic environment; in other words, it quickly becomes part of the organism within which it now resides. Horizontal transfer is just another source (perhaps the biggest source) of the genetic variation that Darwinian evolution requires.

I have an old classic sports car, a 1968 Lotus Super Seven. Over the past 46 years, most of this car has been replaced at one time or another—some parts several times. Yet, according to the North Carolina Department of Motor Vehicles, it is still a 1968 Super Seven. Why? Because every time a new part (or collection of parts) was put in, it had to be incorporated into the existing car. Even replacing the engine (which has been done) or major chunks of the framework (ditto) does not void the overall ancestry of the machine. Replacing the original generator with the alternator from a Jaguar changed the behavior of the system and gave it new capabilities, but it did not make the car specifically Jaguar-like. Likewise, the material from which your body is made cycles in and out on average every 7 years; does this mean you aren't the same individual? Does this mean that adults do not have a meaningful mother or father? Of course not.

The current task, at least in my view, is to not throw the baby out with the bathwater, but to come up with a theory of microbial species and evolutionary history that incorporates both vertical and horizontal inheritance. This is probably the single most important theoretical problem in microbiology today. Remember that *On the Origin of Species* by Charles Darwin, and everything that

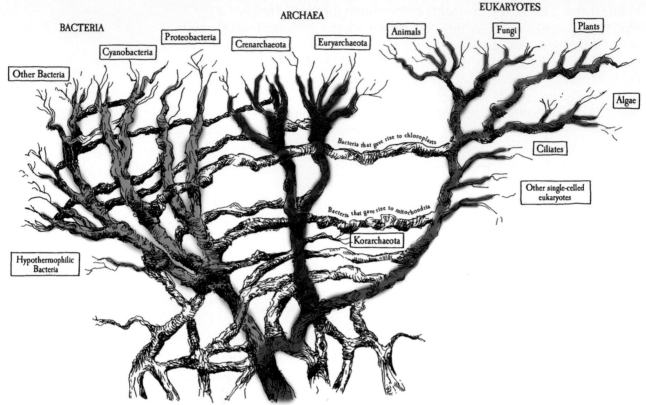

**Figure 7.7** The Doolittle "Tree of Life," with the underlying three-domain tree representing the overall descent of the genome highlighted in color. The colored branches represent the main lines of descent, and they clearly reveal the three-domain tree. The original image was intended to emphasize horizontal transfer, but the cross-branches represent a minority of genes compared to the main lines of descent. If this tree were drawn with the thickness of the branches representing genetic content, it might convey a very different picture. (Adapted from a drawing by Perdita Phillips.) doi:10.1128/9781555818517.ch7.f7.7

has flowed from it, originated from trying to understand what a "species" really is in the macroscopic world. What will we learn from a much-needed understanding of microbial species, the organisms that dominate our world?

## *Horizontal transfer and the origin of the domains from the last common ancestor*

There is reason to believe that horizontal transfer was more common in the deep past, before the emergence of the three domains, than it is today. In fact, the emergence of the *Bacteria*, *Archaea*, and *Eukarya* from their common ancestor may actually represent the emergence of more or less independently evolving lineages from a prior amalgamated gene pool. Perhaps those who think that microbes exist as a continuous gene pool are correct after all, but only for organisms before the emergence of distinct lineages and the three domains. Notice in Fig. 7.6 and 7.7 that the three domains emerge from a network of branches and that ancestry gets harder to distinguish as we look deeper in the tree. The last common ancestor probably was not a specific organism, or even species, but a communal system with very different evolutionary properties from those of modern organisms.

By the way, the same sort of argument from a tree network can be used to show that there was no meaningful "mitochondrial Eve." The last common ancestor of humans, following the mitochondrial genes and therefore the maternal lineage, was no doubt an unexceptional female member of the breeding population of protohumans. Only chance resulted in her, and only her, mitochondrial lineage persisting into the far future. Nothing at the time, or for a long time thereafter, singled out her lineage. It had to be someone.

## Questions for thought

1. Do molecular phylogenetic trees really represent biological diversity? How would you define diversity?

2. In the big tree, most of the deep (early-branching) species of eukaryotes are parasites (*Giardia*, microsporidia, trypanosomes, etc.). Can you think of a reason why this might be?

3. Do microbes actually have meaningful genealogies?

4. Do molecular phylogenies represent how primitive or derived an organism is, or just how primitive or derived the gene studied is?

5. Given that *Bacteria* and *Archaea* (and many members of the *Eukarya*, for that matter) do not reproduce sexually, how would you define "species" in these organisms?

6. How could the properties of the three evolutionary groups be useful in predicting specific properties of the last common ancestor (or gene pool)?

7. It appears that fewer than 1% of living things have been characterized, and in only a very small fraction of these have SSU rRNA sequences been determined. How might the Big Tree change as more and more SSU rRNA sequences become available for analysis?

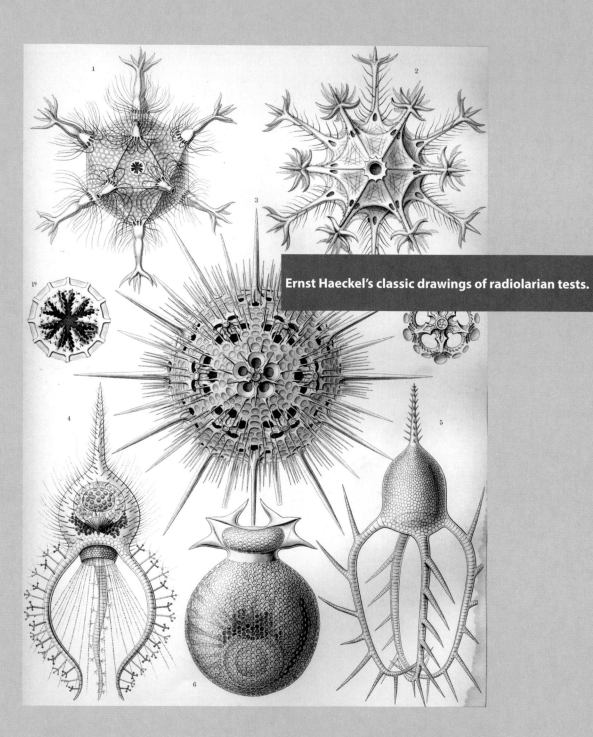

Ernst Haeckel's classic drawings of radiolarian tests.

*The Microbial Zoo*

In this section, we take a stroll through the microbial world, as if it were a zoo or garden. No zoo contains *all* animals, but only representatives; a complete collection would overwhelm both visitor and keeper alike. Thus, we focus on the most prevalent and well understood groups of organisms, and we discuss only a small number of representatives from each. Most of these phylogenetic groups contain a wide range of organisms; there are something like 5,000 described species of *Bacteria* alone, and given any reasonable definition of "species" there must be many orders of magnitude more to be found in nature.

Section II is dominated by a discussion of various groups of *Bacteria* (chapters 8 to 14), followed by shorter discussion of the *Archaea* (chapter 15), *Eukarya* (chapter 16), and lastly viruses and prions (chapter 17). Because so many chapters in this section are about the various groups of *Bacteria*, an introduction to the relationships within this group might be helpful.

## The *Bacteria*

The *Bacteria* can mostly be divided into 13 traditional groups (Kingdoms or Phyla, depending on your nomenclature), related by the tree shown in Fig. 1. Most of these branches of *Bacteria*, and certainly most of the branches with lots of known and abundant species, radiate from a single region of the tree, the "main radiation." There are (perhaps) some earlier branches; these are minor groups with few species that are not generally abundant, and they are primitive organisms (short branch length) and mostly thermophilic.

Early, primitive, thermophilic branches:

> *Aquifex* and relatives (including *Thermocrinus*, the pink filamentous mat former in Yellowstone hot springs)
>
> *Thermotoga* and relatives
>
> Green nonsulfur bacteria (including *Chloroflexus*, the golden mat former in Yellowstone hot springs)

Main radiation:

**Deinococci** (including *Thermus*)

**Spirochetes:** common in the environment and as animal symbionts

**Green sulfur bacteria:** green anaerobic sulfur-oxiding photosynthesizers, not familiar to most people

**Cytophaga/Bacteroids/Flavobacteria:** *Bacteroides* is an abundant gut organism in humans; the aerobes are common but unfamiliar

**Planctomycetes:** common but few are cultivated and so are little understood

**Chlamydiae** (hopefully you know this one, but not too well): a small, closely related cluster of species

**Firmicutes:** the low-G+C gram-positive bacteria (now we have reached the hugely common groups—these are predominant in soil and include lots of pathogens and symbionts)

**Actinobacteria:** the high-G+C gram-positive bacteria (also abundant in soil and as pathogens and symbionts)

**Cyanobacteria:** the blue-green algae, which carry out oxygenic photosynthesis and power the biosphere today

**Proteobacteria:** most gram-negative bacteria you would think of are members of this group

And, of course, there are many other less well-known branches, which are discussed in chapter 14.

**Figure 1** Phylogenetic tree of the major bacterial phyla, based on SSU rRNA sequence alignments. doi:10.1128/9781555818517.sII.f1

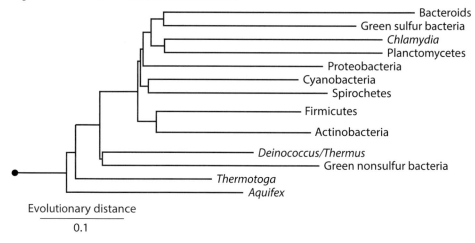

# 8

# Primitive Thermophilic Bacteria

The phyla *Aquificae* and *Thermotogae* represent the best known of several small groups of organisms that are both primitive and deeply branching in the bacterial tree (Fig. 8.1).

They are *primitive* in that they have changed less than other organisms since their common ancestry, at least in terms of small-subunit ribosomal RNA (SSU rRNA) sequences; the distance from the common ancestor to the modern sequences is shorter for them than for other members of the *Bacteria*. This does not mean that they are the ancestors of other *Bacteria* or that they are any less complex.

They are *deeply branching* in that their branches connect to the bacterial tree closer to the root than do the other branches. This statement, that these groups are deeply branching, comes with two important caveats. The first is that it presumes that the bacterial phyla represent distinct "kinds" of creatures regardless of how these phyla are related, in the same way that animal phyla are different "kinds" (evidenced by differences in basic body plan) regardless of the details of how these phyla are related. This is related to the difference between evolutionary rates and modes, and implies that each phylum emerged during a distinct transition. If this is not the case, then "deeply branching" becomes a relative term only; *Aquifex* and *Thermotoga* are deeply branching with respect to, for example, the cyanobacteria, proteobacteria, or gram-positive bacteria, but *Escherichia coli* could be viewed as a deeply branching mesophile from the perspective of *Aquifex* and its relatives.

doi:10.1128/9781555818517.ch8

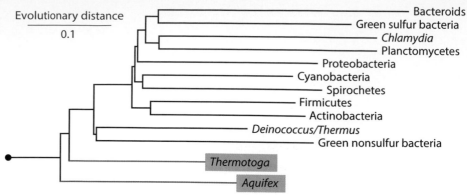

**Figure 8.1** Phylogenetic tree of the bacterial phyla, with the phyla *Thermotogae* and *Aquificae* highlighted. doi:10.1128/9781555818517.ch8.f8.1

The second caveat to the view that these phyla are deeply branching comes from uncertainties inherent in phylogenetic trees, particularly as a result of unequal evolutionary rates. This can result in long-branch attraction, in which long branches (groups with higher evolutionary rates) are artificially "pushed" toward each other as a result (in part) of underestimating the actual evolutionary distance between them. In the case of the bacterial tree, this may result in the artifactual clustering of the longer branches, excluding the shortest branches into the deeper parts of the tree. One observation in favor of this view is that as more and more diverse sequences are included in the bacterial tree (see chapter 14), *all* of the bacterial phyla come closer to seeming to emerge from a single point of radiation.

Even with these caveats in mind, however, it does seem that the shortest branches of the bacterial tree are thermophilic. The fact that they are primitive with respect to 16S rRNA sequences may represent only a lower tolerance at high temperatures for the transient non-Watson-Crick base pairs that are the inevitable intermediates of evolutionary change in RNA structures, and may not represent primitiveness in other features of these organisms.

## Phylum *Aquificae* (*Aquifex* and relatives)

A tree showing relationships between representative members of the *Aquificae* is shown in Fig. 8.2.

**Figure 8.2** Phylogenetic relationships between representative members of the *Aquificae*. doi:10.1128/9781555818517.ch8.f8.2

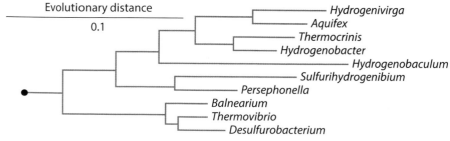

| Phylum | Aquificae | | | | |
|--------|-----------|--------|----------|-------------|--------|
| Class | | Aquificae | | | |
| Order | | | Aquificales | | |
| Family | | | | Aquificaceae | |
| Genus | | | | | Aquifex |
| Genus | | | | | Calderobacterium |
| Genus | | | | | Hydrogenivirga |
| Genus | | | | | Hydrogenobaculum |
| Genus | | | | | Hydrogenobacter |
| Genus | | | | | Hydrogenothermus |
| Genus | | | | | Persephonella |
| Genus | | | | | Sulfurihydrogenibium |
| Genus | | | | | Thermocrinus |
| Genus | | | | | Venenivibrio |
| Incertae sedis | | | | | |
| Genus | | | | | Balnearium |
| Genus | | | | | Desulfurobacterium |
| Genus | | | | | Thermovibrio |

## General characteristics of the Aquificae

### Diversity

Fewer than two dozen species have been described in the *Aquificae* (only one or a few species in each genus), and they are relatively homogenous in phenotype, especially among the *Aquificales* (the genera that are currently *incertae sedis* are more phenotypically distinct).

### Metabolism

These organisms are all thermophilic or extremely thermophilic, and obtain energy by respirative hydrogen oxidation. This reaction is known as the Knallgas reaction. They are obligate aerobes, generally microaerophilic (the solubility of oxygen in water at the temperatures at which these organisms grow is very low in any case). Reduced-sulfur compounds such as sulfide, thiosulfate ($S_2O_3^{2-}$), or elemental sulfur can generally replace hydrogen, but nitrate cannot replace oxygen as the terminal electron acceptor in most species. These organisms are autotrophic, fixing carbon dioxide via the reverse tricarboxylic acid cycle. Few have been grown heterotrophically; they are generally considered to be obligate autotrophs.

### Morphology

These organisms are rod shaped (ca. 0.5 by 2 to 8 μm) to filamentous, with a typical gram-negative-type envelope, some with an external crystalline protein S-layer. Some are motile and flagellated, but other appendages have not been seen. During growth on sulfide or thiosulfate, elemental sulfur granules can appear, but no storage granules or internal membranous structures have been observed. Some produce carotenoid pigments.

## Habitat

These organisms are common inhabitants of near-neutral-pH, high-temperature geothermal springs, including hot springs and submarine vents.

### EXAMPLE SPECIES

#### Aquifex pyrophilus

*Aquifex pyrophilus* (Fig. 8.3) is, like the other members of this group, a thermophilic hydrogen oxidizer. *A. pyrophilus* is an extreme thermophile, growing optimally at 85°C but up to 95°C. This makes it the most extremely thermophilic isolated and characterized bacterium. It is unusual among its relatives in that it can use nitrate as a terminal electron acceptor and so can grow anaerobically. It is also more sensitive to oxygen than are most members of this group, and when grown without nitrate it is an obligate microaerophile. *A. pyrophilus* was isolated in 1992 from a deep-sea hydrothermal vent on the Kolbeinsey Ridge north of Iceland (which is an outcropping of the Mid-Atlantic Ridge).

The phylum is named after this organism because it was the first of its members to be characterized phylogenetically and discovered to represent a distinct branch of the *Bacteria*. It was only later found that previously isolated species (of the genera *Hydrogenobacter* and *Calderobacterium*) were related to *A. pyrophilus*.

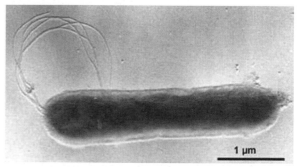

**Figure 8.3** Electron micrograph (shadow cast) of *Aquifex pyrophilus*. (Courtesy of Reinhard Rachel, Harald Huber, Reinhard Wirth, and Michael Thomm, University of Regensburg.) doi:10.1128/9781555818517.ch8.f8.3

#### Thermocrinus ruber

*Thermocrinus ruber* (Fig. 8.4) is also a thermophilic hydrogen oxidizer, growing optimally at 80°C but up to 89°C. It grows either as individual rod-shaped cells

**Figure 8.4** Scanning electron micrograph of *Thermocrinus ruber*. (Reprinted from Huber R, Eder W, Heldwein S, Wanner G, Huber H, Rachel R, Stetter KO, *Appl Environ Microbiol* **64**:3576–3583, 1998, with permission.) doi:10.1128/9781555818517.ch8.f8.4

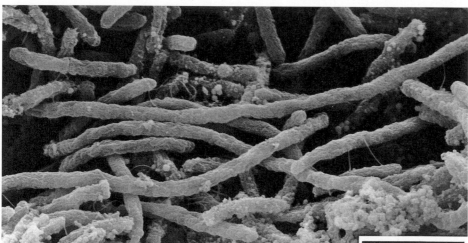

**Figure 8.5** Octopus Spring, a neutral-pH hot spring in Yellowstone National Park.
doi:10.1128/9781555818517.ch8.f8.5

(0.5 by 1 to 3 μm) that are motile by monopolar polytrichous flagella (multiple flagella at one end of the cell only) or as filaments. *T. ruber* can grow heterotrophically using formate or formamide, as well as autotrophically using hydrogen or reduced-sulfur compounds as electron donors, but it cannot replace oxygen with nitrate as an alternative electron acceptor, and so it is an obligate microaerophile.

*T. ruber* was isolated from pink filamentous growth in Octopus Spring, Yellowstone National Park (Fig. 8.5). This pink filamentous growth is common in the 80 to 90°C temperature zones of neutral to slightly alkaline hot springs throughout the park and has been described since the early work of Thomas Brock in the 1960s. Numerous attempts to cultivate the pink filamentous organism failed, although *Thermus aquaticus* (the source of *Taq* polymerase, which made polymerase chain reaction amplification a reasonable technology) was isolated as a by-product of these attempts.

Ultimately, of course, *T. ruber* was isolated from the pink filaments of Octopus Spring, using insight gained by molecular phylogenetic analysis of the organism prior to cultivation. This story is described in detail in chapter 18.

# Phylum *Thermotogae* (*Thermotoga* and relatives)

A tree showing relationships between representative *Thermotogae* is shown in Fig. 8.6.

| Phylum | Thermotogae | | | |
|--------|-------------|---|---|---|
| Order | | *Thermotogales* | | |
| Family | | | *Thermotogaceae* | |
| Genus | | | | *Fervidobacterium* |
| Genus | | | | *Geotoga* |
| Genus | | | | *Marinitoga* |
| Genus | | | | *Petrotoga* |
| Genus | | | | *Thermopallium* |
| Genus | | | | *Thermosipho* |
| Genus | | | | *Thermotoga* |

## *General characteristics of the* Thermotogae

### Diversity

There are fewer than three dozen described species in a single family in the phylum *Thermotogae*.

### Metabolism

The *Thermotogae* are thermophilic (65 to 90°C), anaerobic fermentative organisms that grow on a wide range of organic compounds, using protons or elemental sulfur as the terminal electron acceptor. Ferric ion ($Fe^{3+}$) can serve as an electron acceptor in some isolates. The products of the oxidation of organic material are typically $CO_2$ and organic acids such as acetate and lactate. When they are grown in the absence of sulfur, hydrogen is the waste product of proton reduction and hydrogen is a strong inhibitor of growth; hydrogen removal, often by symbiosis with methanogens, is required to maintain the energetic favorability of this reaction.

### Morphology

The distinguishing feature of *Thermotogae* is the presence of a loose sheath, the "toga," covering the rod-shaped cells. This sheath is typically snug over the sides

**Figure 8.6** Phylogenetic tree of the genera of the phylum *Thermotogae*.
doi:10.1128/9781555818517.ch8.f8.6

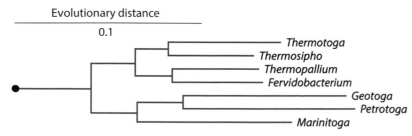

of the cells but balloons out at each end. The toga is the outer membrane of an otherwise typical gram-negative-type envelope, and so the space captured by the toga is periplasm. The toga is rich in porin-like proteins, arrayed in a regular pattern over the entire surface of the toga. The function of this unusual toga structure is unknown. Most members of this group are flagellated.

## Habitat

These organisms have been isolated primarily from geothermally heated soils and sediments, including solfataras, the soil surrounding hot springs, and hot sediments in the vicinity of deep-sea hydrothermal vents.

### EXAMPLE SPECIES

#### *Thermotoga maritima*

By far the best-studied member of the phylum *Thermotogae* is *Thermotoga maritima* (Fig. 8.7), and for many years it was the only known member of the group. It was isolated from a heated submarine sediment off the island of Vulcano, Italy. Cells are rods, measuring ca. 0.6 by 2 to 5 μm, with pronounced togas. This species is an extreme thermophile, growing in a temperature range of 50 to 90°C, optimally at 80°C. It is motile via terminal flagella, but other species of *Thermotoga* are nonmotile. It is able to fix nitrogen but cannot use nitrate or sulfur compounds other than elemental sulfur as terminal electron acceptors. *T. maritima* lacks genes for the electron transport chain and so must generate its proton motive force (proton gradient) using adenosine triphosphatase (ATPase) (run in the reverse of the usual reaction) or by a membrane-associated pyrophosphatase proton pump.

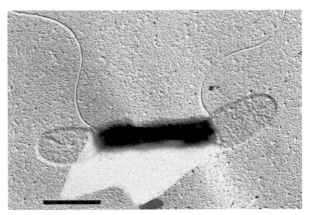

**Figure 8.7** Electron micrograph (thin section) of *Thermotoga maritima*. (Courtesy of Reinhard Rachel, Harald Huber, Reinhard Wirth, and Michael Thomm, University of Regensburg.) doi:10.1128/9781555818517.ch8.f8.7

*T. maritima* is easy to grow, and the complete genome sequence was one of the earliest available from a thermophilic bacterium, following only that of *Aquifex aeolicus*, which is notoriously difficult to handle. This, in combination with the fact that proteins from extreme thermophiles often form high-quality crystals more readily than do those of mesophiles (perhaps because they are more rigid and so more uniform in structure than are the homologous proteins from mesophilies), and are usually much easier to express in recombinant form in *E. coli* than are the proteins of *Archaea*, makes this organism an attractive source of proteins for structural examination by X-ray diffraction.

The genome sequence of *T. maritima* provided the best evidence available for the large-scale horizontal transfer of genes across large phylogenetic distances (discussed in chapter 7). It has been argued that up to one-quarter of the genome of *T. maritima* may have been acquired from an archaeal source recently enough that it can still be identified as foreign; others would revise this figure downward to about 5%. Either way, it is clear that this organism has acquired a great deal of genetic potential from outside sources.

### Thermosipho africanus

*Thermosipho africanus* was isolated from hot, sandy sediment off the coast of Obock, Djibouti. This species is morphologically much like *Thermotoga* but grows in chains up to 12 cells long. Cells within chains are connected and separated by their togas. *T. africanus* is also a thermophile but grows at lower temperatures than other members of this phylum: 35 to 77°C, optimally at 75°C. Figure 8.8 shows a close relative.

### Fervidobacterium islandicum

*Fervidobacterium islandicum* (Fig. 8.9) was the second member of this phylum to be isolated, after *T. maritima*, but is much less well characterized. It was isolated from a solfatara in the Hveragerði geothermal fields of Iceland. In this species, and its close relative *F. nodosum*, the toga forms a large spheroid "nodule" at one end of the cell. *F. islandicum* grows at temperatures ranging from 40 to 80°C, optimally at 65 to 70°C.

## Other primitive thermophiles

There are a number of other groups of primitive thermophiles that are even less well understood than are the *Aquificae* and *Thermotogae*. Most are known only from SSU rRNA sequences obtained directly from thermal environments and have not been cultivated (see chapter 14). Others are known only from single isolates, their placement in the tree is uncertain, or both.

## Thermodesulfobacterium

The phylum *Thermodesulfobacteriales* consists of only six named isolates in three genera: *Geothermobacterium*, *Thermodesulfatator*, and *Thermodesulfobacterium*. The best (but still poorly) characterized member of this group is *Thermodesulfobacterium hydrogenophilum*. *T. hydrogenophilum* is a thermophilic (50 to 80°C, optimum 75°C) anaerobic sulfate reducer isolated from a deep-sea hydrothermal vent in the Guaymas Basin. It is a small (0.4 to 0.5 by 0.5 to 0.8 μm) motile rod with a single, polar flagellum. In stationary-phase culture, cells tend to form short chains of longer cells, and sometimes cyst-like cells appear. This organism grows by sulfate reduction to sulfide, using hydrogen ($H_2$) as the only electron donor (other species in this genus use organic acids). Unlike other species of this genus, *T. hydrogenophilum* cannot use sulfite or thiosulfate as an electron acceptor. It is an obligate chemolithoautotroph.

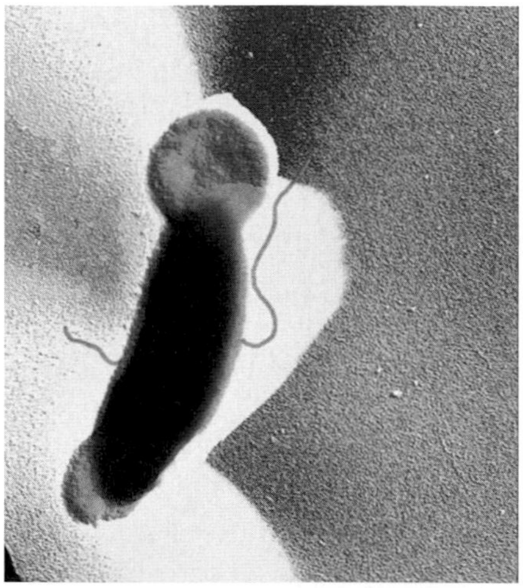

## Thermomicrobium

The phylum *Thermomicrobia* consists of only a single characterized species, *Thermomicrobium roseum*, a potential second uncharacterized

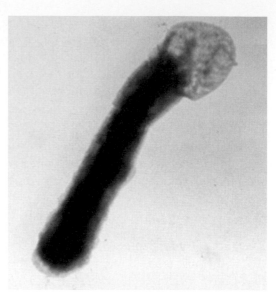

**Figure 8.8** Electron micrograph of *Thermosipho melanesiense*, a close relative of *Thermosipho africanus*. (From the Joint Genome Institute-U.S. Department of Energy.)
doi:10.1128/9781555818517.ch8.f8.8

**Figure 8.9** Electron micrograph (thin section) of *Fervidobacterium islandicum*. (Courtesy of Reinhard Rachel, Harald Huber, Reinhard Wirth, Michael Thomm, and Veronika Heinz, University of Regensburg.)
doi:10.1128/9781555818517.ch8.f8.9

isolate, and some environmental sequences. This phylum may be a distant but specific relative of the phylum *Chloroflexi*. *T. roseum* was isolated from Toadstool Spring in Yellowstone National Park; it grows optimally at 75°C. It forms irregular to pleomorphic short rods of ca. 1.5 by 3 to 6 μm, is nonmotile, and forms pink colonies on plates. It is an obligately aerobic heterotroph, but its respiratory pathway is unknown and, as it is cyanide resistant, probably unique.

### Chloroflexi *and* Deinococci

Although the phyla *Chloroflexi* and *Deinococci* are discussed in chapter 13, they are also largely thermophilic and perhaps deeply branching as well. However, these phyla do contain many mesophilic species and, except perhaps for *Thermus* and its relatives, are not strikingly primitive. It is not clear whether they should be included among the primitive thermophiles and/or the ancestral thermophiles (see below).

### Thermophilic ancestry of *Bacteria*

The fact that the (perhaps) deepest branches in the bacterial tree are thermophilic, and often extremely thermophilic, would imply that the ancestors of the *Bacteria* were also thermophilic and that mesophily arose later in the tree, approximately at the main evolutionary radiation where most of the nonthermophilic phyla originate. The possibility of a thermophilic ancestry for *Bacteria* is strengthened by the fact that the deeply branching and primitive *Archaea* are likewise thermophilic, and so probably the *Bacteria* and *Archaea* share a thermophilic ancestry.

But not all bacterial thermophiles are primitive and related only to other thermophiles; many are close relatives of otherwise mesophilic species. This means that there are two types of fundamentally different thermophiles: ancestral thermophiles and adapted thermophiles.

Ancestral thermophiles (e.g., *Thermotoga* and *Aquifex*) are thermophiles for which there is no evidence of mesophilic ancestors. These organisms have been thermophiles as far back as we can observe phylogenetically, and so seem never to have adapted to mesophily; they evolved from scratch in thermophilic environments and thermophily is built into them from the basics up. In contrast, adapted thermophiles (e.g., thermophilic *Bacillus* and *Clostridium* species) are related to and evolved from mesophiles, (re)adapting to thermophily more recently.

This is a critical difference for those interested in examining how life is possible at high temperatures.

### Life at high temperatures

How are these organisms capable of growth at such high temperatures, such as those found in hot springs (Fig. 8.10)? What is the upper temperature limit for life? The answers to these questions are not entirely clear, and high temperature

**Figure 8.10** Obsidian Pool, a slightly acidic boiling hot spring in Yellowstone National Park. The black color of the water is from metal sulfides, primarily pyrite, which combined with silica precipitate as obsidian sand. doi:10.1128/9781555818517.ch8.f8.10

may not really be an extreme condition except from our anthropocentric point of view. Nevertheless, there are some characteristics of life at very high temperatures, which are discussed here.

## Membrane fluidity and integrity

As temperature increases, the fluidity of the lipoprotein membranes of the cells increases. This fluidity must be balanced. In general, the membrane lipids of thermophiles have a higher melting point than those of mesophiles, so that the fluidity of the membranes of these organisms is appropriate at their optimum growth temperature. Organisms can also change their mix of membrane lipids in response to changes in temperature.

## DNA structure

The two strands of typical DNA in solution separate at about 70°C, depending on ionic strength and which ions are present. Although increasing the fraction of G-C pairs increases this melting point somewhat, there is no correlation between the growth temperature of an organism and the G+C content of its DNA. It turns out that in the cell, the DNA is inherently resistant to denaturation because of the high ionic strength and low water activity of the cytoplasm (i.e., most of the water is already tied up in hydration shells). Most organisms negatively supercoil their DNA, which makes it more easily denatured, but extremely thermophilic *Archaea* and *Bacteria* positively supercoil their DNA. DNA-binding proteins such as histones or histone-like proteins also stabilize DNA to thermal denaturation.

## RNA structure

The folded structure of non-messenger RNAs (mRNAs) (e.g., rRNA, transfer RNA [tRNA], ribonuclease [RNase] P RNAs, small nuclear RNAs [snRNAs], small nucleolar RNAs [snoRNAs], and transfer-messenger RNA [tmRNA]) can be denatured just like the strands of DNA. However, modest changes in the sequences and structures of RNAs can stabilize them against denaturation. Some thermophilic RNAs are rich in G-C base pairs and, more importantly, are very low in G-U pairs, mismatches, bulges, and other irregularities that might lead to excessive flexibility in the RNA. The RNAs are also usually short, with no extra sequences; shorter sequences have fewer nonfunctional folding possibilities. In addition, base modifications and changes in protein binding can stabilize RNAs.

## Protein structure

The denaturation of proteins from mesophiles at high temperature is dramatic; a good example is what happens to an egg when you boil it. However, stabilizing a protein for function at high temperatures seems to be a relatively easy evolutionary task, although our understanding of these changes is poor at best.

## Enzymatic function

The function of an enzyme is tuned to the growth temperature of the organism. Mesophilic enzymes work best at 20 to 40°C and denature at higher temperatures, but enzymes from thermophiles work best at their own growth temperature and denature when heated further. Thermophilic enzymes work slowly if at all at mesophilic temperatures; they are too rigid at these temperatures and are essentially frozen. In other words, enzymes are tuned for optimal flexibility at the temperature at which they function (Fig. 8.11).

A more complex issue is balancing catalytic function in the cell. Different reactions increase in speed at different rates as the temperature goes up, and these must be kept in balance.

## Small-molecule stability

Small-molecule stability may be a more difficult problem for thermophiles. The half-life of guanosine triphosphate (GTP) at 100°C is measured in seconds, and yet even organisms that grow at this temperature use GTP for translation, RNA synthesis, and many other processes. Many other small molecules are not very heat resistant, e.g., uridine triphosphate (UTP), nicotinamide adenine dinucleotide (NAD), and flavin adenine dinucleotide (FAD). These may be synthesized on a just-in-time basis so that their degradation is not too great a loss

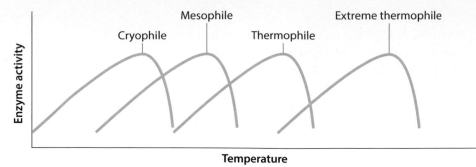

**Figure 8.11** Representation of enzymatic activities from different kinds of organisms as affected by temperature. doi:10.1128/9781555818517.ch8.f8.11

for the cell. Another way to put this is that the flux of reactions through these intermediates may be very high even though their steady-state concentration may be low.

The highest growth temperature of cultivated species is about 120°C, but there is good evidence for life growing at up to about 135 to 140°C in hydrothermal environments. This may represent the upper limit for life, because at this temperature amino acids become racemized (flip from L to D) at significant rates. This is independent of any of the normally stabilizing mechanisms, and the flipping of an amino acid in a protein would potentially lead to its irreversible denaturation. However, one temperature after another has previously been suggested to be the limit for life, for seemingly very good reasons, and been proven wrong by biology. So who knows?

## Questions for thought

**1.** Going back to the description in chapter 4 of the neighbor-joining algorithm for generating phylogenetic trees, how do you think long-branch attraction might affect the bacterial tree?

**2.** The *Aquificae* and *Thermotogae* are both primitive and (probably) deep-branching phyla. Can you draw a tree with deep branches that are not primitive, and with primitive branches that are not deep? Do you know of any examples of either?

**3.** Organisms such as *Thermotoga maritima* that reduce protons to generate hydrogen can do so only if very low ambient hydrogen concentrations make it energetically favorable. How do you suppose this hydrogen is gotten rid of so efficiently in their native environment?

**4.** What do you think the absolute limiting issue for high-temperature growth might be? What do you think is the highest temperature at which any familiar type of organism could grow?

**5.** Can you think of any opportunities that extreme thermophiles might have that are not available to mesophiles? In other words, what might be some advantages of life in high-temperature environments?

6. *Thermocrinus ruber* is found in Yellowstone hot springs separated by many miles of inhospitably cold (for them) territory. How do you suppose they colonize new hot springs when they emerge? Would you predict that the same organisms exist in similar hot springs in other parts of the world?

7. How many continuously or intermittently high-temperature microbial environments can you identify close to where you are right now? Do you think you could isolate thermophiles from these environments? How would you go about it?

# 9

# Green Phototrophic Bacteria

The green phototrophic bacteria include three phylogenetic groups: the green sulfur bacteria (*Chlorobi*), the green nonsulfur bacteria (*Chloroflexi*), and the cyanobacteria (blue-green algae) (Fig. 9.1). They contain bacteriochlorophyll *c*, *d*, or *e* or chlorophyll *a* (sometimes with chlorophyll *b* as well). These give many of these organisms a green color, but color is not a consistent trait; the green color is often masked by the abundant accessory pigments. These

**Figure 9.1** Phylogenetic tree of the bacterial phyla, with the phyla *Chlorobi* (green sulfur bacteria), *Chloroflexi* (green nonsulfur bacteria), and *Cyanobacteria* highlighted. doi:10.1128/9781555818517.ch9.f9.1

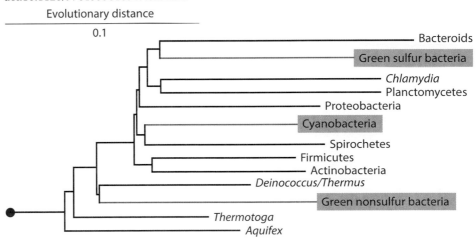

organisms also usually contain their antenna complex in discrete membranous organelles, the chlorosomes (green sulfur or nonsulfur bacteria) or thylakoids (cyanobacteria). Purple phototrophic bacteria use bacteriochlorophylls *a* and *b* and house their antenna complexes in invaginations of the cytoplasmic membrane.

These three phylogenetic groups are, as a whole, predominantly phototrophic (at least among the familiar species), whereas the purple phototrophic bacteria and heliobacteria are members of phylogenetic groups (the *Proteobacteria* and *Firmicutes*, respectively) that are predominantly nonphototrophic.

## Phylum *Chloroflexi* (green nonsulfur bacteria)

A phylogenetic tree of the genera of *Chloroflexi* is shown in Fig. 9.2.

| Phylum | *Chloroflexi* | | | | |
|---|---|---|---|---|---|
| Class | | *Chloroflexi* | | | |
| Order | | | *Chloroflexales* | | |
| Family | | | | *Chloroflexaceae* | |
| Genus | | | | | *Chloroflexus* |
| Genus | | | | | *Chloronema* |
| Genus | | | | | *Chlorothrix* |
| Genus | | | | | *Heliothrix* |
| Genus | | | | | *Roseiflexus* |
| Genus | | | | | *Kouleothrix* |
| Family | | | | *Oscillochloridaceae* | |
| Genus | | | | | *Oscillochloris* |
| Order | | | *Herpetosiphonales* | | |
| Family | | | | *Herpetosiphonaceae* | |
| Genus | | | | | *Herpetosiphon* |
| Class | | *Anaerolineae* | | | |
| Order | | | *Anaerolinaeles* | | |
| Family | | | | *Anaerolinaeceae* | |
| Genus | | | | | *Anaerolinea* |
| Order | | | *Caldilineales* | | |
| Family | | | | *Caldilineae* | |
| Genus | | | | | *Bellilinea* |
| Genus | | | | | *Caldilinea* |
| Genus | | | | | *Leptolinea* |
| Genus | | | | | *Levilinea* |
| Genus | | | | | *Longilinea* |

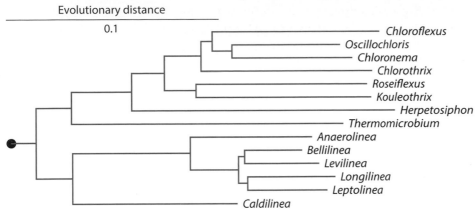

**Figure 9.2** Phylogenetic tree of the genera of the *Chloroflexi*.
doi:10.1128/9781555818517.ch9.f9.2

## *General characteristics of the* Chloroflexi

### Diversity

The phylum *Chloroflexi* consists of just over a dozen genera, with only one or sometimes two species in each. *Thermomicrobium roseum* is sometimes included in this group (and so is included in the tree above), or it may constitute a phylum of its own.

### Metabolism

The familiar members of this phylum fall into two general metabolic types: the thermophilic phototrophs (*Chloroflexales*) and the heterotrophs, some of which are thermophiles (*Anaerolineae*) and some of which are mesophiles (*Herpetosiphon*).

The phototrophic *Chloroflexi* carry out anoxygenic photosynthesis; they have a single type of photosystem and carry out cyclic photophosphorylation. They grow best photoheterotrophically; however, most can fix carbon if need be, but do not use either the Calvin cycle or reverse tricarboxylic acid (TCA) cycle. Instead they use an unusual pathway found otherwise only in a few of the *Archaea*, the hydroxypropionate pathway. The electrons used to reduce $CO_2$ to glyoxylate are obtained from reduced nicotinamide adenine dinucleotide phosphate (NADPH) derived by reverse electron transport from the oxidation of sulfide or hydrogen. Most of these organisms can also grow chemoheterotrophically by aerobic respiration.

### Morphology

Members of this phylum are flexible unbranched filaments, usually thin (<2 μm) with a uniform diameter, and motile by gliding. Septa are present but not usually visible. Phototrophic species are green or orange en masse, the color depending on which photopigments are produced. Most contain membranous "chlorosomes" resembling individual thylakoids directly beneath the cell membrane; these contain very high concentrations of accessory photopigments (carotinoids and bacteriochlorophyll *c*), whereas the reaction centers and bacteriochlorophyll *a* are found primarily in the cytoplasmic membrane. These chlorosomes resemble those of the *Chlorobi* (see below), but phototrophy in *Chloroflexi* otherwise resembles that of typical purple-bacterium photosynthesis. Phototrophic species grown aerobically become etiolated and resemble the nonphototrophic species.

The presence of an outer membrane makes these organisms formally gram negative, but they generally seem to lack the lipopolysaccharide typical of the gram-negative envelope. Some are said to have thin sheaths, visible at the ends of filaments, but these may represent cell envelope material left over after breakage of the filaments at dead cells rather than true sheaths.

## Habitat

Phototrophic members of this phylum are common and conspicuous in moderate-pH hot springs at temperatures from 35 to 70°C or more, but they are most abundant at about 60°C. Mesophilic heterotrophs are commonly abundant in wastewater sludge. They are also commonly seen in sediments, soil, and freshwater.

### EXAMPLE SPECIES

### *Chloroflexus aurantiacus*

*Chloroflexus aurantiacus* (Fig. 9.3) is the best-studied phototrophic member of this phylum, most of whose members have not been grown in pure culture. *C. aurantiacus* was originally isolated from photosynthetic mats surrounding hot springs in Yellowstone National Park (Fig. 9.4). It seems to be the primary producer of the distinctive orange mats found in these hot springs (but see the discussion of *Roseiflexus*, below), and is also abundant in an orange layer directly beneath the light green cyanobacterial mats at somewhat higher temperatures. Filaments are thin (0.5 to 1.2 μm) and glide slowly, ca. 1 μm per minute.

**Figure 9.3** Phase-contrast micrograph of *Chloroflexus aurantiacus* (Source: Wikimedia Commons Chlorofl.jpg, attributed to the Joint Genome Institute of the U.S. Department of Energy.) doi:10.1128/9781555818517.ch9.f9.3

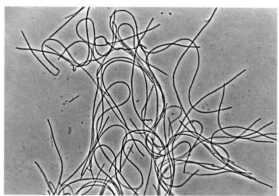

**Figure 9.4** *Chloroflexus*-rich microbial mat in the outflow of Excelsior Geyser, Yellowstone National Park. doi:10.1128/9781555818517.ch9.f9.4

*Chloroflexus* contains chlorosomes and bacteriochlorophyll *c*, whereas other members of the *Chloroflexi* do not. These are traits otherwise found only in the green sulfur bacteria (*Chlorobi*), and it is likely that the genes encoding these were acquired by *Chloroflexus* by horizontal transfer.

Although *C. aurantiacus* is capable of carbon fixation, in most mats it is found in association with cyanobacteria. In these communities, the cyanobacteria are the primary producers and the *Chloroflexi* grow photoheterotrophically from organic carbon produced by the cyanobacteria. They are not capable of fixing nitrogen but can acquire their sulfur from sulfate.

### Roseiflexus castenholzii

*Roseiflexus* (Fig. 9.5) very closely resembles *Chloroflexus* except that it lacks chlorosomes and the associated accessory bacteriochlorophyll *c*. Although *Chloroflexus* is usually thought to be the primary member of this phylum in hot-spring mats and is readily cultivated from these mats, both molecular phylogenetic analysis (fluorescent in situ hybridization and ribosomal RNA [rRNA]-based surveys) and spectral analysis of photopigments suggest that *Roseiflexus* is more abundant.

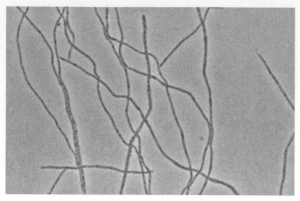

**Figure 9.5** Phase-contrast micrograph of *Roseiflexus castenholzii*. (Source: Marcel van der Meer.) doi:10.1128/9781555818517.ch9.f9.5

## Herpetosiphon aurantiacus

*Herpetosiphon aurantiacus* (Fig. 9.6) is a mesophilic heterotrophic member of the phylum *Chloroflexi*, isolated from the polysaccharide matrix of the eukaryotic alga *Chara* from Birch lake in Minnesota. *Herpetosiphon* and morphologically similar organisms are commonly seen in soils and sediments, and especially activated sludge from the wastewater treatment process, but have rarely been cultivated. Although they are always present in activated sludge, their overgrowth is thought to cause the bacterial flocs they reside in to retain bulk water and not settle into compact sludge. In some cases, these flocs trap gas bubbles and float, producing a thick foam. Microbiologists at wastewater treatment facilities therefore carefully monitor these organisms in the aerobic digestion process, guided by morphological keys to their categorization rather than by cultivation.

**Figure 9.6** Phase-contrast micrograph of *Herpetosiphon aurantiacus*, showing its flexibility; notice the small overhand knot in the filament at the bottom left and the complex "fisherman's" knot at the top. (Source: Irina Arkhipova and Michael Shribak.) doi:10.1128/9781555818517.ch9.f9.6

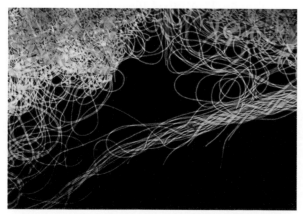

### Anaerolinea thermophila

*Anaerolinea thermophila* (Fig. 9.7) was isolated from an industrial thermophilic anaerobic digester treating wastewater from fried soybean curd. It represents a second branch of the *Chloroflexi* that is poorly understood, composed of thermophilic heterotrophic filaments. *A. thermophila* grows at 45 to 65°C (55°C optimum) and is an obligate anaerobe; other members of this group are facultatively aerobic. Protons are the terminal electron acceptor, generating hydrogen. Growth is strongly inhibited by the accumulation of hydrogen, and so is promoted by cocultivation with hydrogen-consuming methanogens. Unlike other members of this phylum, *A. thermophila* is nonmotile.

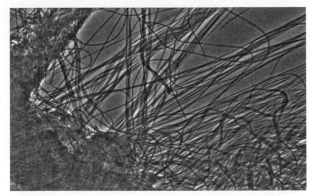

**Figure 9.7** Phase-contrast micrograph of *Anaerolinea thermophila*. (Reprinted from Sekiguchi Y, et al., *Int J Syst Evol Microbiol* **53**:1843–1851, 2003, with permission.) doi:10.1128/9781555818517.ch9.f9.7

## Phylum *Chlorobi* (green sulfur bacteria)

A phylogenetic tree of the genera of *Chlorobi* is shown in Fig. 9.8.

| Phylum | *Chlorobi* | | | | |
|---|---|---|---|---|---|
| Class | | *Chlorobia* | | | |
| Order | | | *Chorobiales* | | |
| Family | | | | *Chlorobiaceae* | |
| Genus | | | | | *Chlorobium* |
| Genus | | | | | *Anacalochloris* |
| Genus | | | | | *Chlorobaculum* |
| Genus | | | | | *Chloroherpeton* |
| Genus | | | | | *Pelodictyon* |
| Genus | | | | | *Prosthecochloris* |

### *General characteristics of the* Chlorobi

#### Diversity

The phylum *Chlorobi* contains only a small number of relatively closely related species. Most are poorly characterized, and the relationships between them are uncertain. This phylum is probably related to the phylum *Bacteroidetes* and is often grouped with it.

**Figure 9.8** Phylogenetic tree of the genera of the phylum *Chlorobi*. doi:10.1128/9781555818517.ch9.f9.8

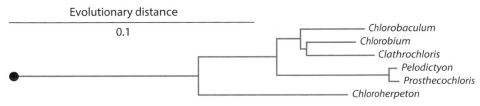

## Metabolism

These organisms are strict photolithoautotrophs, using hydrogen or sulfide as the electron donor for reverse electron flow to generate reduced nicotinamide adenine dinucleotide (NADH) for carbon fixation via the reverse TCA cycle. Energy is generated by cyclic photophosphorylation. Most are also capable of fixing nitrogen but cannot assimilate sulfate, and so they require sulfide as a sulfur source even when using hydrogen as the electron donor for $CO_2$ fixation. Elemental sulfur is the product of sulfide oxidation and accumulates as extracellular globules.

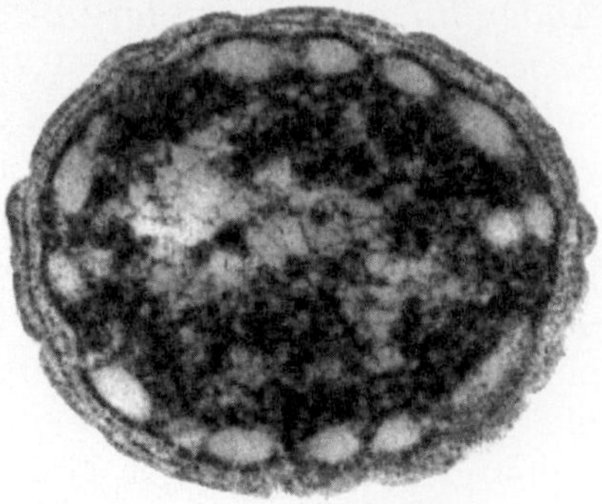

**Figure 9.9** *Chlorobium tepidum* chlorosomes. (Reprinted from Frigaard NU, Voigt GD, Bryant DA, *J Bacteriol* **184:**3368–3376, 2002, with permission.) doi:10.1128/9781555818517.ch9.f9.9

## Morphology

Cellular morphology is diverse in this group, with short chains of rods being the most common. Some produce gas vacuoles, but they are otherwise nonmotile except *Chloroherpeton*, which is a unicellular gliding filament. Cells contain internal membranous chlorosomes along the inside of the cytoplasmic membrane (Fig. 9.9); these house high concentrations of the accessory photopigments bacteriochlorophyll *c*, *d*, or *e* and carotenoids. Unlike cyanobacterial antenna complexes, in which the pigments are associated with proteins, these pigments in *Chlorobi* and *Cloroflexi* are not bound by protein, existing instead nearly in a solid state, and transfer light energy very efficiently. Reaction center chlorophyll is bacteriochlorophyll *a*, which is located in the cytoplasmic membrane. Sulfur granules are produced externally.

## Habitat

Members of the *Chlorobi* are commonly found in anaerobic, sulfide-rich freshwater and marine sediments. Because they require less light than other phototrophic organisms (about a quarter as much as typical green or purple bacteria), they can live deeper in the more anoxic zones of their environments and so are less conspicuous than the other green bacteria. They can be conspicuous, however, in meromictic (permanently stratified) lakes, in which they form a brown or green layer in the water column beneath the red or purple layer of purple bacteria at the chemocline (the interface between the oxygenated surface water and the denser anaerobic sulfide-rich deep water).

## Symbiotic consortia

Some members of this phylum participate in a symbiosis with an uncultivated rod-shaped heterotroph (betaproteobacteria in known cases). The motile heterotroph is bound by many (about a dozen) nonmotile *Chlorobi* organisms; the heterotroph is provided resources by the *Chlorobi*, which in turn are provided motility by the heterotroph (Fig. 9.10). The two cell types divide synchronously, and the heterotroph swims phototactically, attracted by light at the absorption maximum of bacteriochlorophyll *c* (740 nm) of the phototroph; the cells are in close communication, but the details of this interaction are unknown. Unlike the heterotrophs, the *Chlorobi* can often be cultivated alone (although they are distinct from the free-living species). These consortia are common in meromictic lakes and provide one of the best examples of specific bacterium-bacterium symbiosis.

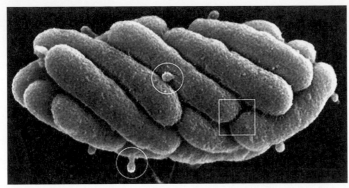

**Figure 9.10** *Chlorobium* symbiotic consortium. The visible cells are nonmotile *Chlorobium*, and the encapsulated cell is a flagellated betaproteobacterium. (Reprinted from Wanner G, Vogl K, Overmann J, *J Bacteriol* **190:**3721–3730, 2008, with permission.) doi:10.1128/9781555818517.ch9.f9.10

## EXAMPLE SPECIES

### *Chlorobium limicola*

*Chlorobium limicola* (Fig. 9.11) strains are sometimes found in symbiosis with motile betaproteobacteria, as described above, but most often they are free-living sediment dwellers. They grow as short chains of rod-shaped cells, each 0.7 to 1.1 by 0.9 to 1.5 μm, and generate large sulfur globules. This species does not produce gas vacuoles. Cultures are green, and the predominant photopigment is bacteriochlorophyll *c* or *d*.

### *Pelodictyon phaeoclathratiforme*

*Pelodictyon phaeoclathratiforme* (Fig. 9.12) is common in meromictic lakes and is composed of trapezoidal rod-shaped cells of ca. 1 by 2 μm that form branched chains, circles, and three-dimensional networks. Although generally

**Figure 9.11** Phase-contrast micrograph of *Chlorobium limicola*. Notice the extracellular phase-bright sulfur globules. (Unattributed source from the Joint Genome Institute of the U.S. Department of Energy.) doi:10.1128/9781555818517.ch9.f9.11

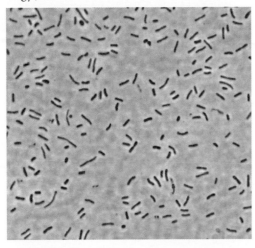

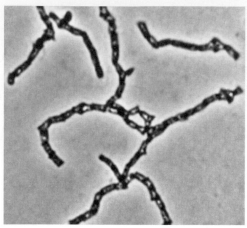

**Figure 9.12** Phase-contrast micrograph of *Pelodictyon phaeclathratiforme*. (Unattributed source from the Joint Genome Institute of the U.S. Department of Energy.) doi:10.1128/9781555818517.ch9.f9.12

considered to be nonmotile, this species produces gas vacuoles that allow it to control its vertical location in the water column. The accessory pigment is primarily bacteriochlorophyll *e*. The species requires very highly reduced conditions for growth but can use elemental sulfur or thiosulfate as well as sulfide for reducing power for carbon fixation. Unlike most members of this phylum, it is capable of using acetate to grow photoheterotrophically.

## Phylum *Cyanobacteria* (blue-green algae)

A phylogenetic tree of the genera of *Cyanobacteria* is shown in Fig. 9.13.

### *Taxonomy*

Because of the large number of genera in the phylum, only examples are given below.

| Phylum | *Cyanobacteria* | | |
|--------|-----------------|---|---|
| Class | | *Cyanobacteria* | |
| Order | | | *Chroococcales* (e.g., *Microcystis, Synechococcus*) |
| Order | | | *Pleurocapsales* (e.g., *Dermocarpa, Pleurocapsa*) |
| Order | | | *Oscillatoriales* (e.g., *Lyngbya, Oscillatoria, Spirulina*) |
| Order | | | *Nostocales* (e.g., *Anabaena, Nostoc, Calothrix*) |
| Order | | | *Stigonematales* (e.g., *Fischerella, Stigonema*) |
| Order | | | *Prochlorales* (e.g., *Prochloron, Prochlorothrix*) |

### *General characteristics of the* Cyanobacteria

#### Diversity

The members of the phylum *Cyanobacteria* are incredibly diverse phenotypically but represent a relatively small phylogenetic range. Traditionally mistakenly classified with the eukaryotic "algae," this group has a problematic

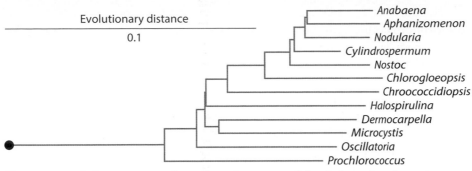

**Figure 9.13** Phylogenetic tree of representative genera of the phylum *Cyanobacteria*.
doi:10.1128/9781555818517.ch9.f9.13

taxonomy. Most species do not have formal names and are instead referred to by their genus name and number in the Pasteur Culture Collection (PCC), e.g., *Anabaena* PCC 6309, rather than *Anabaena variabilis*.

The prochlorophytes (*Prochloron* and *Prochlorothrix*) are often considered separately from the remainder of the cyanobacteria, although they are phylogenetically members of the *Chroococcales*. Photosynthesis in prochlorophytes is different in some ways from that in other cyanobacteria, resembling that of the plastids of eukaryotic phototrophs, which are also members of this group.

## Metabolism

Despite the morphological diversity of cyanobacteria, they are physiologically much alike. They carry out oxygenic photosynthesis, using two photosystems in the traditional "Z-scheme" to obtain both energy and reducing power for fixing $CO_2$ via the Calvin cycle. Water is the electron donor for $CO_2$ reduction, generating oxygen. Cyanobacteria use chlorophyll *a* as their only chlorophyll, and they use phycobilins as accessory photopigments. One form of phycobilin, phycocyanin, in combination with chlorophyll *a* produces the blue-green color from which these organisms get their name. However, most familiar members of this group produce phycoerythrin instead, resulting in a rust-red or brown color.

Most cyanobacteria can fix atmospheric nitrogen, but this creates a dilemma for them. Nitrogenases, the enzymes that reduce $N_2$ to $NH_4^+$, are strongly inhibited by $O_2$, the product of oxygenic photosynthesis. As a result, unicellular species generally fix nitrogen at night and photosynthesize by day, separating these mutually exclusive processes in time. Some filamentous species separate the process physically, by producing specialized cells called heterocysts for fixing nitrogen. Heterocysts are terminally differentiated cells (they cannot reproduce or revert to vegetative cells) that do not carry out oxygenic photosynthesis; they fix nitrogen and carry out cyclic photophosphorylation. Heterocysts provide fixed nitrogen (in the form of glutamine) to the nearby cells in the filament in exchange for energy (in the form of glutamic acid or sugar). This represents a differentiation between somatic cell lines and germ cell lines; in other words, these filaments are true multicellular organisms.

## Morphology

Cyanobacteria are very diverse morphologically, being found in all shapes and sizes. Unicellular forms are typically rods or cocci and are found in distinctive organized clusters (see the discussion of the *Chroococcales*, below). Filamentous forms are generally motile by gliding and may have sheaths. Filamentous species can have complex life cycles with a variety of specialized cell types, including nitrogen-fixing heterocysts and resting akinetes. Only the *Stigonematales* contain filaments with branches. Cyanobacteria range widely in size, from typical bacterial sizes (about 1 μm) to macroscopic; many types of cyanobacterial filaments can be seen distinctly without magnification.

Cells typically contain thylakoids flattened against the cytoplasmic membrane. These can be discrete disks, like the chlorosomes of *Chlorobi*, or concentric layers of thylakoid membrane around the periphery of the cell. Many planktonic species produce gas vacuoles to regulate buoyancy and position in the water column.

## Habitat

Cyanobacteria are common in any almost environment in which there is sunlight and water. They are especially common in freshwater and marine environments, in soils, and on and in the surface of rocks. They are the predominant primary producers in environments that are inhospitable for eukaryotic algae. Cyanobacteria are found in acid springs and soda lakes, hot springs and permanently frozen rocks, and the ocean surface layer and crusty desert soils.

Many species of cyanobacteria form symbioses with fungi, animals, plants, various protists, and other bacteria. Lichens are composite organisms, composed of a fungus and an alga, and these algae are very often cyanobacteria. Cyanobacteria form endosymbiotic relationships with a wide variety of unicellular eukaryotes, for example, diatoms (this is in addition to their usual chloroplasts). The water fern *Azolla* remains associated with its cyanobacterial symbiont (an *Anabaena*) throughout its life cycle; specialized lobes on the leaves house the symbionts, which provide their host with fixed nitrogen. Cyanobacteria are commonly associated with freshwater and marine sponges, in the gill arches of crustaceans, and in tropical-reef clams and corals. The most extreme examples of symbiosis of cyanobacteria with eukaryotes are the plastids (chloroplasts) of photosynthetic eukaryotes.

## *Family* Chroococcales

Members of the family *Chroococcales* are unicellular but usually grow as aggregates embedded in sheaths, capsules, or slime. The form of these aggregates depends on the type of covering (if any) the cells have, and how the cells divide. Division is by budding, by binary fission in one, two, or three dimensions, or irregularly. Fission in one plane produces only strings of cells, fission in two dimensions (cleavage planes alternate) produce sheets of cells in rows and columns, and division in three dimensions produces three-dimensional arrays of cells. Irregular cleavage produces irregular cell masses.

## EXAMPLE *Microcystis*

*Microcystis* (Fig. 9.14) is a common freshwater and estuarine species. Cells are coccoid, 3 to 5 μm in diameter, and are clustered irregularly in gelatinous masses. *Microcystis* flourishes in the summer months, and in polluted waters it can form blooms that look like bright green latex paint floating on and in the water. It may produce toxins (microcystins) that cause skin irritation or gastrointestinal discomfort (in the short term) or liver damage (in the long term) if ingested.

## *Family* Pleurocapsales

Members of the family *Pleurocapsales* reproduce by multiple fission, with a single large mother cell producing a number of small spore-like daughter cells known as baeocytes. The simplest forms are unicellular, with a single baeocyte growing into a large vegetative mother cell which divides into many baeocytes. If the cell divisions are in alternating planes, the baeocytes are arranged in orderly three-dimensional cubical arrays. In some species, an early single asymmetric binary fission of the baeocyte produces a large cell that continues to grow and then divides by multiple fission. The smaller "mother" cell then grows and divides again in an asymmetric binary fission, regenerating both the mother cell and a cell destined for multiple fission. In the genus *Pleurocapsa* and its relatives, the growing baeocyte also divides early into two cells; one of these cells expands and ultimately undergoes multiple fission. The other cell goes through a series of asymmetric divisions, creating branched pseudohyphae of vegetative cells. These can be simple masses of cells or complex three-dimensional structures. Eventually, cells within these pseudohyphae undergo multiple fission, releasing baeocytes.

**Figure 9.14** Bright-field micrograph of *Microcystis*. (Source: Jason Oyadomari.) doi:10.1128/9781555818517.ch9.f9.14

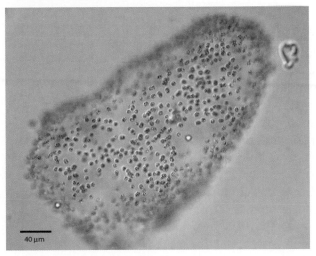

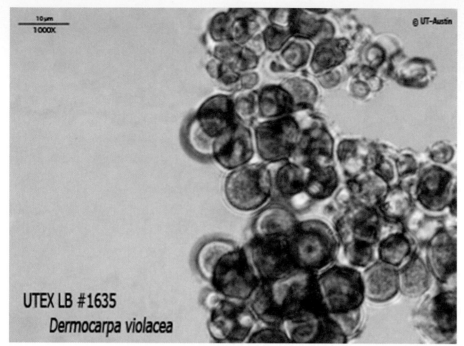

**Figure 9.15** Bright-field micrograph of *Dermocarpa violacea*. (Photo credit: UTEX Culture Collection of Algae, The University of Texas at Austin.) doi:10.1128/9781555818517.ch9.f9.15

**EXAMPLE** *Dermocarpa*

*Dermocarpa* (Fig. 9.15) has a relatively simple life cycle, in which the baeocytes (which are transiently motile, by gliding) grow into large vegetative mother cells, which divide by multiple fission into many baeocytes. These baocytes are initially contained in the spherical husk of the mother cell, which splits open to release the baeocytes into the environment.

## *Family* Oscillatoriales

These very common and conspicuous cyanobacteria are filamentous and divide only by binary fission. The only cellular differentiation in these filaments appears in the terminal cells in some cases, which can be rounded, tapered, or pointed. The form of the ends of filaments is important in the identification of these species. Filaments can be straight, loosely coiled, or tightly helical. The individual cells of the filament can be obvious, or the septa can be difficult to see. Sheaths are common, and most of these organisms are motile by gliding.

**EXAMPLE** *Oscillatoria*

*Oscillatoria* (Fig. 9.16) is a common member of this group and is ubiquitous in freshwater environments. These organisms are often large enough to be mistaken for filamentous eukaryotic green algae. The filaments typically have a light sheath, if any, and the individual cells are cylindrical disks. Filaments are rigid and rotate during gliding; this can make them appear to "oscillate" as they rotate, because of irregularities in the filament.

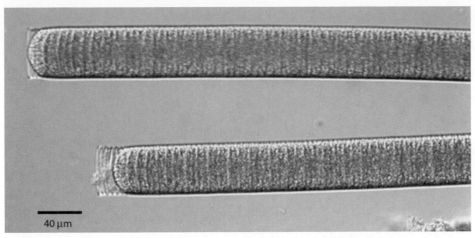

**Figure 9.16** Bright-field micrograph of *Oscillatoria*. Notice that these individuals have a sheath, visibly protruding past the rounded terminal cells of the filaments. (Source: Jason Oyadomari.) doi:10.1128/9781555818517.ch9.f9.16

### *Family* Nostocales

The *Nostocales* are morphologically similar to the *Oscillatoriales*, growing in linear filaments, but they have complex life cycles and cellular differentiation. In nitrogen-limiting conditions, some cells differentiate into nitrogen-fixing heterocysts. This is a terminal differentiation; heterocysts can neither divide nor develop back into vegetative cells. Heterocysts have a heavy cell wall and are much lighter in color than are vegetative cells, and so they are readily identified. They provide fixed nitrogen to the nearby cells. As the distance between heterocysts increases while the vegetative cells between them continue to grow and divide, the cells midway between the heterocysts become starved for fixed nitrogen, and one therefore develops into a new heterocyst.

Most members of this group also produce "resting"-stage cells called akinetes when filaments are nutritionally limited. Akinetes are usually larger than vegetative cells, contain good reserves of storage granules, and are depleted for photopigments. It is common, however, for akinetes to accumulate other pigments, making them appear dark or brown. Akinetes are resistant to a variety of environmental insults but not to heat. Filaments are prone to breakage at the junctions between akinetes or heterocysts and the adjacent vegetative cells, and so it is common to see filaments with these specialized cells at one or both ends. When provided with a fresh supply of nutrients, akinetes germinate to produce a new filament.

Some members of this group, especially the plant symbionts, also produce specialized filaments called hormogonia. Hormogonia are short filaments composed of small cells that glide rapidly. They are produced in response to detection by the vegetative filament of a soluble factor produced by the plant. The hormogonia glide rapidly toward the source of this "hormone" and develop heterocysts, and then the rest of the cells of the hormogonia develop into vegetative cells.

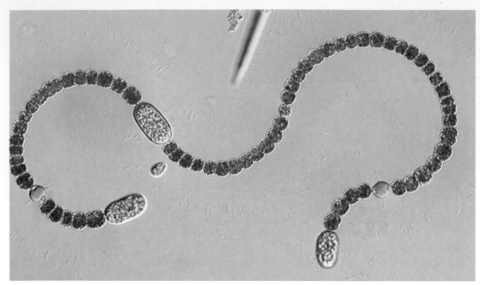

**Figure 9.17** Bright-field micrograph of *Anabaena*. Dark green cells are vegetative; the small, round lighter cells interspersed in the filament are heterocysts; and the large, ovoid light cells are akinetes. (Source: Jason Oyadomari.) doi:10.1128/9781555818517.ch9.f9.17

EXAMPLE *Anabaena*

*Anabaena* (Fig. 9.17) is a common freshwater species. Cells are in the shape of flattened beads to short barrel-like cylinders. Filaments are not covered in slime or sheaths, and akinetes are not produced adjacent to heterocysts. The cells do not produce hormogonia. Filaments are not tapered.

## *Family* Stigonematales

The *Stigonematales* divide in more than one plane, producing branched filaments or filaments composed of clusters of cells (multiseriate). Side branches are often morphologically distinct from the main filament. In some species the filaments are easily disrupted, producing clusters of cells that are easily confused with *Chroococcales*, but unlike these nonfilamentous forms, *Stigonematales* clusters often contain heterocysts or akinetes. Hormogonia are common.

EXAMPLE *Fischerella*

*Fischerella* (Fig. 9.18) is composed of multiceriate main filaments and uniseriate side branches. The cells grow to become large, dense mats of filaments that can glide along the substrate. Heterocysts form in both the main filament and side branches, whereas akinetes form only on the main filaments. The main filament, but not the side branches, is usually covered in a dense sheath.

## *Family* Prochlorales

Members of the *Prochlorales* resemble the chloroplasts of eukaryotic algae and plants more than any other cyanobacteria do. Chloroplasts are, in fact, specific

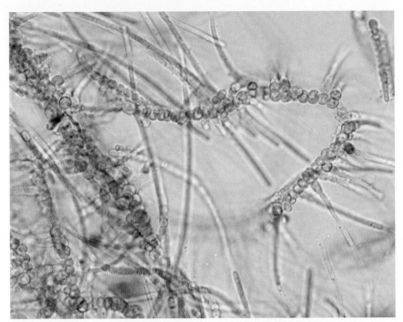

**Figure 9.18** Bright-field micrograph of *Fischerella*. Notice the main filaments composed of multiple layers of round cells, and the elongated single-width branches. (Source: Peter A. Siver and Hannah A. Shayler.) doi:10.1128/9781555818517.ch9.f9.18

relatives of the *Prochlorales* (hence the name) and so are formally members of this family as well.

The *Prochlorales* utilize both chlorophyll *a* and *b* as primary photopigments for oxygenic photosynthesis. Cells contain distinct thylakoids, but these lack phycobilins or phycobilisomes. Thylakoids are generally stacked, reminiscent of those of chloroplasts.

These organisms are sometimes considered separately from the other cyanobacteria, and in this context they are referred to as the *Oxychlorobacteria*. Nevertheless, they are phylogenetic members of the cyanobacteria. Species of the genus *Prochlorococcus*, which also contain chlorophyll *b* instead of phycobilin, were originally thought to be members of this group, but they seem to have evolved this trait independently or by horizontal transfer and are in reality members of the *Chroococcales*.

EXAMPLE *Prochloron*

*Prochloron* (Fig. 9.19) is a symbiont of ascidians (sea squirts); these organisms reside on the surface and embedded in the surface test of the animals, and especially in the communal cloacal cavity (these are colonial species), where they form visible green patches. Although usually considered exosymbionts, they are also found in vacuoles of the host cells living as an endosymbiont. They are especially common in sea squirts living in shady zones of the coral reefs. *Prochloron* provides up to half of the organic carbon requirements of the host animal. The cells are individual spheres or ovals 10 to 30 μm in diameter.

**Figure 9.19** Photograph of an ascidian (sea squirt) with *Prochloron* symbionts in Kimbe Bay, Papua New Guinea. (Source: Antonio Baeza, http://cfb.unh.edu/phycokey/Choices/ Chloroxybacteria/PROCHLORON/Prochloron_Image_page.html, with permission.) doi:10.1128/9781555818517.ch9.f9.19

## Other green phototrophs

The heliobacteria are green, anoxygenic phototrophs that use bacteriochlorophyll *g*. However, these organisms are members of the phylum *Firmicutes*, and so are described in chapter 11.

## Bacterial photosynthesis

Phototrophism is the biological process of converting (capturing) light energy into chemical energy. The process by which green bacteria (this chapter) and purple bacteria (see chapter 10) carry out phototrophism shares a common basic mechanism, usually referred to as photophosphorylation, which is only briefly reviewed here.

### Cyclic photophosphorylation

Photosynthetic bacteria generally perform cyclic photophosphorylation (Fig. 9.20). Light energy, captured by antenna chlorophylls and accessory pigments, is channeled to reaction center complexes. This light energy excites an electron in the reaction center chlorophyll and makes the reaction center a strong electron donor (reductant). This electron is transferred from the reaction center (which becomes oxidized) to the electron transport chain. The electron traverses the electron transport chain, resulting in the pumping of protons from the cytoplasm to the periplasm or lumen of the thylakoid. At the end of the electron transport chain, the electron is transferred back to the oxidized reaction center

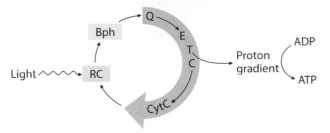

**Figure 9.20** Diagrammatic view of the flow of electrons during cyclic photosynthesis. RC, reaction center; Bph, bacteriopheophytin; Q, quinone; ETC, electron transport chain; CytC, cytochrome *c*. Notice that all electrons are recycled; the absorption of light results in the establishment of a proton gradient, the energy of which can be used for ATP synthesis, but there are no net reactants or products. doi:10.1128/9781555818517.ch9.f9.20

in the ground state. The reduced reaction center is now ready to accept another photon of light energy and repeat the cycle. With each cycle comes an increase in the proton motive force, which is used by adenosine triphosphatase (ATPase) to generate adenosine triphosphate (ATP) from adenosine diphosphate (ADP) and phosphate. The net reaction of cyclic photophosphorylation is light + ADP + phosphate → ATP. All of the components of the cycle are regenerated within the cycle; no other inputs are required.

## *Obtaining reducing power for carbon fixation*

Cyclic photophosphorylation generates only ATP. Photosynthetic organisms (i.e., those that also fix carbon) have several approaches to obtain NADH: from organic compounds (photoheterotrophs), reverse electron flow (purple photoautotrophs and *Chloroflexi*), from ferridoxin from the electron transport chain (*Chlorobi* and heliobacteria), or oxygenic photosynthesis (cyanobacteria).

Photoheterotrophs use light only for their energy needs; reducing power and carbon are obtained from organic compounds. This is independent of cyclic photophosphorylation.

Purple photoautotrophic bacteria and *Chloroflexi* rely on a source of a strong chemical reductant, such as sulfide, thiosulfate, elemental sulfur, ferrous cation, or hydrogen as a source of electrons (Fig. 9.21). These electrons are used to reduce cytochrome *c*; they traverse the electron transport chain in reverse, at the expense of the proton motive force (and so ultimately at the expense of ATP), and are used to reduce NAD$^+$ to NADH used for carbon fixation. This uses the same electron transport chain as photophosphorylation but is otherwise an independent process.

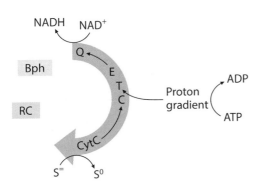

**Figure 9.21** Diagrammatic view of the flow of electrons through the electron transport chain during reverse electron flow. RC, reaction center; Bph, bacteriopheophytin; Q, quinone; ETC, electron transport chain; CytC, cytochrome *c*. Notice that the chemical half-reactions are separated at each end of the chain, and the reaction is driven by energy from the proton gradient at the expense (usually indirectly) of ATP. doi:10.1128/9781555818517.ch9.f9.21

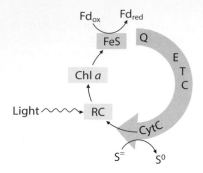

**Figure 9.22** Diagrammatic view of the flow of electrons during sulfur-dependent photosynthesis. RC, reaction center; Chl *a*, chlorophyll *a*; Q, quinone; ETC, electron transport chain; CytC, cytochrome *c*; Fd, ferredoxin. Notice that electrons are siphoned from the iron-sulfur protein (FeS) by the reduction of ferredoxin. The reaction center is recycled using electrons from cytochrome *c*, taken from the oxidation of sulfide. doi:10.1128/9781555818517.ch9.f9.22

Members of the *Chlorobi* have more strongly reducing reaction centers than do other organisms (Fig. 9.22). As a result, the electrons transferred to the electron transport chain pass first through iron-sulfur proteins that can be used to generate reduced ferredoxin as a source of reducing power for carbon fixation. Electrons removed from the electron transport chain to reduce ferredoxin are replaced from external sources of reductant, typically sulfide, sulfur, thiosulfate, ferrous cation, or hydrogen. Although these electrons are transferred to cytochrome *c*, as in the purple photoautotrophs, they subsequently are transferred to the oxidized reaction center, rather than undergoing reverse electron flow. Therefore, photosynthesis in these organisms generates both ATP and reducing power for carbon fixation but requires a chemical reductant as well as light.

Cyanobacteria, including chloroplasts, carry out oxygenic photosynthesis; the traditional "Z-scheme" of photosynthesis (Fig. 9.23). This process is based on the same cyclic photophosphorylation used by other photosynthetic bacteria. As in the other green bacteria, reducing power (NADH or NADPH) is siphoned out of the system at the stage of ferredoxin reduction. These electrons must be replaced for cyclic photophosphorylation to continue. Rather than directly use an external chemical reductant, however, cyanobacteria use a second reaction center (photosystem II; photosystem I is the reaction center used for cyclic photophosphorylation). This reaction center transfers electrons to the electron transport chain via reduced quinone. These electrons must be replaced in turn, but photosystem II is so strongly oxidizing that it can accept electrons from water, generating molecular oxygen in the process.

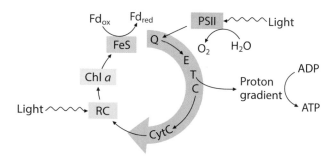

**Figure 9.23** Diagrammatic view of the flow of electrons during oxygenic photosynthesis. RC, reaction center; Chl *a*, chlorophyll *a*; Q, quinone; ETC, electron transport chain; CytC, cytochrome *c*; Fd, ferredoxin; PSII, photosystem II. As in sulfur-dependent photosynthesis, electrons are siphoned from the iron-sulfur protein (FeS) via reduction of ferredoxin. The reaction center of photosystem I is recycled by electrons from the electron transport chain originating in photosystem II. Photosystem II is, in turn, recycled using electrons split from water, generating molecular oxygen as a waste product. doi:10.1128/9781555818517.ch9.f9.23

## Rhodopsin phototrophy

It has recently been discovered that a wide range of bacteria can capture light energy for ATP production by using rhodopsin, a simple light-driven proton pump composed of a single protein (rhodopsin) and photopigment (retinal). This is apparently a significant form of phototrophy in the ocean, and perhaps other ecosystems, but is best known only from genes found in uncultivated organisms. This form of phototrophy seems to have spread widely, by horizontal transfer, from its presumptive origin in the halophilic *Archaea*. The mechanism of rhodopsin phototrophy is therefore discussed in chapter 15, and the discovery of these genes in uncultivated bacteria is discussed in chapter 22.

## Carbon fixation

Most phototrophic organisms can fix carbon for growth; they are photoautotrophic or photosynthetic. These three phyla of green phototrophic bacteria each use a different pathway for carbon ($CO_2$) fixation: the Calvin cycle (*Cyanobacteria*), the reverse TCA cycle (*Chlorobi*), and the hydroxypropionate pathway (*Chloroflexi*). Although each is an entirely independent pathway, they share several fundamental features imposed by chemistry: they are cyclical and require reductant and energy. Another, noncyclical pathway for carbon fixation is known from nonphototrophic acetogenic bacteria and some methanogens; the reductive acetyl coenzyme A (acetyl-CoA) pathway.

### Calvin cycle

The Calvin cycle is the most familiar and common carbon fixation pathway (Fig. 9.24). The key enzyme in this pathway is ribulose *bis*-phosphate carboxylase/oxidase ("rubisco"), which carboxylates ribulose 1,5-*bis*-phosphate

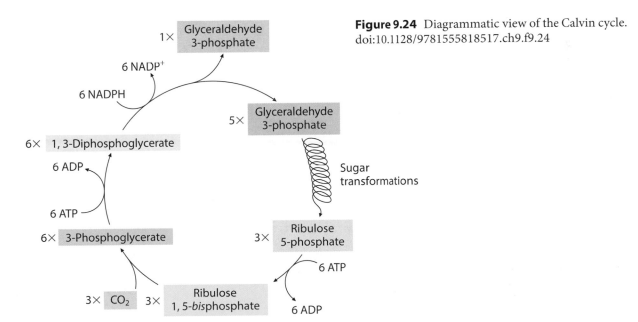

**Figure 9.24** Diagrammatic view of the Calvin cycle. doi:10.1128/9781555818517.ch9.f9.24

to form two molecules of 3-phosphoglycerate; this is the step in which $CO_2$ is incorporated into organic carbon. After phosphorylation, the resulting 1,3-diphosphoglycerate is reduced to glyceraldehyde 3-phosphate. The remainder of the steps are sugar rearrangements, using enzymes common in carbohydrate metabolism in most organisms, ending in the phosphorylation of ribulose 5-phosphate to regenerate ribulose 1,5-*bis*-phosphate. For each three molecules of ribulose 1,5-*bis*-phosphate that enter the cycle, three molecules of $CO_2$ are fixed, generating a net of one molecule of glyceraldehyde 3-phosphate that can be siphoned off for metabolism after regenerating three more molecules of ribulose 1,5-*bis*-phosphate to close the cycle. This glyceraldehyde 3-phosphate feeds directly into the general carbohydrate metabolism.

## Reverse TCA cycle

Many autotrophic organisms that do not use the Calvin cycle to fix carbon (e.g., the *Chlorobi*) use the reverse TCA cycle, also known as the reductive TCA cycle (Fig. 9.25). This is the same pathway as the familiar TCA cycle, but all of the reactions are run in the reverse direction. The TCA cycle usually consumes pyruvate, generating ATP, NADH or NADPH, reduced ferredoxin, and waste $CO_2$. The reverse TCA cycle therefore consumes $CO_2$, ATP, NADH/NADPH, and reduced ferredoxin to produce pyruvate. In the *Chlorobi*, two key steps use the oxidation of ferredoxin rather than NADH to drive the reactions in the reverse direction. The acetyl-CoA generated by this pathway can be used directly or carboxylated further to produce pyruvate in a reverse of the "transition reaction."

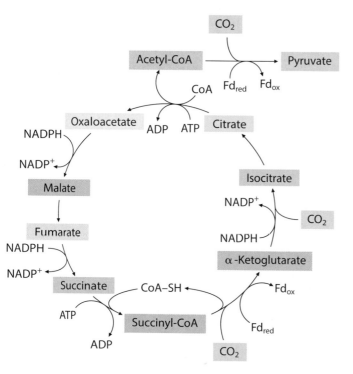

**Figure 9.25** Diagrammatic view of the reductive TCA cycle. Fd, ferredoxin.
doi:10.1128/9781555818517.ch9.f9.25

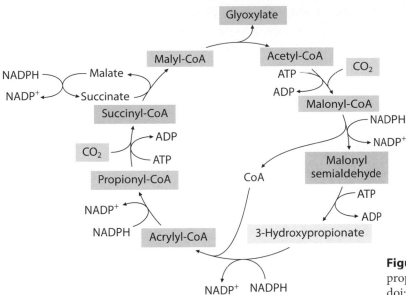

**Figure 9.26** Diagrammatic view of the hydroxy-propionate pathway.
doi:10.1128/9781555818517.ch9.f9.26

## Hydroxypropionate pathway

The *Chloroflexi* and many crenarchaea use a third carbon fixation method, the hydroxypropionate pathway (Fig. 9.26). Some of the reactions in this pathway are the same as in the TCA cycle, but it is a unique pathway. The hydroxypropionate generates glyoxylate, which can feed into the central metabolism after amidation to glycine.

## Reductive acetyl-CoA pathway

Acetogenic bacteria, as well as some sulfate reducers and some methanogenic members of the *Archaea*, fix carbon by using molecular hydrogen in the Wood reaction, also known as the reductive acetyl-CoA pathway (Fig. 9.27). Unlike the other pathways for carbon fixation, the reductive acetyl-CoA pathway is not cyclic; it has two branches, one being the formation of a methyl-corrinoid and the other being the generation of carbon monoxide, which come together to produce acetyl-CoA. Unlike in the other pathways, the carbon from $CO_2$ is carried through the reductive transformations on one-carbon-carrier cofactors; tetrahydrofolate (in the case of *Bacteria*) or tetrahydromethanopterin (in the case of *Archaea*). This pathway is also unusual in that the electrons for reduction of $CO_2$ come directly from hydrogenase rather than from NADH or other reduced cofactors. As in the reductive TCA pathway, acetyl-CoA produced can be further carboxylated (fixing another $CO_2$ molecule) to pyruvate by the reverse of the transition reaction. Although acetogenic bacteria fix carbon using this pathway, they also generate energy (in the form of ATP) via this same pathway, by cleaving acetate from the acetyl-CoA coupled to phosphorylation of ADP to ATP. This acetate is excreted as a waste product.

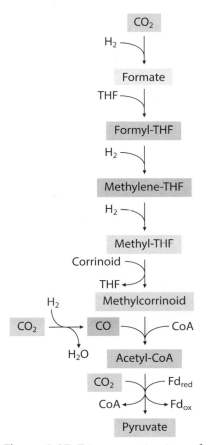

**Figure 9.27** Diagrammatic view of the Wood reaction. THF, tetrahydrofolate; Fd, ferredoxin.
doi:10.1128/9781555818517.ch9.f9.27

This pathway shares many of the reactions of methanogenesis (see chapter 15), in which the methyl group from methanopterin or the corrinoid protein would otherwise be transferred to coenzyme M, reduced one step further, and released as methane.

## Questions for thought

**1.** Can you think of any bacterium-bacterium symbioses other than that of *Chlorobium*? What about bacterium-eukaryote symbioses?

**2.** Some cyanobacteria can grow photoheterotrophically rather than photosynthetically. What does this mean? What effect would this have on photosystems I and II?

**3.** Some cyanobacteria can grow by using sulfide instead of water as the donor of electrons for reducing power for $CO_2$ fixation. Can you develop some hypotheses about how they might be doing this? How would you test these?

**4.** *Chloroflexus* contains chlorosomes and chlorophyll *c*, unlike other members of the *Chloroflexi*, and it has been suggested that these were acquired by horizontal transfer from the *Chlorobi*. How would you test this hypothesis?

**5.** Most phototrophic bacteria are capable of fixing carbon (i.e., they are photoautotrophic). Why do you suppose this is? Why are so many also able to fix atmospheric nitrogen?

**6.** Why do you suppose it is that most phototrophic bacteria except the cyanobacteria are anaerobic?

**7.** The final proof that chloroplasts originated from cyanobacteria came from phylogenetic analysis of chloroplast genes, especially the rRNA genes. Given that chloroplasts did arise from the cyanobacteria (and in particular the *Prochlorales*), do you see chloroplasts as *descendants* of cyanobacteria or as a *kind* of cyanobacteria?

**8.** The chloroplasts of various types of algae (green, golden-brown, red, etc.) are quite distinct, and it remains controversial whether they all descend from a single chloroplast ancestor or whether each represents an independent symbiotic event. How would you test these opposing hypotheses? What are the complications that might cloud your answer(s)?

# 10 Proteobacteria

### Phylum *Proteobacteria* (purple bacteria and relatives)

Most of the familiar gram-negative bacteria are members of the *Proteobacteria* (Fig. 10.1). The members of this very successful phylum predominate in many environments. They have a very broad range of phenotypes scattered around the phylogenetic tree. This implies that these organisms can change phenotype readily, at least in terms of evolutionary history; for this reason they have been named after the shape-shifting sea god Proteus. Unlike most of the other bacterial phyla, there is no phenotypic trait that unites the *Proteobacteria*. The purple phototrophic bacteria are members of this group, and so this phylum is also sometimes referred to as the purple bacteria and relatives. In addition to purple phototrophs, members of this phylum include many heterotrophs (including some important symbionts and pathogens) and key players in the sulfur and nitrogen cycles.

doi:10.1128/9781555818517.ch10

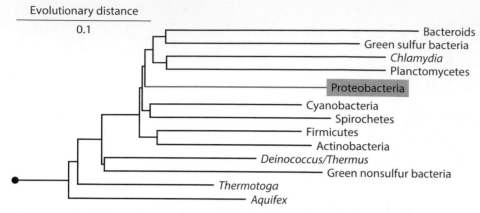

**Figure 10.1** Phylogenetic tree of the bacterial phyla, with the phylum *Proteobacteria* highlighted. doi:10.1128/9781555818517.ch10.f10.1

Because this is such a broad phylum, with thousands of described species, the proteobacteria are usually divided into classes, which are given the generic labels α-, β-, γ-, δ-, and ε-proteobacteria (Fig. 10.2). Perhaps this generic labeling has persisted because, like the phylum, within each of these classes there is a range of phenotypes: heterotrophs, phototrophs, and chemoautotrophs of various kinds, without any unifying themes.

Given the incredible diversity of organisms in the phylum *Proteobacteria*, this chapter touches on only a few examples.

**Figure 10.2** Phylogenetic tree of representative proteobacteria, with each of the five major classes indicated. doi:10.1128/9781555818517.ch10.f10.2

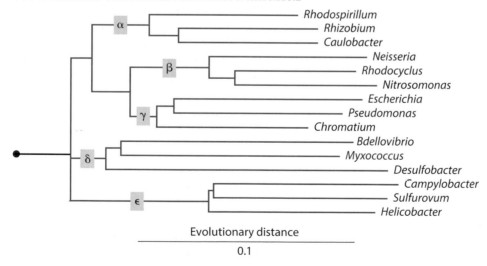

# Class *Alphaproteobacteria*

Figure 10.3 is a phylogenetic tree of representative *Alphaproteobacteria*.

| Class | Alphaproteobacteria | | |
|---|---|---|---|
| Order | | Caulobacterales | |
| Family | | | Caulobacteraceae (e.g., *Caulobacter*) |
| Order | | Parvularculales | |
| Family | | | Parvularculaceae (*Parvularcula*) |
| Order | | Rhodobacterales | |
| Family | | | Rhodobacteraceae (e.g., *Rhodobacter, Paracoccus, Roseobacter*) |
| Order | | Rhodospirillales | |
| Family | | | Acetobacteraceae (e.g., *Acetobacter, Gluconobacter, Stella*) |
| Family | | | Rhodospirillaceae (e.g., *Rhodospirillum, Magnetospirillum*) |
| Order | | Rhizobiales | |
| Family | | | Aurantimonadaceae (e.g., *Aurantimonas*) |
| Family | | | Bartonellaceae (e.g., *Bartonella, Rochalimaea*) |
| Family | | | Beijerinckiaceae (e.g., *Beijerinckia, Methylocella*) |
| Family | | | Bradyrhizobiaceae (e.g., *Bradyrhizobium, Rhodopseudomonas*) |
| Family | | | Brucellaceae (e.g., *Brucella, Ochrobactrum*) |
| Family | | | Hyphomicrobiaceae (e.g., *Hyphomicrobium, Ancylobacter*) |
| Family | | | Methylobacteriaceae (e.g., *Methylobacter, Microvirga, Meganema*) |
| Family | | | Methylocystaceae (e.g., *Methylocystis, Pleomorphomonas*) |
| Family | | | Phyllobacteriaceae (e.g., *Phyllobacterium, Mesorhizobium*) |
| Family | | | Rhizobiaceae (e.g., *Rhizobium, Agrobacterium*) |
| Family | | | Rhodobiaceae (e.g., *Rhodobium, Roseospirillum*) |
| Order | | Rickettsiales | |
| Family | | | Anaplasmataceae (e.g., *Anaplasma, Ehrlichia, Wolbachia*) |
| Family | | | Holosporaceae (*Holospora*) |
| Family | | | Rickettsiaceae (e.g., *Rickettsia, Orientia*) |
| Order | | Sphingomonadales | |
| Family | | | Sphingomonadaceae (e.g., *Sphingomonas, Erythrobacter*) |

## *General characteristics of the* Alphaproteobacteria

### Diversity

The class *Alphaproteobacteria* is a large group of organisms encompassing 7 orders, 20 families, and more than 200 genera. Also in this group are the mitochondria, probably as relatives of the *Rickettsiales*, although this is obscured by the very high evolutionary rates in the gene sequences of mitochondria.

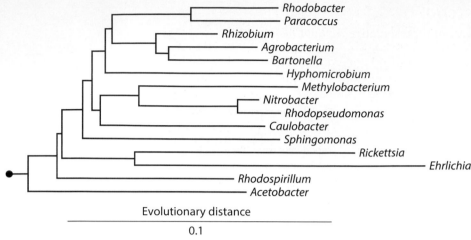

**Figure 10.3** Phylogenetic tree of representative *Alphaproteobacteria*.
doi:10.1128/9781555818517.ch10.f10.3

## Metabolism

Most of the purple nonsulfur phototrophic bacteria are members of the α-proteobacteria, but there are also a wide range of heterotrophic species, including some important pathogens. Also in this class are autotrophic methane oxidizers (methylotrophs) and organisms that are capable of nitrogen fixation, usually as symbionts of plant hosts. Autotrophic members of this class fix $CO_2$ via the Calvin cycle. The sphingomonads lack lipopolysaccharide in their cell envelopes, instead having glycosphingolipids similar to those found in the central nervous systems of animals.

## Morphology

Although most members of this class are generic rods and cocci, and a few spirilla, many of the appendaged bacteria are members of this group. These appendages are extensions of the cell, surrounded by the cell envelope, unlike the stalks of the *Planctomycetes* (see chapter 13) in which they are fibrous extracellular structures. Appendages can be single stalks, as in *Caulobacter*, or multiple "arms," as in *Ancalomicrobium*. In the hyphomicrobia, including the purple phototroph *Rhodomicrobium*, reproduction is by budding from the ends of appendages, often resulting in chains and networks of cells connected by thin bridging appendages.

## Habitat

Members of this class are abundant in nearly all habitats and are predominant members of both aerobic and anaerobic aquatic environments.

## *Purple nonsulfur phototrophs*

With the exception of the genus *Rhodocyclus* and a couple of close relatives, all purple nonsulfur bacteria are members of the *Alphaproteobacteria*. These organisms carry out anoxygenic photosynthesis via a single photosystem, i.e., cyclic photophosphorylation (see chapter 9 for a discussion of photosynthesis). Although all can grow photoheterotrophically, using light for energy and organic compounds for carbon growth requirements, most are also capable of photoautotrophy, using hydrogen or reduced sulfur compounds as a source of reducing power for carbon fixation; this is captured in the form of reduced nicotinamide adenine dinucleotide (NADH) as the product of reverse electron flow and is used to fix carbon via the Calvin cycle. A few can grow fermentatively, producing organic acids, $CO_2$, and hydrogen, or by anaerobic respiration using nitrate or nitrite as terminal electron acceptors. Most are capable of nitrogen fixation. Photosynthesis takes place in internal membranes, which can take the form of vesicles lining the cytoplasmic membrane, flattened stacks resembling thylakoids, or concentric lamellae lining the cytoplasmic membrane.

### EXAMPLE *Rhodomicrobium vannielii*

*Rhodomicrobium vannielii* (Fig. 10.4) is a commonly isolated purple phototroph, producing deep red growth in liquid culture or colonies. It is an appendaged ovoid or rod-shaped cell that reproduces by budding from the ends of appendages. It prefers to grow photoheterotrophically and is capable of utilizing a wide range of organic compounds for growth, but in the absence of these it grows photoautotrophically, using sulfide or hydrogen as the source of reducing power for carbon fixation. Thiosulfate, not elemental sulfur or sulfate, is the product of sulfide oxidation. Unlike some other purple nonsulfur bacteria, *R. vannielii* is not capable of nitrogen fixation. It is also unique among these organisms in having peritrichous flagella and in forming moderately heat-resistant resting cysts (exospores). Photosynthetic membranes are lamellar.

**Figure 10.4** Phase-contrast micrograph of *Rhodomicrobium vannielii*, showing polar stalks and budding division. (Courtesy of Jobst H. Klemme.) doi:10.1128/9781555818517.ch10.f10.4

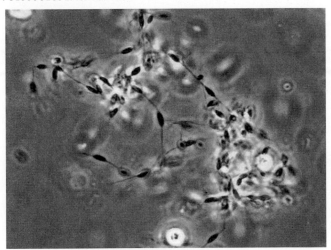

## Appendaged bacteria

The appendaged, or "prosthecate," *Alphaproteobacteria* contain cytoplasmic extensions. Most are aerobic heterotrophs and live in oligotrophic environments. If only a single appendage is present, it is referred to as a stalk. These stalks are often used to attach to surfaces, with terminal holdfasts containing powerful adhesives. Some species have multiple appendages, which apparently serve to increase the surface area of the cells and may also make the organisms resistant to grazing by some protists. Most divide by binary fission, but a few divide by budding from the ends of appendages (e.g., *Hyphomicrobium*).

Some stalked *Alphaproteobacteria* are dimorphic, reproducing by an asymmetrical binary fission in which the original nonmotile stalked "mother" cell remains but the other offspring is flagellated and lacks a stalk. These motile "swarmer" cells disperse in the environment, attach to a surface, and develop into sessile stalked cells. This dispersal of offspring prevents the accumulation of nonmotile cells in one place, where they would compete with each other for resources in their usually oligotrophic environment; this is the same reason most sessile reef animals have planktonic larvae.

### EXAMPLE *Caulobacter crescentus*

*Caulobacter crescentus* (Fig. 10.5) is a well-studied dimorphic appendaged bacterium. Cells are vibrioid or fusiform. Mature cells have a single thin terminal stalk. Division is by binary fission; before division is complete, a single flagellum

**Figure 10.5** Life cycle of *Caulobacter crescentus*. Flagellated swarmer cells transform into stalked mother cells, which produce swarmer cell daughters by asymmetric division. (Reprinted with permission from Brun YV, Janakiraman R. 2000. The dimorphic life cycle of *Caulobacter* and stalked bacteria, p 297–317. *In* Brun YV, Shimkets LJ [ed], *Prokaryotic Development*. ASM Press, Washington, DC.) doi:10.1128/9781555818517.ch10.f10.5

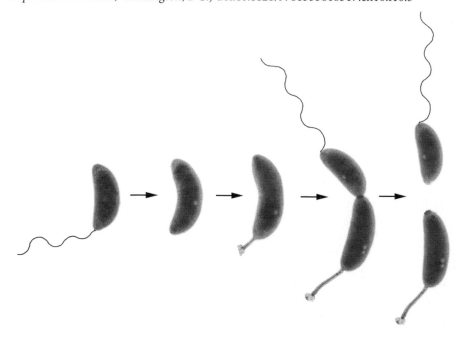

is created at the pole opposite the stalk. Although the mother and swarmer cells produced are approximately the same size, the DNA of the swarmer cell is condensed and transcriptionally inactive, and metabolism in the swarmer is much reduced compared to that in the mother cell. When the swarmer cell comes in contact with a solid substrate, it adheres, loses its flagellum, and develops a stalk (with a terminal holdfast) from the same end of the cell that previously had the flagellum. *Caulobacter* is ubiquitous in aquatic samples and is most readily observed attached to diatoms and eukaryotic algae. *Caulobacter* and other appendaged bacteria are readily isolated from the surfaces inside laboratory distilled-water containers.

## Nitrogen-fixing plant symbionts

Nitrogen-fixing plant symbionts, known as rhizobia, form intimate symbioses with leguminous plants. The bacteria enter the plant via the root hairs, enter the body of the root through an infection thread, and induce the formation of root nodules. In these nodules the bacterial cells reproduce and then develop into the symbiotic forms (bacteroids) that fix atmospheric nitrogen both for the bacterial symbionts and for the host plant. These bacteroids are terminally differentiated; they cannot revert to vegetative bacteria, nor can they reproduce. The host provides the bacteroids with nutrients and vitamins for growth and metabolism in return. The growth of these leguminous plants, with their nitrogen-fixing rhizobia, is commonly used to replenish impoverished soil.

As described previously in the discussion of nitrogen fixation by cyanobacteria (chapter 9), nitrogenases (the enzymes that reduce $N_2$ to $NH_4^+$) are strongly inhibited by oxygen. Rhizobia, however, are obligate aerobes. In culture, rhizobia can be grown and fix nitrogen microaerophilically. In nodules, the bacteroids are provided with plenty of oxygen bound to a legume-hemoglobin, *leghemoglobin*, which is readily available for respiration but maintains a very low concentration of free oxygen that would inhibit nitrogenase.

The production of effective (nitrogen-fixing) nodules is a complex process requiring a matching pair of compatible host and bacterial symbiont. Mismatched host-symbiont pairs often can produce ineffective nodules that fix little or no nitrogen.

When the host plant dies, the small numbers of vegetative (nonbacteroid) rhizobia escape from the decaying nodule and persist in the soil, capable of infecting emerging host plants.

### EXAMPLE *Rhizobium etli*

*Rhizobium etli* (Fig. 10.6), previously known as *Rhizobium leguminosarum* biovar phaseoli type I, is specific for the host *Phaseolus vulgaris*, the Latin American common bean. The genome of *R. etli* comprises a single large chromosome (4.38 Mbp) and six smaller chromosomes ranging from 0.18 to 0.64 Mbp. Because the smaller chromosomes contain mostly nonessential genes, they would usually be considered plasmids, but they contain a wide range of important genes, including genes required for normal cell cycling and the *nod* genes required for their normal symbiotic life cycle. The largest of these so-called plasmids are as big as the entire genomes of some obligately parasitic bacteria!

**Figure 10.6** Alfalfa root nitrogen-fixing nodules. (Courtesy of Ninjatacoshell, http://en
.wikipedia.org/wiki/Root_nodule, http://creativecommons.org/licenses/by-sa/3.0/deed.en.)
doi:10.1128/9781555818517.ch10.f10.6

## Obligate intracellular parasites

These include a wide range of obligately intracellular symbionts or parasites of
eukaryotes, especially animals. Best known are the insect symbionts (e.g., *Wolbachia*) and insect-borne human and mammal pathogens (e.g., *Rickettsia*). Also
a member of this group phylogenetically, *Bartonella* (previously *Rochalimaea*) is
an obligate parasite but lives on the outside surface of the host cell rather than
intracellularly. These organisms are often compared to viruses, but although
they are small and obligately intracellular, they are otherwise typically bacterial.

Also members of this group are the mitochondria. These seem
to be specifically related to *Rickettsia*, the causative agents of spotted
fevers and typhus. Both groups lack enzymes for glycolysis (which is
carried out in the cytoplasm of eukaryotes) but contain the enzymes
of the complete tricarboxylic acid cycle and electron transport. Amino
acids and nucleotide precursors are provided by the host in either case.
The sequences encoding these proteins in *Rickettsia* resemble those of
mitochondria, even in cases where these genes now reside in the nuclear genome rather than in the mitochondrial genome.

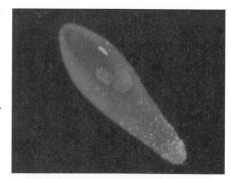

**Figure 10.7** Egg of the wasp *Tricho-
gramma kaykai*. The brightly stained dots
concentrated at the bottom right end of
this egg (which will develop into the germ
line of the animal) are cells of the symbiotic bacterium *Wolbachia*. (Courtesy of
Merijn Salverda and Richard Stouthamer.)
doi:10.1128/9781555818517.ch10.f10.7

### EXAMPLE *Wolbachia pipientis*

*Wolbachia pipientis* (Fig. 10.7) is the only formally described species
of this genus. *Wolbachia* is a very common intracellular symbiont of
arthropods and nematodes. A majority of insect species are susceptible to infection, and perhaps 15 to 20% of individuals are infected.
Although infection with *Wolbachia* is not associated with outright disease, it does cause a range of phenotypes in the reproductive biology

of the host. This is because infected females transmit the symbiont directly to their eggs, but infected males do not transmit the symbionts to their offspring via infected sperm. As a result, in an effort to maximize the number of infected offspring generated by the host, *Wolbachia* manipulates the host in favor of producing females at the expense of males. Often this means inducing parthenogenesis, feminization (causing genetically male eggs to develop into females), or embryonic death in males (son-killing). When infected males are viable, they can usually successfully fertilize the eggs only from females infected with the same strain of *Wolbachia*, resulting in the phenomenon of "cytoplasmic incompatibility"; this seems to be mediated by imprinting of the host chromosomes in both sperm and eggs. This can create a reproductive barrier favoring host speciation in the absence of geographic isolation.

## Class *Betaproteobacteria*

Figure 10.8 is a phylogenetic tree of representative *Betaproteobacteria*.

| Class | *Betaproteobacteria* | |
|---|---|---|
| Order | | *Burkholderiales* |
| Family | | | *Alcaligenaceae* (e.g., *Alcaligenes, Bordetella, Achromobacter*) |
| Family | | | *Burkholderiaceae* (e.g., *Burkholderia, Ralstonia*) |
| Family | | | *Comamonadaceae* (e.g., *Comamonas, Acidovorax*) |
| Family | | | *Oxalobacteraceae* (e.g., *Oxalobacter, Janthinobacterium*) |
| Order | | *Hydrogenophilales* |
| Family | | | *Hydrogenophilaceae* (e.g., *Hydrogenophilus, Thiobacillus*) |
| Order | | *Methylophilales* |
| Family | | | *Methylophilaceae* (e.g., *Methylophilus, Methyobacillus*) |
| Order | | *Neisseriales* |
| Family | | | *Neisseriaceae* (e.g., *Neisseria, Aquaspirillum, Vitreoscilla*) |
| Order | | *Nitrosomonadales* |
| Family | | | *Gallionellaceae* (*Gallionella*) |
| Family | | | *Nitrosomonadaceae* (e.g., *Nitrosomonas, Nitrospira*) |
| Family | | | *Spirillaceae* (*Spirillum*) |
| Order | | *Procabacterales* |
| Family | | | *Procabacteraceae* (*Procabacter*) |
| Order | | *Rhodocyclales* |
| Family | | | *Rhodocylaceae* (e.g., *Rhodocyclus, Azoarcus, Zoogloea*) |

### *About this class*

#### Diversity

Although less diverse phylogenetically, and perhaps less abundant in the environment, than either *Alphaproteobacteria* or *Gammaproteobacteria*, *Betaproteobacteria* nevertheless is a diverse and abundant class of bacteria, consisting of at least 125 characterized genera.

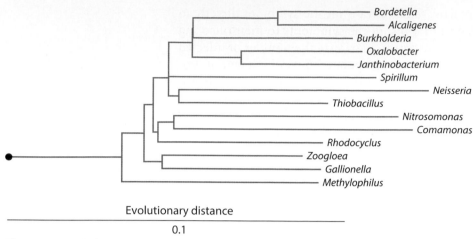

Evolutionary distance

0.1

**Figure 10.8** Phylogenetic tree of representative *Betaproteobacteria*.
doi:10.1128/9781555818517.ch10.f10.8

Included in this class are a number of organisms that were previously considered to be members of the genus *Pseudomonas*. This genus is now reserved for organisms specifically related to *Pseudomonas aeruginosa* and *P. fluorescens*; other phylogenetic groups that had been lumped into this genus on the basis of generic phenotypic criteria have been divided into a number of new genera and reclassified on the basis of their phylogenetic relationships.

## Metabolism

Like the other classes of proteobacteria, the *Betaproteobacteria* are very diverse metabolically and phenotypically. Most of these organisms are either heterotrophs (including some important pathogens, although mostly opportunistically so) or chemolithoautotrophs. They are generally aerobic or facultatively anaerobic.

Heterotrophic members of this class are able to utilize a wide range of substrates for growth, including compounds important in waste management, such as phenol and lignin. Lithotrophic members are key players in the nitrogen cycle, primarily in nitrification. Many are also sulfur, ferrous iron, or manganese oxidizers, and those that grow autotrophically use the Calvin cycle to fix $CO_2$. Also in this class are the methylotrophs, which are methane oxidizers.

Although the bulk of the purple nonsulfur phototrophs are members of the α-proteobacteria, one main genus (*Rhodocyclus*) and a few less well-known genera (e.g., *Rubrivivax* and *Roseateles*) are members of the *Betaproteobacteria*. Morphologically, *Rhodocyclus* spp. are very tightly wound short helices, resembling old-fashioned lock washers.

## Morphology

Although organisms in this class are typically rods and cocci, there are also a number of conspicuous spirilla and the noncyanobacterial filamentous sheathed bacteria. Filamentous iron- and manganese-oxidizing organisms typically become coated with granules of insoluble metal salts. *Simonsiella* and its relatives are gliding flattened multicellular filaments that inhabit the oral epithelium of mammals, often in great numbers.

## Habitat

These organisms are abundant in most environments, particularly in organic-rich soils, sediments, wastewater, and eutrophic aquatic systems. They are also frequently found in association with plant and animal surfaces.

## Heterotrophs and pathogens

The β-proteobacteria contain numerous heterotrophic species that are common in the environment and are capable of utilizing a wide range of organic compounds. Most of these heterotrophs are aerobic rod-shaped organisms, and many are at least opportunistic pathogens of plants and animals. The most important animal pathogens include *Bordetella*, *Burkholderia*, and *Neisseria*.

### EXAMPLE *Ralstonia solanacearum*

*Ralstonia solanacearum* (Fig. 10.9) is an important plant pathogen, causing southern bacterial wilt in a wide range of crop plants worldwide, including tobacco, potato (brown rot), tomato, pepper, and bananas (Moko disease). Species now in the genus *Ralstonia* were previously classified as members of the genus *Pseudomonas* and are obligately aerobic motile rods. The pathogen enters the plant through the root hairs, grows, and is transported throughout the plant in the xylem. The organism grows to such large numbers in the plant that one of the standard diagnostic tests is to touch the cut end of an infected stem to a container of water; if the infection is *R. solanacearum*, cells and exopolysaccharide can been easily seen as a milky stream flowing out of the xylem. *R. solanacearum* can overwinter or persist for even longer in the soil or water. The genome of *R. solanacearum* is in two chromosomes, a 3-Mbp circle containing most of the essential genes and a 2-Mbp "megaplasmid" that also contains essential genes and genes required for pathogenicity.

**Figure 10.9** Diagnostic test for southern bacterial wilt. Cut off a section of the diseased plant stem, and suspend it in the end of a water-filled tube. If milky-white material flows out, the test is positive. (Courtesy of David B. Langston, University of Georgia, Bugwood.org. http://www.forestryimages.org/browse/detail.cfm?imgnum=5077045#sthash.oxZMZ18i.dpuf; http://creativecommons.org/licenses/by/3.0/us/.) doi:10.1128/9781555818517.ch10.f10.9

## Chemolithoautotrophs

Most of the chemolithoautotrophs among the β-proteobacteria fall into four categories: sulfur oxidizers (which can sometimes oxidize metal ions as well), methylotrophs, ammonia oxidizers, and hydrogen oxidizers. None of these phenotypes are unique to the β-proteobacteria; all are also found at least in the α- and γ-proteobacteria. Many of these organisms are obligate autotrophs. The Calvin cycle is used for $CO_2$ fixation.

The most prevalent group of chemoautotrophs in the β-proteobacteria is the sulfur oxidizers. These organisms use reduced-sulfur compounds (sulfide, thiosulfate, thiocyanate, elemental sulfur) as the reductant for electron transport; oxygen is the common terminal electron acceptor, although many species can use nitrate or nitrite as alternative terminal electron acceptors. Sulfuric acid is the product of sulfur oxidation, although some organisms accumulate elemental sulfur as long as more reduced sulfur compounds are available. These organisms are ubiquitous in environments at the interface between aerobic and anaerobic zones, where sulfides and oxygen coexist. Many are capable of iron oxidation (ferrous to ferric); this, combined with their production of sulfuric acid from sulfur oxidation, makes them important contributors to the corrosion of iron plumbing. Sulfur oxidizers are used routinely in the mining process to extract metals from ores by leaching.

EXAMPLE *Thiobacillus thioparus*

*Thiobacillus thioparus* (Fig. 10.10) is an obligate chemolithotroph, oxidizing reduced-sulfur compounds by using oxygen or nitrate as the terminal electron acceptor. Nitrate is reduced to nitrite. This rod-shaped motile organism is a neutrophile, in contrast to the acidophilic γ-proteobacterial *Athiobacillus* species that were previously in this genus. Carboxysomes are present when the organisms are grown under conditions of $CO_2$ limitation; these are small proteinaceous organelles that concentrate $CO_2$ and contain the enzymes of the Calvin cycle. Many organisms that use the Calvin cycle contain carboxysomes.

**Figure 10.10** A river polluted by acid mine drainage and the resulting overgrowth of *Thiobacillus*, pollution's "worst-case scenario." (Source: Wikimedia Commons, http://commons.wikimedia.org/wiki/File:Riotintoagua.jpg.) doi:10.1128/9781555818517.ch10.f10.10

## Bacteria with sheathed filaments

The sheathed filamentous β-proteobacteria are commonly seen in polluted streams and wastewater. They are distinguished from other filamentous bacteria, which are common in these environments, by their sheath, which is a tubular structure surrounding the cells of the filament. Sheathed β-proteobacteria usually "strengthen" their sheaths with a precipitate of iron hydroxide or manganese dioxide, which the organisms produce by oxidation of soluble reduced metal ions. These organisms are obligately aerobic heterotrophs; it is unlikely that energy from the oxidation of metals contributes to the energy needs of the cells. Sheaths often are surrounded by a slime layer and are attached to the substrate by a terminal holdfast. In addition to attachment, the sheath protects the filaments from predation.

### EXAMPLE *Sphaerotilus natans*

*Sphaerotilus natans* (Fig. 10.11) is an iron-accumulating organism, perhaps the most common member of this group. Filaments contain false branches, breaks in the sheath from which filaments can protrude. Sheaths are thin and smooth; they are brown because of their impregnation with iron hydroxide. Filaments can sometimes leave their sheath and create new ones, or individual cells can leave to create new filaments. Empty space in sheaths is common. Filaments are 1.2 to 2.5 μm in diameter, and individual cells are 2 to 10 μm in length. *S. natans* is commonly found in aerated wastewater, where it can be a nuisance, contributing to bulking (poor settling of sludge).

**Figure 10.11** Phase-contrast micrograph of *Sphaerotilus natans*. Notice the individual rod-shaped cells contained within the sheath. (Reprinted from Pellegrin V, Juretschko S, Wagner M, Cottenceau G, *Appl Environ Microbiol* **65:**156–162, 1999, with permission.) doi:10.1128/9781555818517.ch10.f10.11

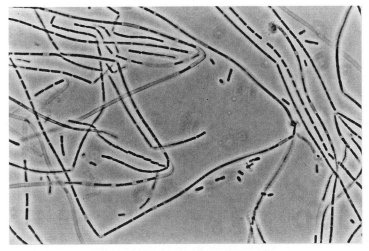

# Class *Gammaproteobacteria*

Figure 10.12 is a phylogenetic tree of representative γ-proteobacteria.

| Class | Gammaproteobacteria | | |
|---|---|---|---|
| Order | | Acidithiobacillales | |
| Family | | | Acidithiobacillaceae (Acidothiobacillus) |
| Family | | | Thermithiobacillaceae (Thermothiobacillus) |
| Order | | Aeromonadales | |
| Family | | | Aeromonadaceae (e.g., Aeromonas, Oceanomonas) |
| Family | | | Succinivibrionaceae (e.g., Ruminobacter, Succinivibrio) |
| Order | | Alteromonadales | |
| Family | | | Alteromonadaceae (e.g., Alteromonas, Glaciecola) |
| Family | | | Colwelliaceae (Colwellia, Thalassomonas) |
| Family | | | Ferrimonidaceae (Ferrimonas) |
| Family | | | Idiomarinaceae (e.g., Idomarina) |
| Family | | | Moritellaceae (e.g., Moritella) |
| Family | | | Pseudoalteromonadaceae (Algicola, Pseudoalteromonas) |
| Family | | | Psychromonadaceae (Psychromonas) |
| Family | | | Shewanellaceae (Shewanella) |
| Order | | Cardiobacteriales | |
| Family | | | Cardiobacteriaceae (e.g., Cardiobacter) |
| Order | | Chromatiales | |
| Family | | | Chromatiaceae (e.g., Chromatium, Thiocapsa, Thiocystis) |
| Family | | | Ectothiorhodospiraceae (e.g., Ectothiorhodospira, Nitrococcus) |
| Family | | | Halothiobacillaceae (e.g., Halobacillus, Thiovirga) |
| Order | | Enterobacteriales | |
| Family | | | Enterobacteriaceae (e.g., Buchnera, Escherichia, Yersina) |
| Order | | Legionellales | |
| Family | | | Coxiellaceae (e.g., Coxiella, Rickettsiella) |
| Family | | | Legionellaceae (e.g., Legionella) |
| Order | | Methylococcales | |
| Family | | | Methylococcaceae (e.g., Methylobacter, Methylococcus) |
| Order | | Oceanospirillales | |
| Family | | | Alcanivoraceae (e.g., Alcanivorax) |
| Family | | | Hahellaceae (e.g., Hahella, Halospina) |
| Family | | | Halomonadaceae (e.g., Chromohalobacter, Halomonas) |
| Family | | | Oceanospirillaceae (e.g., Marinomonas, Oceanospirillum) |
| Family | | | Oleiphilaceae (Oleiphilus) |
| Family | | | Saccharospirillaceae (Saccharospirillum) |
| Order | | Pasteurellales | |
| Family | | | Pasteurellaceae (e.g., Actinobacillus, Haemophilus, Pasteurella) |

| Order | | Pseudomonadales | |
|---|---|---|---|
| Family | | | Moraxellaceae (e.g., Acinetobacter, Psycrobacter, Moraxella) |
| Family | | | Pseudomonadaceae (e.g., Azotobacter, Pseudomonas) |
| Order | | Thiotrichales | |
| Family | | | Francisellaceae (Francisella) |
| Family | | | Piscirickettsiaceae (e.g., Thiomicrospira, Methylophaga) |
| Family | | | Thiotrichaceae (e.g., Beggiatoa, Thiothrix, Thiomargita) |
| Order | | Salinisphaerales | |
| Family | | | Salinisphaeraceae (Salinsphaera) |
| Order | | Vibrionales | |
| Family | | | Vibrionaceae (e.g., Photobacterium, Vibrio) |
| Order | | Xanthomonadales | |
| Family | | | Xanthomonadaceae (e.g., Stenotrophomonas, Xanthomonas) |

## About this class

### Diversity

*Gammaproteobacteria* is a very large and diverse class. As mentioned at the beginning of this chapter, most of the familiar gram-negative bacteria are members of the *Proteobacteria*, but even within this large and diverse phylum, a large fraction of familiar gram-negative organisms are specifically members of the class *Gammaproteobacteria*. This class contains 15 orders, 35 recognized families, and about 250 genera.

### Metabolism

Members of this class span the metabolic gamut. There are obligate aerobes, facultative anaerobes, microaerophiles, and obligate anaerobes; heterotrophs,

**Figure 10.12** Phylogenetic tree of representative γ-proteobacteria.
doi:10.1128/9781555818517.ch10.f10.12

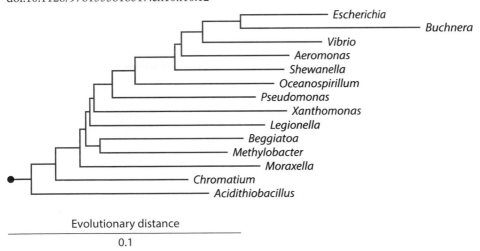

chemoautotrophs, and photoautotrophs; deadly pathogens, opportunistic pathogens, and life-sustaining symbionts; and cryophiles, mesophiles, and moderate thermophiles. Most metabolic possibilities have examples in this class. This being said, most familiar γ-proteobacteria are chemoheterotrophic.

## Morphology

Most of the organisms in this group fall into the stereotypical bacterial morphologies and sizes: rods, cocci, and some spirilla. Curved rods (vibrios) are also common, and a few are filamentous. Organisms occasionally fall outside of this range, an extreme example being the sulfide- and nitrate-oxidizing *Thiomargarita*, a spherical organism with cells nearly 1 mm in diameter.

## Habitat

The γ-proteobacteria are ubiquitous in nature, very often making up the largest fraction of the population. They are not known to exist in hyperthermophilic or strongly alkaline environments but can be found almost anywhere except the fringes of the limits of life.

## *Enterics*

The enterobacteria (enterics) are probably the best-studied group of bacteria. They are easy to isolate, grow, and manipulate, are very common, and are important symbionts (and pathogens) of plants and animals. Enterics are mesophilic facultatively anaerobic heterotrophs, growing respiratively under aerobic conditions and fermentatively under anaerobic conditions. Many can also grow by anaerobic respiration, using nitrate as the terminal electron acceptor and producing nitrite. Motility is by peritrichous flagella; a few are nonmotile. Acid and gas are commonly produced from carbohydrate fermentation. They are oxidase negative but catalase positive with few exceptions.

Most or all of the enterics are at least opportunistically pathogenic, and this group contains many well-known important pathogens; *Salmonella*, *Shigella*, *Escherichia*, *Yersinia*, and *Klebsiella* are good examples. Plant pathogens in this group are responsible for a wide range of plant wilts and blights in commercial and wild plants.

### EXAMPLE SPECIES

### *Escherichia coli*

*Escherichia coli* (Fig. 10.13) is certainly the best known and understood of bacteria. It is generally a commensal inhabitant of mammalian colon. Because it is much easier to grow than are the far more abundant obligately anaerobic *Clostridium* and *Bacteroides*, it is routinely used as an indicator of fecal contamination of food and water. It is the standard workhorse of molecular biology because it is easy to grow, safe, and readily manipulated genetically. Nevertheless, it is an opportunistic pathogen; even otherwise benign strains can cause urinary tract infections. More virulent strains cause gastroenteritis ranging from mild to life-threatening.

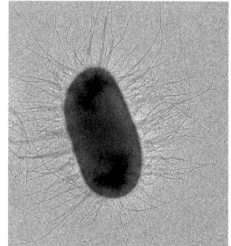

**Figure 10.13** Electron micrograph of the familiar bacterium *Escherichia coli*. Short straight pili and longer fimbriae protrude in all directions. (Courtesy of Esther Bullit.) doi:10.1128/9781555818517.ch10.f10.13

### Buchnera aphidicola

*Buchnera aphidicola* is an endosymbiont of aphids (Fig. 10.14). These animals are sap-sucking parasites of plants, and so their diet is rich in minerals and plant sugars but essentially devoid of essential vitamins and amino acids. They have a unique organ, the bacteriome, consisting of about 70 bacteriocytes. These are specialized host cells containing vacuoles filled with *B. aphidicola*. These endosymbiotic bacteria provide the insect with the vitamins and essential amino acids it needs for survival; aphids fed antibiotics stop growing and reproducing, and die prematurely. The endosymbionts are transmitted to offspring in utero and, like *Wolbachia* (see the above discussion of α-proteobacteria), have a dramatic impact on the reproductive biology of the insect. Aphids are parthenogenic, producing live pregnant offspring without the need for fertilization.

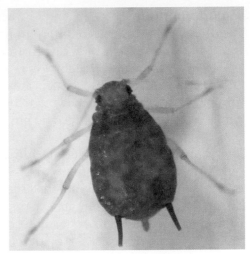

**Figure 10.14** A typical aphid, host of the bacterial symbiont *Buchnera aphidicola*. (Courtesy of Mary Harris.) doi:10.1128/9781555818517.ch10.f10.14

### Thiotrichs

Thiotrichs are gliding filamentous sulfur oxidizers, commonly found in marine and freshwater sediments and cold sulfur springs. Sulfide is oxidized first to elemental sulfur, which is stored in intracellular granules. When sulfide is depleted, these sulfur granules are oxidized to sulfate. Thiotrichs are microaerophilic, facultative autotrophs, and most are capable of nitrogen fixation. Few have been grown in pure culture, and species identification is based on morphology.

EXAMPLE  *Beggiatoa alba*

*Beggiatoa alba* (Fig. 10.15) is found on the surface of freshwater and marine sediments but is most conspicuous in freshwater sulfide-rich springs, where it can form spectacular white fuzzy mats covering all submerged surfaces. *B. alba* is the only formally recognized species of this genus, but a wide range of species have been defined informally. The primary distinction between species is filament diameter; *B. alba* is 2 to 3 μm is diameter. Unlike other filamentous colorless sulfur bacteria, no holdfasts or sheaths are present.

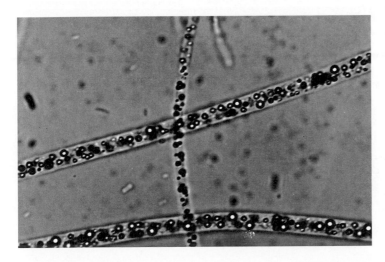

**Figure 10.15** Two filaments of *Beggiatoa alba* (the thinner vertical filament is its relative, *B. minima*), from Pluto Spring, French Lick, Indiana. doi:10.1128/9781555818517.ch10.f10.15

## Pseudomonads

Pseudomonads are very common in most aerobic environments. They are obligately aerobic, except for a few that can grow anaerobically by nitrate reduction, and are straight or slightly curved (not helical) rods with polar flagella. "Pseudo-monad," or "false unit," is an apt name—until recently, this was a huge group with species that turned out to be unrelated γ- and β-proteobacteria. The genus has recently been divided up phylogenetically, and only organisms related to the "fluorescent" *Pseudomonas* species remain. "Fluorescent" refers to diffusible pigments that are produced by these organisms and are siderophores; they bind iron with very high affinity and are used by the organism to scavenge trace quantities of this essential mineral. These organisms are common in laboratory distilled water because they are experts at extracting trace nutrients from sparse environments. Some are opportunistic pathogens (*P. aeruginosa* is usually the proximal cause of death for cystic fibrosis and burn patients), but most are free-living oligotrophic aquatic species. This group also contains the free-living nitrogen fixers, the azotobacteria.

### EXAMPLE  *Azotobacter vinelandii*

The azotobacteria are free-living nitrogen fixers, distinguished from most species of the genus *Pseudomonas* only by the ability to fix nitrogen, and distinguished from the rhizobia by the fact that they do not infect plants (although some are external symbionts of roots). Like other members of the genus *Azotobacter*, *A. vinelandii* differentiates into resting spore-like microcysts in stationary phase (Fig. 10.16). *A. vinelandii* is a common soil, freshwater, and marine inhabitant, preferring slightly alkaline (pH 7.5 to 8.0) conditions. Unlike other motile pseudomonads, which have polar flagella, *A. vinelandii* has peritrichous flagella.

**Figure 10.16** Electron micrograph of *Azotobacter vinelandii* cysts. (Reprinted from Segura D, Vite O, Romero Y, Moreno S, Castañeda M, Espín G, *J Bacteriol* **191**:3142–3148, 2009, with permission.) doi:10.1128/9781555818517.ch10.f10.16

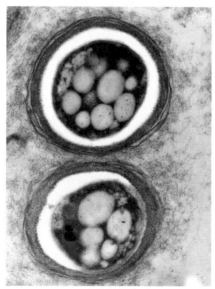

## Purple sulfur bacteria

The purple sulfur bacteria (chromatia) are all phylogenetically and metabolically much alike. They are anaerobic photosynthetizers that require sulfide for growth, and so in some ways they resemble the green sulfur bacteria (*Chlorobi*) and are often found in the same environment (see chapter 9 for a discussion of photosynthesis and the *Chlorobi*). In these environments, the purple sulfur bacteria are often found overlying the green sulfur bacteria because they require more light and less sulfide. Photosynthesis is by cyclic photophosphorylation, and reducing power (reduced nicotinamide adenine dinucleotide [NADH]) for autotrophic carbon fixation is generated by reverse electron flow using sulfide as the electron donor. Elemental sulfur or polysulfide generated by sulfide oxidation is stored in intracellular globules; when environmental sulfide is depleted, these globules are oxidized first to sulfite and then sulfate. Most purple sulfur bacteria are also capable of heterotrophic growth in the absence of light. These organisms appear in Winogradsky columns as pastel purple blotches in the sulfide-rich anaerobic regions of the column.

EXAMPLE **Chromatium vinosum**

*Chromatium vinosum* (Fig. 10.17) is a large (ca. 2- by 3- to 6-$\mu$m) rod-shaped species that accumulates many small sulfur globules per cell. The cells are motile by way of polar flagella, and they grow individually, not in clumps like most purple bacteria of other genera. Species in this genus are usually distinguished by cell size and absorption spectra. Unlike most other purple sulfur bacteria, *C. vinosum* can use hydrogen in place of sulfide as an electron donor for reverse electron flow.

**Figure 10.17** Bright-field micrograph of *Chromatium okenii* (a close relative of *C. vinosum*). Notice the internal elemental sulfur granules. (Courtesy of Michael Plewka, www.plingfactory.de.) doi:10.1128/9781555818517.ch10.f10.17

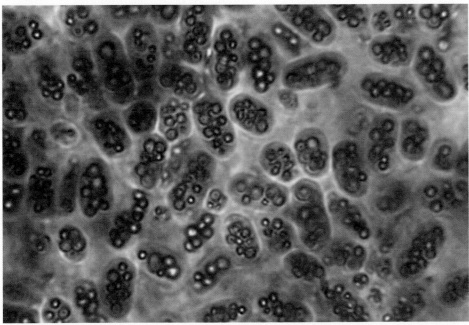

# Class *Deltaproteobacteria*

Figure 10.18 is a phylogenetic tree of representative members of the *Deltaproteobacteria*.

| Class | Deltaproteobacteria | | |
|---|---|---|---|
| Order | | Bdellovibrionales | |
| Family | | | Bdellovibrionaceae (e.g., Bdellovibrio, Vampirovibrio) |
| Family | | | Bacteriovoracaceae (e.g., Bacteriovorax, Peredibacter) |
| Order | | Desulfarculales | |
| Family | | | Desulfarculaceae (Desulfarculus) |
| Order | | Desulfovibrionales | |
| Family | | | Desulfovibrionaceae (e.g., Desulfovibrio) |
| Family | | | Desulfomicrobiaceae (e.g., Desulfomicrobium) |
| Family | | | Desulfohalobiaceae (e.g., Desulfohalobium, Desulfomonas) |
| Family | | | Desulfonatronumaceae (Desulfonatronum) |
| Order | | Desulfobacterales | |
| Family | | | Desulfobacteraceae (e.g., Desulfobacter, Desulfonema) |
| Family | | | Desulfobulbaceae (e.g., Desulfobulbus, Desulfocapsa) |
| Family | | | Nitrospinaceae (Nitrospina) |
| Order | | Desulfurellales | |
| Family | | | Desulfurellaceae (Desulfurella, Hippea) |
| Order | | Desulfuromonadales | |
| Family | | | Desulfuromonadaceae (e.g., Desulfuromonas, Pelobacter) |
| Family | | | Geobacteraceae (e.g., Geobacter, Trichlorobacter) |
| Order | | Myxococcales | |
| Family | | | Cystobacteraceae (e.g., Cystobacter, Stigmatella) |
| Family | | | Myxococcaceae (e.g., Angiococcus, Myxococcus) |
| Family | | | Polyangiaceae (e.g., Chondromyces, Polyangium) |
| Family | | | Nannocystaceae (e.g., Nannocyctis) |
| Family | | | Haliangiaceae (Haliangium) |
| Family | | | Kofleriaceae (Kofleria) |
| Order | | Syntrophobacterales | |
| Family | | | Syntrophobacteraceae (e.g., Desulforhabdus, Syntrophobacter) |
| Family | | | Syntrophaceae (e.g., Desulfomonile, Syntrophus) |

## *About this class*

### Diversity

Most of the δ-proteobacteria are sulfate reducers, and this is probably the primitive phenotype of the group. This is also true of the ε-proteobacteria, suggesting that these two groups share a common ancestral sulfate-reducing phenotype, either because they are specifically related (i.e., they share a common branch off the remainder of the proteobacterial tree) or because sulfate reduction is the primitive phenotype of the proteobacteria in general.

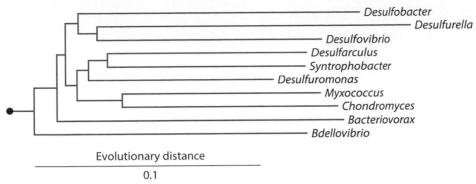

**Figure 10.18** Phylogenetic tree of representative δ-proteobacteria.
doi:10.1128/9781555818517.ch10.f10.18

## Metabolism

Although a much smaller group than the α-, β-, or γ-proteobacteria, like these other groups the δ-proteobacteria have a very broad range of metabolic pathways. Characterized members of this group fall into three general classes of metabolism: anaerobic sulfate reducers and syntrophic hydrogen-generating heterotrophs, aerobic heterotrophs, and parasites of other bacteria.

## Morphology

Most of the organisms in this group are straight or slightly curved rods.

## Habitat

The habitats of these organisms vary with their phenotypes. The sulfate reducers are the most common and are sometimes predominant members of sulfate-containing anaerobic environments.

### Sulfate reducers and hydrogenic syntrophs

Sulfate reducers are common in many anaerobic environments but are predominant in marine and estuary sediments (saltwater is rich in sulfate). Metabolism is anaerobic respiration, using organics (or sometimes $H_2$) as the electron donor for electron transport. Sulfate is the terminal electron acceptor. In the environment, this sulfide often reacts with metal cations (chiefly $Fe^{3+}$) to produce insoluble black metal sulfides. This is the black color typical of marine, estuary, and nutrient-rich freshwater sediments and muds.

Metabolism by the hydrogenic syntrophs is related to that of the sulfate reducers, except that protons are used as the terminal electron acceptor in place of sulfate to generate molecular hydrogen. This process is energetically favorable only if the ambient concentration of hydrogen is vanishingly low, and so they must live in symbiosis with hydrogen-utilizing organisms, such as methanogens, that consume the hydrogen as quickly as it is generated.

EXAMPLE *Desulfovibrio desulfuricans*
*Desulfovibrio desulfuricans* cells (Fig. 10.19) are slightly curved rods, motile by a single polar flagellum. They oxidize hydrogen and a wide range of organics,

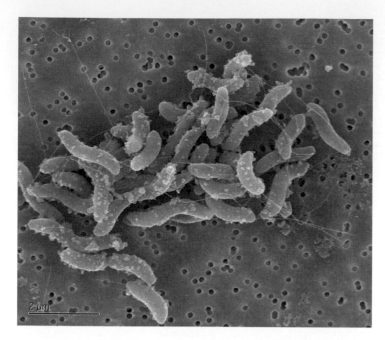

**Figure 10.19** Scanning electron micrograph of *Desulfovibrio desulfuricans*. (Source: M. Auer and M. Zemia, Lawrence Berkeley National Laboratory.) doi:10.1128/9781555818517.ch10.f10.19

including glycerol (which most anaerobes cannot utilize). Organics are incompletely oxidized by *D. desulfuricans*, generating acetate as the primary waste product. When oxidizing hydrogen, they require acetate as a source of carbon for growth; they are not autotrophic. *D. desulfuricans* can reduce sulfate or sulfite, or even protons; hydrogenic growth requires removal of this hydrogen product, either by thorough flushing of the medium or by cocultivation with methanogens (i.e., syntrophically).

## Myxobacteria

The myxobacteria are unicellular aerobic gliders (twitching motility is also used) with complex life cycles, usually found in terrestrial organic-rich environments, especially on bark or decomposing leaves and wood. They grow individually in thin swarming sheets, excreting lytic and digestive enzymes that lyse other bacteria, on which the myxobacteria feed. When starved, myxobacteria aggregate and develop into fruiting bodies, with base, stalk, and spore cells. This is a terminal differentiation; only the spore cells have a future, and so the fruiting body is a true multicellular organism. Spores are released into the environment, and those that migrate to a better environment germinate to produce a new crop of free-living cells.

### EXAMPLE *Myxococcus xanthus*

*Myxococcus xanthus* (Fig. 10.20) is the best-studied member of this group, being the easiest to grow to high density in liquid cultures and being genetically manipulatable. As a result, aggregation and sporangium formation in *M. xanthus* are model systems for bacterial cell-cell communication, self-organization, and development. This species produces simple spheroid fruiting bodies on short stalks.

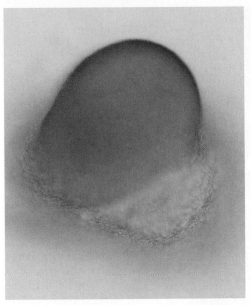

**Figure 10.20** Close-up photograph of simple *Myxococcus xanthus* fruiting bodies. (Courtesy of John R. Kirby.) doi:10.1128/9781555818517.ch10.f10.20

### *Parasites of other bacteria*

Although many bacteria are predatory, i.e., can kill others and feed on the nutrients released, the bdellovibrios actually invade or attach to the surface of other bacterial cells and parasitize them. As a result, they grow as plaques on lawns of host bacteria, much like the plaques produced by viruses. Members of the genus *Bdellovibrio* can parasitize a wide range of gram-negative bacteria (and a few gram-positive bacteria) and take up residence in the periplasmic space of the host. Others, e.g., *Vampirococcus*, are very host specific (infecting only *Chromatium*, in this example) and are epibiotic, attaching to the surface but not entering into the host.

#### EXAMPLE  *Bdellovibrio bacteriovorans*

*Bdellovibrio bacteriovorans* (Fig. 10.21) is probably the best-studied member of this group. Commonly found in soil and freshwater environments, the swarmer "attack phase" of this organism is a small (0.25 to 0.4 by 1 to 2 μm) vibrio, motile by a single polar flagellum. This flagellum is unusual in being ensheathed by the outer membrane. After attachment of the attack-phase parasite to a host cell, the parasite loses the flagellum, passes through the outer membrane of the host, and resides in the periplasmic space. In some cases the parasite can reside quiescently in the host, but more often it immediately begins to extract nutrients from the host cell for growth. The parasitic cell grows by elongation; the length of the resulting spiral-shaped parasite depends primarily on the initial size of the host. When the host is spent, the parasite divides into a number of attack-phase cells, which are released by lysis of the host.

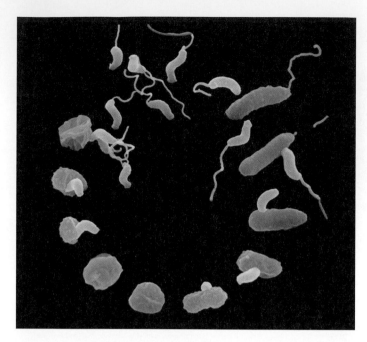

**Figure 10.21** Life cycle of *Bdellovibrio bacteriovorans*, assembled from scanning electron micrographs. Host cells are false-colored blue, *B. bacteriovorans* cells are false-colored yellow. (© Max Planck Institute for Developmental Biology/Rendulic, Berger, and Schuster.) doi:10.1128/9781555818517.ch10.f10.21

## Class *Epsilonproteobacteria*

Figure 10.22 is a phylogenetic tree of representative members of the *Epsilonproteobacteria*.

| Class | Epsilonproteobacteria | | |
|---|---|---|---|
| Order | | Campylobacterales | |
| Family | | | Campylobacteraceae (e.g., Campylobacter, Sulfurospirillum) |
| Family | | | Helicobacteraceae (e.g., Helicobacter, Sulfuricurvum, Wolinella) |
| Family | | | Hydrogenimonaceae (Hydrogenimonas) |
| Order | | Nautiliales | |
| Family | | | Nautiliaceae (e.g., Caminibacter, Nautilia, Thioreductor) |

**Figure 10.22** Phylogenetic tree of representative ε-proteobacteria.
doi:10.1128/9781555818517.ch10.f10.22

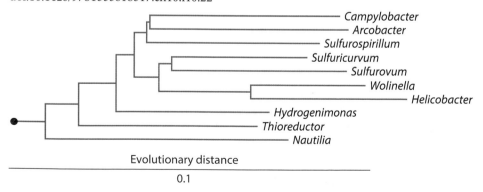

## About this class

### Diversity

Familiar ε-proteobacteria form a relatively narrow phylogenetic group of intestinal symbionts, but the phylogenetic range of environmental members of this group, represented by ribosomal RNA (rRNA) sequences extracted primarily from deep-sea environments, suggests that this is in actuality a very large and phylogenetically diverse class.

### Metabolism

These organisms are microaerophilic or anaerobic heterotrophs or chemoautotrophs. Metabolism is by anaerobic respiration using a wide range of organic and inorganic electron donors and acceptors, but generally not carbohydrates. A common feature of the ε-proteobacteria is the ability to use hydrogen as an electron donor for electron transport. Sulfate reduction seems also to be common.

### Morphology

Most of the cultivated ε-proteobacteria are helically curved rods and are motile by polar flagella. The deep-sea environmental ε-proteobacteria have a wide range of morphologies, including vibrio and helical forms but also rods, cocci, and filaments of all types.

### Habitat

The best-known ε-proteobacteria are intestinal symbionts, but in deep-sea environments, particularly marine hydrothermal zones, the ε-proteobacteria are predominant. Deep-sea hydrothermal ε-proteobacteria are abundant not just in water and sediment but also as external and endosymbiotic inhabitants of vent-associated animals.

## Intestinal symbionts

Members of the genera *Campylobacter*, *Helicobacter*, and *Wolinella* are inhabitants of the upper gastrointestinal epithelium of mammals and birds. Most are commensalistic, at least in their natural host, but some are pathogens and many cause zoonotic disease. For example, *Campylobacter* is a common commensal in birds, but in humans it is perhaps the single most common cause of food-borne disease.

### EXAMPLE *Helicobacter pylori*

*Helicobacter pylori* (Fig. 10.23) is a microaerophilic curved rod with several unipolar flagella that are sheathed, with a distinctive bulb at the distal ends. It is a common symbiont of the stomach and duodenal lining, colonizing about 70% of humans. In most cases no symptoms occur and the symbiosis persists for life. In some cases, however, colonization by *H. pylori* results in gastritis or peptic ulcers and is a contributing factor for stomach cancer. However, there is evidence that *H. pylori* may also help modulate stomach acidity and reduce acid reflux.

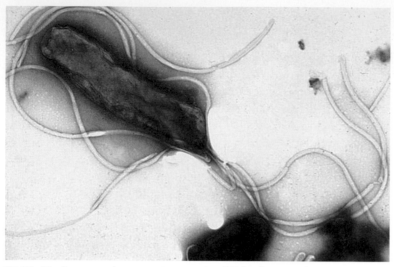

**Figure 10.23** Shadow-cast electron micrograph of *Helicobacter pylori*. Notice the polar flagella with a bulbous end. (Source: Y. Tsutsumi, Fujita Health University School of Medicine.) doi:10.1128/9781555818517.ch10.f10.23

## *Deep-sea hydrothermal vent-associated species*

Although few have been cultivated, molecular surveys of environmental samples (see chapter 19) show that ε-proteobacteria are very common, even predominant, in many marine hydrothermal vent environments. These environments usually bring to mind the hyperthermophilic *Archaea*, but the cold area surrounding the hot vents is an oasis of both macroscopic and microscopic life. In fact, it is the mixture of hot, reduced geothermal water and cold, oxygenated seawater, each by itself more or less at equilibrium, that creates the chemical disequilibria that provide the chemical potential energy that can be harvested by lithotrophic organisms. Except in the hottest regions of these vent zones, ε-proteobacteria are very abundant. This includes the sediments, surfaces, and waters of these regions but also the symbionts of the animals that inhabit the regions.

EXAMPLE **The endosymbiont of the scaly snail *Crysomallon squamiferum***

The scaly snail (*Crysomallon squamiferum* [Fig. 10.24]) is a unique animal found only in Indian Ocean hydrothermal vent fields. Instead of an operculum (the other half of the shell that other snails usually use to cover themselves when retracted), the body of the scaly snail is covered in tough, iron-sulfide-reinforced scale-like plates. More amazing, the scaly snail has only a vestigial digestive tract and radulus (a scraping tongue). Instead, the scaly snail has greatly enlarged esophogeal glands filled with ε-proteobacterial endosymbionts. The animal probably absorbs sulfides from the environment through the crawling surface of its foot; these sulfides are brought to the esophageal glands along with oxygen absorbed in the gills. Here the sulfide-oxidizing ε-proteobacterial endosymbionts use the sulfides to generate energy and fix carbon from $CO_2$. In return for a

**Figure 10.24** The deep-sea hydrothermal vent scaly snail *Crysomallon squamiferum*. (Courtesy of Cindy Van Dover.) doi:10.1128/9781555818517.ch10.f10.24

place to live and a supply of resources, the bacteria provide the snail with some form of nutrition. The snail, then, is a chemoautotrophic animal. This symbiosis is analogous to that of the giant vent tubeworm (*Riftia*) and several other hydrothermal vent animals. The symbiotic biofilm covering the snail is by a wide range of ε-proteobacteria, and these presumably participate in the iron sulfide mineralization of the host scales and shell surface.

## The concept of "proteobacteria"

### *Review of the basics of electron transport*

Electron transport is carried out by a series of electron carriers in an "electron transport chain" in the cell membrane (Fig. 10.25). This allows the oxidation half-reaction (electron donation) and the reduction half-reaction (electron acceptance) to be physically separated, so that the energy released by the reaction can be captured.

Some electron carriers in the electron transport chain really do carry just electrons, while others carry hydrogen atoms (electrons plus protons). When an electron carrier transfers an electron to a hydrogen carrier, the hydrogen carrier must capture a proton from solution; when it then transfers the hydrogen to an electron carrier, it releases the proton into solution.

The carriers in the electron transport chain are physically organized in such a way that when a carrier needs to capture a proton (when an electron carrier is donating to a hydrogen acceptor), it gets it from the cytoplasmic side of the membrane. When a carrier needs to release a proton (when a hydrogen carrier is donating to an electron acceptor), it does so at the outside surface of the membrane. The result, then, of electron transport down the chain is that protons are collected from inside and released to the outside of the membrane. In this way, the chemical energy of the oxidation-reduction reaction mediated by

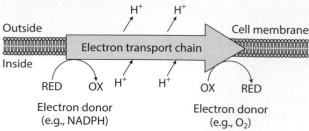

**Figure 10.25** Simplified representation of the electron transport chain. The oxidation half-reaction occurs at the upstream end of the electron transport chain, while the reduction half-reaction occurs at the downstream end. The physical separation of the oxidation-reduction reaction into its component half-reactions allows the energy released by the transfer of electrons to be captured in the form of a proton gradient. doi:10.1128/9781555818517.ch10.f10.25

the chain is captured in the form of a proton gradient, rather than being entirely lost to heat as would occur if the reaction occurred in solution.

The energy in this proton gradient is in two forms: a chemical gradient (higher concentration of protons outside than inside) and an electrical potential (positive outside, negative inside). The energy in the proton gradient is converted back to chemical energy, which the cell can use, by adenosine triphosphatase (ATPase). ATPase (named for the reverse reaction) is a membrane protein that leaks the protons back into the cell, using the energy of the proton gradient to phosphorylate adenosine diphosphate (ADP) to make adenosine triphosphate (ATP).

## The proteobacteria

The purple bacteria and relatives really are "proteobacteria"; they seem to have the ability to change readily (in evolutionary terms) between sulfur oxidation or reduction, photosynthesis, heterotrophy, nitrogen oxidation or reduction, and many other phenotypes. These different lifestyles are based on the same electron transport chain; what have changed are the inputs and outputs, i.e., the electron donors and acceptors of the oxidation-reduction reaction from which they derive energy.

Most heterotrophs oxidize organic compounds into $CO_2$ to generate NADH, which serves as the electron donor for electron transport (and ultimately ATP synthesis), using $O_2$ as the terminal electron acceptor. Sulfur-oxidizing autotrophs use $H_2S$ as an electron donor (converting it to sulfate, if completely oxidized) and oxygen as the electron acceptor (generating water). Other electron donors are thiosulfate, elemental sulfur, activated photosystem chlorophylls, hydrogen, methane, and ammonia. Other electron acceptors are sulfate, sulfite, sulfur, nitrite, nitrate, ferric ion (and many other oxidized metal ions), and reduced photosystem chlorophylls. For the production of a proton gradient for ATP synthesis, the electron flow is in the "forward" direction. Reverse electron flow is generally reserved for the synthesis of NADH by autotrophs (which do not use organic carbon to make NADH) for reducing $CO_2$ to organic carbon (see chapter 9).

Because all of these metabolisms are based on the same electron transport chain, all that is needed to change metabolic phenotype is to change either the enzyme that catalyzes the oxidation reaction (feeding electrons into the beginning of the electron transport chain) or the reduction reaction (transferring electrons from the end of the electron transport chain to the terminal electron acceptor) or both. For example, the acquisition of a single polypeptide, nitrate reductase, could convert a heterotroph from obligate aerobe into an anaerobic nitrate reducer.

The proteobacteria seem to be particularly good at changing the inputs and outputs of the electron transport chain, hence the Latin name, meaning "changeable bacteria." The genes for these enzymes feeding into and out of the electron transport chain are most frequently acquired by horizontal transfer. The proteobacteria may be particularly good at acquiring such useful genes from other sources, and perhaps they are more readily able to integrate such foreign enzymes into electron transport chain function.

In any case, the metabolic phenotype of an organism can clearly be a superficial trait, and it is certainly not a reliable guide to phylogenetic relationships.

## Questions for thought

1. Make as long a list of gram-negative bacteria as you can from your memory or old notes. Now look them up; what fraction of them are proteobacteria? What fraction are α-, β-, γ-, δ-, or ε-proteobacteria?

2. What complex bacterial life cycles are you familiar with other than the ones described in this chapter?

3. The cells of *Caulobacter* have a specified polarity; each end is distinct. Swarmer cells have a flagellum at only one end, and the receptors for chemotaxis are also localized to this end. Stalked cells have a stalk at only one end, and daughter cells are produced by fission from the other end. What other bacteria are you familiar with that have defined polarity? Which ones do not have it? Are you sure? If bacteria have terminal (end-to-end) polarity, what about dorsal/ventral and lateral polarity?

4. What do you think you would see if you made a time-lapse close-up video of growing films of *Myxococcus*? What about growing colonies of *E. coli*? *Bacillus*?

5. Do you know of any examples where the entire ecological niche of an organism is defined by one or a few genes it acquired horizontally? How might you recognize these cases?

6. Imagine that a segment of DNA was recently acquired by an organism and integrated into its genome. One of the genes in this segment conferred a useful trait on its new host, while the rest did not. How might a scientist studying this organism recognize that this is a foreign gene? How might this gene change in the time shortly after being acquired? What might happen to the other, unhelpful acquired genes?

# 11

# Gram-Positive Bacteria

Traditional taxonomy divides all bacteria into two groups, gram positive and gram negative. In phylogenetic terms, this division is false. As can be seen in the tree in Fig. 11.1, the gram-positive members of the domain *Bacteria* are actually two phyla that may be related, the *Firmicutes* (the low-G+C gram-positive bacteria) and the *Actinobacteria* (the high-G+C gram-positive bacteria). The remaining bacterial phyla have a generally gram-negative cell envelope. Therefore, like the terms "prokaryote" (see chapter 2) and "invertebrate," the term "gram negative" tells you what an organism *is not* but not what an organism *is*. Whether the *Firmicutes* and *Actinobacteria* are specifically related is doubtful; these two branches lost their outer membranes in independent evolutionary events (as did the *Planctomycetes*). In other words, the "gram-positive" bacteria are unlikely to be a coherent phylogenetic group.

Nevertheless, the use of the Gram stain as a starting point in the identification of bacteria has been an incredibly important tool. Perhaps this is because, although the gram-positive bacteria represent only a fraction of the immense phylogenetic diversity of bacteria in nature, this group contains more than its fair share of the most important human and animal pathogens.

The original names for the *Firmicutes* (low-G+C gram-positive bacteria) and *Actinobacteria* (high-G+C gram-positive bacteria) came from the fact that many of the most familiar members of the *Firmicutes* have relatively low genomic G+C contents whereas many of the familiar members of the *Actinobacteria* have relatively high genomic G+C contents. This rule is by no means universal, nor particularly meaningful.

doi:10.1128/9781555818517.ch11

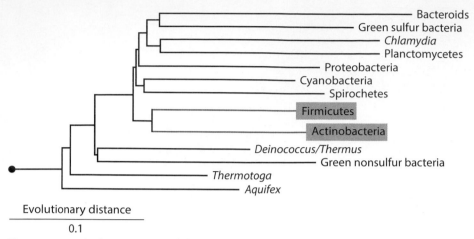

**Figure 11.1** Phylogenetic tree of the bacterial phyla, with the two phyla of (predominantly) gram-positive bacteria highlighted. doi:10.1128/9781555818517.ch11.f11.1

Given the incredible diversity of organisms in the phyla *Firmicutes* and *Actinobacteria*, this chapter touches on only a few examples, focusing on less familiar representatives rather than the well-known pathogens.

## What does being gram positive mean?

The structural distinction between gram-positive and gram-negative bacteria lies in the structure of the cell envelope. Gram-positive bacteria have no phospholipid outer membrane; they are bound by the cytoplasmic membrane and often a very thick cell wall (and so stain gram positive). There is therefore no periplasmic space, which gram-negative bacteria use as an environmental buffer. But they do have some control of the conditions in the spongy thick wall, and so this may serve a purpose similar to that of the periplasmic space. Most other bacteria (gram-negative bacteria) have both an outer and inner membrane, sandwiching a thin cell wall, and therefore stain gram negative. Like many other members of the *Bacteria* and *Archaea*, some gram-positive bacteria also have an outer protein coat, the S-layer.

However, a wide range of factors, in addition to the outer membrane and cell wall thickness, determine whether a culture stains gram positive or negative. Even in the same species, cells may stain differently in different stages of the growth cycle or when grown under different conditions.

In addition, some members of the gram-positive phylogenetic group do have an outer membrane. Some members of the *Actinobacteria* have an outer membrane composed of mycolic acids rather than phospholipids. These organisms are sometimes considered neither gram positive nor gram negative, but "acid fast." Some members of the *Firmicutes* have a traditional gram-negative envelope, complete with phospholipid outer membrane and lipopolysaccharide. It is very important, therefore, when using the terms "gram positive" and "gram negative," to be clear about whether this refers to how the cells stain, the structure of their envelope, or a phylogenetic group.

## An alternative view of gram-positive bacteria

Some molecular phylogenists argue, primarily on the basis of conserved insertions and deletions (indels) in protein-coding genes, that the gram-positive members of the *Bacteria* are more closely related to *Archaea* and *Eukarya* than are any other members of the *Bacteria*. This view is not supported by analyses of conserved protein or RNA gene sequences, but it does separate organisms into those with two-membrane envelopes (gram-negative bacteria), which in this scheme are termed "diderms," and those with single-membrane envelopes (gram-positive bacteria, archaea, and eukaryotes), termed "monoderms." This issue remains a matter of contention.

## Phylum *Firmicutes* (low-G+C gram-positive bacteria)

Figure 11.2 is a phylogenetic tree of representative members of the *Firmicutes*.

| Phylum | *Firmicutes* | | |
|--------|-----------|---|---|
| Class | | *Bacilli* | |
| Order | | | *Bacillales* | |
| Family | | | | *Bacillaceae* (e.g., *Bacillus, Geobacillus, Halobacillus*) |
| Family | | | | *Alicyclobacillaceae* (e.g., *Alicyclobacillus*) |
| Family | | | | *Listeriaceae* (e.g., *Listeria*) |
| Family | | | | *Paenibacillaceae* (e.g., *Paenibacillus, Brevibacillus*) |
| Family | | | | *Pasteuriaceae* (*Pasteuria*) |
| Family | | | | *Planococcaceae* (e.g., *Planococcus, Kurthia*) |
| Family | | | | *Sporolactobacillaceae* (e.g., *Sporolactobacillus*) |
| Family | | | | *Staphylococcaceae* (e.g., *Staphylococcus, Salinicoccus*) |
| Family | | | | *Thermoactinomycetaceae* (e.g., *Thermoactinomyces*) |
| Order | | | *Lactobacillales* | |
| Family | | | | *Lactobacillaceae* (e.g., *Lactobacillus, Pediococcus*) |
| Family | | | | *Aerococcaceae* (e.g., *Aerococcus, Facklamia*) |
| Family | | | | *Carnobacteriaceae* (e.g., *Carnobacterium, Atopobacter*) |
| Family | | | | *Enterococcaceae* (e.g., *Enterococcus, Vagococcus*) |
| Family | | | | *Leuconostocaceae* (e.g., *Leuconostoc, Weissella*) |
| Family | | | | *Streptococcaceae* (e.g., *Streptococcus, Lactococcus*) |
| Class | | *Clostridia* | |
| Order | | | *Clostridiales* | |
| Family | | | | *Clostridiaceae* (e.g., *Clostridium, Sarcina*) |
| Family | | | | *Eubacteriaceae* (e.g., *Eubacterium, Acetobacterium*) |
| Family | | | | *Gracilibacteraceae* (*Gracilibacter*) |
| Family | | | | *Heliobacteriaceae* (e.g., *Heliobacterium, Heliophilum*) |
| Family | | | | *Lachnospiraceae* (e.g., *Butyrovibrio, Coprococcus*) |
| Family | | | | *Peptococcaceae* (e.g., *Peptococcus, Desulfotomaculum*) |
| Family | | | | *Peptostreptococcaceae* (e.g., *Peptostreptococcus*) |

*(continued)*

| Phylum | Firmicutes (continued) | | | |
|--------|-----------|---|---|---|
| Family | | | | Ruminococcaceae (e.g., Ruminococcus, Acetivibrio) |
| Family | | | | Syntrophomonadaceae (e.g., Syntrophomonas) |
| Family | | | | Veillonellaceae (e.g., Sporomusa, Megasphaera) |
| Order | | | Halanaerobiales | |
| Family | | | | Halanaerobiaceae (e.g., Halanaerobium, Halocella) |
| Family | | | | Halobacteroidaceae (e.g., Halanaerobacter, Orenia) |
| Order | | | Thermoanaerobacterales | |
| Family | | | | Thermoanaerobacteraceae (e.g., Thermoanaerobacter) |
| Family | | | | Thermodesulfobiaceae (e.g., Coprothermobacter) |
| Class | | Erysipelotrichia | | |
| Order | | | Erysipelotrichales | |
| Family | | | | Erysipelotrichaceae (e.g., Erysipelothrix) |
| Class | | Mollicutes | | |
| Order | | | Mycoplasmatales | |
| Family | | | | Mycoplasmataceae (e.g., Mycoplasma, Ureaplasma) |
| Order | | | Entomoplasmatales | |
| Family | | | | Entomoplasmataceae (Entomoplasma, Mesoplasma) |
| Family | | | | Spiroplasmataceae (Spiroplasma) |
| Order | | | Acholeplasmatales | |
| Family | | | | Acholeplasmataceae (Acholeplasma, Phytoplasma) |
| Order | | | Anaeroplasmatales | |
| Family | | | | Anaeroplasmataceae (Anaeroplasma, Asteroleplasma) |

**Figure 11.2** Phylogenetic tree of representative members of the *Firmicutes*.
doi:10.1128/9781555818517.ch11.f11.2

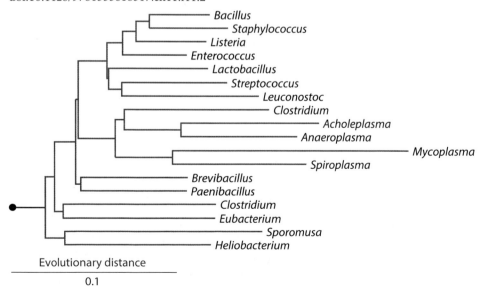

## About this phylum

### Diversity

*Firmicutes* is a large and diverse group of organisms, encompassing four classes, 11 orders, 35 families, and more than 240 species. Included in the *Firmicutes* are the *Mollicutes* (also known as *Tenericutes*), the genus *Mycoplasma*, and relatives. Although not usually considered gram positive (they entirely lack the peptidoglycan cell wall), they are members of this phylogenetic group and, like other gram-positive bacteria, lack the outer membrane. Also among the *Firmicutes* are the members of the family *Veillonellaceae* (*Veillonella*, *Dialister*, *Megasphaera*, and *Sporomusa*) that have traditional gram-negative cell envelopes, complete with outer membrane, despite being members of a gram-positive phylogenetic group. Anaerobic, endospore-forming rods ("clostridia") form several deep lineages in the *Firmicutes*, and so this probably represents the primitive phenotype of this group.

### Metabolism

These organisms are heterotrophic, except for the photosynthetic heliobacteria, but have a wide range of heterotrophic lifestyles. The *Bacilli* are generally obligate aerobes, while other *Firmicutes* are usually anaerobic, although often aerotolerant. Anaerobic metabolism is usually by substrate-level phosphorylation rather than anaerobic respiration; these organisms often lack a complete electron transport chain. Most are mesophilic, although a few psychrophilic or moderately thermophilic species exist. A wide range of carbon and energy sources are used by members of this group and result in a similarly wide range of fermentation products.

### Morphology

Familiar members of this group are either rods or cocci; these sometimes form nearly filamentous chains, as in some species of *Bacillus* and *Streptococcus*. Individual cells or pairs are also very common. Endospores are a common unique feature of this group. Morphology varies more widely in the *Mollicutes*, but their small size and lack of peptidoglycan means that these morphologies are less easily observed.

### Habitat

Both *Firmicutes* and *Actinobacteria* are abundant in most soil and sediment communities. Members of the *Firmicutes* are also predominant symbionts of the skin, mucous membranes, and gut of animals; perhaps this helps explain the large number of human and animal pathogens that also are members of this group.

### *Aerobic endospore-forming rods* (Bacillus *and relatives*)

These are generally aerobic (some grow anaerobically by nitrate reduction), rod-shaped, endospore-forming heterotrophs. They are common soil inhabitants. A few are opportunistic pathogenic to humans, and a very few are bona fide pathogens, *Bacillus anthracis* being the most dangerous. *Listeria* and *Staphylococcus*, although not endospore-forming rods, are members of this phylogenetic group. Most of the endospore-forming aerobic rods were originally considered to be species of the genus *Bacillus*, but the size of this genus and phylogenetic considerations led to its division into a number of new genera, most of which retain the *-bacillus* suffix (e.g., *Paenibacillus*, *Brevibacillus*, and *Geobacillus*).

### EXAMPLE *Bacillus cereus*

*Bacillus cereus* is a very close relative of the human pathogen *B. anthracis*, the insect pathogen *B. thuringiensis* (the source of the widely used insecticide Bt), *B. mycoides*, *B. pseudomycoides*, and *B. weihenstephanensis*; these may more appropriately be considered different strains of the same species. *B. cereus* is easily isolated from soil, and soil has traditionally been considered its natural habitat. However, *B. cereus* is abundant in the hindguts of a wide range of arthropods, in which it is usually filamentous rather than existing as the individual cells or short chains seen in cultivation (Fig. 11.3). Cells shed in the arthropod feces

**Figure 11.3** Phase-contrast micrographs of *Bacillus cereus* (a.k.a. *Arthromitis*) growing as filaments in the hindgut of insects. Endospores are the phase-bright ovals within cells. (Reprinted from Margulis L, Jorgensen JZ, Dolan S, Kolchinsky R, Rainey FA, Lo S-C, *Proc Natl Acad Sci USA* **95**:1236–1241, 1998, with permission. Copyright 1998 National Academy of Sciences, USA.) doi:10.1128/9781555818517.ch11.f11.3

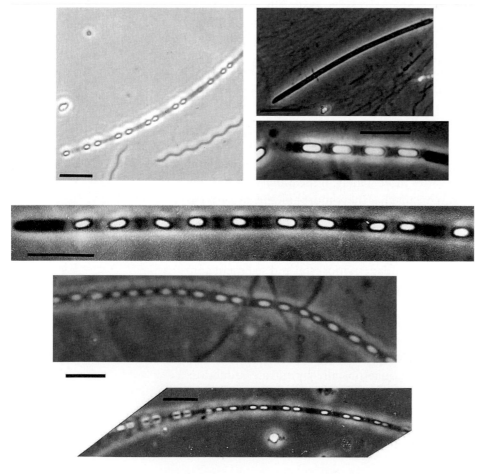

sporulate, awaiting ingestion by a new arthropod host. This life cycle is very similar to that of *B. anthracis* and *B. thuringiensis*, except that the host is not (apparently) harmed. *B. cereus* is often considered to be a food-born opportunistic pathogen, although it is unclear whether these are typically infections or reactions to a Bt-like toxin.

## *Anaerobic endospore-forming rods* (Clostridium *and relatives*)

These organisms and their non-spore-forming relatives (*Eubacterium*) are abundant in anaerobic soils and sediments. They constitute the bulk of the bacteria in the gut contents (and feces) of humans and other large animals and are particularly important in animal decay. Many species are opportunistically pathogenic, and some are well-known pathogens, for example, *C. tetani* (the causative agent of tetanus) and *C. perfringens* (the causative agent of gas gangrene). Clostridia lack the electron transport chain and obtain energy from a wide variety of substrate-level phosphorylation reactions. The proton gradient, which is required to drive many active transport pumps, seems to be maintained by a traditional adenosine triphosphatase (ATPase) run in reverse of the usual ATP synthesis direction; protons are pumped from inside to outside at the expense of the hydrolysis of adenosine triphosphate (ATP).

### EXAMPLE *Clostridium botulinum*

*Clostridium botulinum* (Fig. 11.4) produces a potent neurotoxin (sometimes said to be the most deadly toxin known); it is this neurotoxin, rather than the organism itself, that produces the disease botulism. The most common form of botulism is infantile botulism, in which *C. botulinum* colonizes the gut among the wide range of other normal *Clostridium* species; the toxin is then

**Figure 11.4** Light micrograph of Gram-stained *Clostridium botulinum*. Notice the terminal endospores that swell the end of the mother cell. (Source: Centers for Disease Control and Prevention.) doi:10.1128/9781555818517.ch11.f11.4

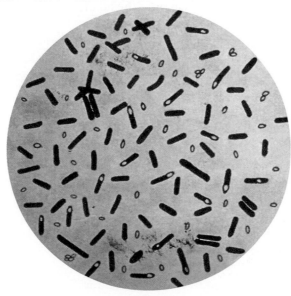

absorbed through the gut mucosa. Sudden infant death syndrome is sometimes (and probably incorrectly) attributed to *C. botulinum* acquired from honey. In wound botulism, *C. botulinum* grows in a gangrenous wound; the toxin is absorbed directly into the surrounding tissue and circulation. This is in many ways analogous to tetanus, caused by a related species, *Clostridium tetani*. The best-known form of botulism, however, is the food-borne type. The organism grows in anaerobic canned food; the toxin is absorbed through the gut mucosa upon ingestion. The botulism toxin inhibits neurotransmission, resulting in flaccid paralysis. The botulism toxin (Botox) has important medical uses in a wide variety of conditions in which muscles contract inappropriately.

## Lactic acid bacteria

Lactic acid bacteria (commonly known by the acronym LAB) are acid-tolerant, nonsporulating relatives of the bacilli and clostridia. Like the clostridia, the LAB lack the electron transport chain and so generate ATP by substrate-level phosphorylation. Unlike clostridia, most are aerotolerant. In nature, these organisms are usually associated with the decomposition of plant material, but they are also important human and animal symbionts. Most grow on simple carbohydrates (sugars), producing either lactic acid alone via glycolysis (homofermentation) or lactic acid, ethanol, and $CO_2$ via the pentose-phosphate pathway (heterofermentation). The LAB are widely used in the food industry for the production of fermented vegetables (pickles and sauerkraut), dairy products (yogurt and cheese), and meat (fermented sausages). Although generally nonpathogenic and very often beneficial to humans, some members of the genus *Streptococcus* are harmful (*S. mutans* causes tooth decay) or pathogenic (*S. pyogenes* causes strep throat and scarlet fever).

### EXAMPLE *Leuconostoc mesenteroides*

*Leuconostoc* species are strictly heterofermentative LAB with distinctly oval cells that grow in chains. *L. mesenteroides* (Fig. 11.5) produces dextran slime when grown on sucrose; colonies become so slimy that they drip onto the lid of culture plates. *L. mesenteroides* predominates in most lactic fermentations in the initial stages; it is succeeded in later stages of these fermentations by the more acidophilic *Lactobacillus* species.

## Mollicutes (Tenericutes; Mycoplasma *and relatives*)

The *Mollicutes* are not often considered to be gram-positive bacteria; lacking peptidoglycan, they stain gram negative. Historically, they were often grouped with the viruses because of their small size and obligately parasitic lifestyle. However, they are phylogenetic members of the *Firmicutes*, being an offshoot of the *Clostridium innocuum* branch, and like other gram-positive bacteria they lack an outer membrane. Mollicutes are obligate extracellular symbionts or parasites of plants and animals. Most are unusual in that they have sterols in their membrane in addition to the usual fatty acid esters; they do not make sterols themselves but acquire them from their eukaryotic host. Mollicutes are very small (typically about 0.25 μm in diameter), among the smallest known cellular organisms with the smallest genomes. Their genomes lack the genes for

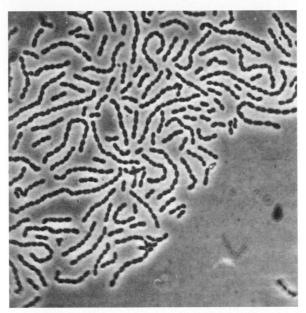

**Figure 11.5** Phase-contrast micrograph of *Leuconostoc mesenteroides*. (Source: U.S. Department of Energy Joint Genome Institute.) doi:10.1128/9781555818517.ch11.f11.5

the Krebs cycle, amino acid biosynthesis, purine and pyrimidine biosynthesis, and many other metabolic pathways. Most do contain the genes for glycolysis, which they use for energy production by substrate-level phosphorylation. They are often motile by gliding (Fig. 11.6). Members of this group have a wide range of complex morphologies but are usually described as amorphous or pleomorphous because of their size (at or below the resolving power of light microscopy) and the fact that their relatively nonrigid envelope cannot withstand traditional fixing treatments.

**Figure 11.6** *Mycoplasma mobile*, showing the typical pear-shaped morphology of these organisms, with appendages used for gliding motility on the "neck." (Reprinted from Hiratsuka Y, Miyata M, Uyeda TQP. 2005. Living microtransporter by uni-directional gliding of *Mycoplasma* along microtracks. *Biochem Biophys Res Commun* **331**:318–324, with permission from Elsevier.) doi:10.1128/9781555818517.ch11.f11.6

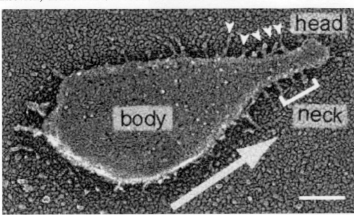

## EXAMPLE *Mycoplasma hominis*

*Mycoplasma* species are obligate parasites of animals, usually infecting the lung (e.g., *M. pneumoniae*) or urogenital tract (e.g., *M. genitalium*). *M. hominis* (Fig. 11.7) is an opportunistic pathogen; it seems to reside asymptomatically in the vagina of healthy women but is also one of many causes of bacterial vaginitis and pelvic inflammatory disease. Disease is usually associated either with invasive surgical procedures or with coinfection with the obligately parasitic protist *Trichomonas vaginalis*. In the former case, the parasite proliferates in deep surgical wounds and can be difficult to identify or treat; lacking a peptidoglycan cell wall, it is resistant to many commonly used postsurgical broad-spectrum antibiotics. In the latter case, there is a symbiosis between *T. vaginalis* and *M. hominis*, and most infected women are coinfected. Although typically an extracellular parasite, *M. hominis* can reside and replicate in the cytoplasm of *T. vaginalis*; this may facilitate both the transfer of *M. hominis* to new human hosts and its resistance to antibiotics. In both men and women, *M. hominis* is associated with reduced fertility. *M. hominis* cells can attach to and invade sperm cells, suggesting that the ability to persist intracellularly may be a general mechanism for infecting new hosts. *M. hominis* is generally spherical, lacking the elongated "flask" shape of most other *Mycoplasma* species. Also unlike other *Mycoplasma* species, *M. hominis* is not saccharolytic but instead uses only arginine for both carbon and energy.

**Figure 11.7** Fluorescent antibody of the parasitic flagellate *Trichomonas vaginalis* (stained red) with intracellular *M. hominis* (stained green). (Reprinted from Dessì D, Delogu G, Emonte E, Catania MR, Fiori PL, Rappelli P, *Infect Immun* **73**:1180–1186, 2005, with permission.) doi:10.1128/9781555818517.ch11.f11.7

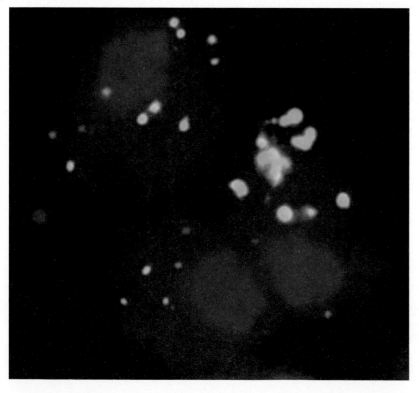

## Heliobacteriaceae (*green photosynthetic gram-positive bacteria*)

The family *Heliobacteriaceae* contains the only known phototrophic gram-positive bacteria. They carry out cyclic photophosphorylation, growing photoheterotrophically using pyruvate or similar organic acids as carbon sources. Heliobacteria use chlorophyll *g*, which absorbs wavelengths in the range of 790 nm not utilized by other photopigments, and so can avoid competition for light with other phototrophs. Unlike other phototrophs, heliobacteria house their photosynthetic complexes in the cytoplasmic membrane; no internal membranous structures or membranous invaginations are present. However, the photoreaction cycle of the *Heliobacteriaceae* is very similar to that of the *Chlorobi* (discussed in chapter 9). Heliobacteria are rod-shaped (sometimes slightly helical) anaerobic endospore formers, motile by gliding or flagella. Cells lack the typical thick gram-positive cell wall; the thin peptidoglycan layer is covered in a regular array of 11-nm protein beads. Heliobacteria are commonly found in soil and are very efficient nitrogen fixers.

EXAMPLE *Heliobacterium chlorum*

*Heliobacterium chlorum* is a long rod-shaped organism (1 by 6 to 10 μm), forming green-brown to emerald green colonies due to the presence of both chlorophyll *g* and the green accessory carotenoid neurosporene. *H. chlorum* is motile by gliding rather than by the flagella present in other heliobacteria. The only carbon sources known to support its photoheterotrophic lifestyle are pyruvate and lactate. Sporulation is rarely observed; older cultures usually degenerate into spheroplasts and lyse.

## Veillonellaceae (Firmicutes *with gram-negative envelopes*)

This little-known group of organisms is unique among the gram-positive phylogenetic bacteria; they have gram-negative envelopes, including an outer membrane and (generally) lipopolysaccharide layer. It is likely that this branch separated prior to the loss of the outer membrane in the remainder of the group. Members of this group are anaerobic heterotrophs and are common symbionts of the gastrointestinal tract, except for *Pectinatus* and *Megasphaera*, which are found in spoiled unpasteurized beer. *Pectinatus* and *Selenomonas* are rod-shaped organisms, motile by flagella that occur only on one side of the cell; they swim sideways in a distinct tumbling X shape. *Sporomusa* produces endospores similar to those of other members of the *Firmicutes*.

EXAMPLE *Veillonella atypica*

*Veillonella* spp. are the most numerous cultivable anaerobes in human saliva. These small (ca. 0.5-μm) cocci grow by decarboxylation of lactate and other organic acids produced by primary oral colonizers such as *Streptococcus salivarius*, fusobacteria, and actinomycetes, to which they adhere. Species of *Veillonella* are specialized to adhere to the primary colonizers of specific portions of the oral cavity; *V. atypica* adheres specifically to *S. salivarius* and *Fusobacterium nucleatum*, inhabitants of the saliva and upper surface of the tongue.

## Phylum *Actinobacteria* (high-G+C gram-positive bacteria)

Figure 11.8 is a phylogenetic tree of representative actinobacteria.

| Phylum | *Actinobacteria* | | | |
|---|---|---|---|---|
| Class | | *Actinobacteria* | | |
| Order | | | *Acidimicrobiales* | |
| Family | | | | *Acidimicrobiaceae (Acidimicrobium)* |
| Order | | | *Rubrobacterales* | |
| Family | | | | *Rubrobacteraceae (e.g., Rubrobacter, Thermoleophilum)* |
| Family | | | | *Patulibacteraceae (Patulibacter)* |
| Order | | | *Coriobacteriales* | |
| Family | | | | *Coriobacteriaceae (e.g., Atopobium, Slackia)* |
| Order | | | *Sphaerobacterales* | |
| Family | | | | *Sphaerobacteraceae (e.g., Sphaerobacter)* |
| Order | | | *Actinomycetales* | |
| Family | | | | *Actinomycetaceae (e.g., Actinomyces, Arcanobacterium)* |
| Family | | | | *Micrococcaceae (e.g., Micrococcus, Arthrobacter)* |
| Family | | | | *Bogoriellaceae (Bogoriella)* |
| Family | | | | *Rarobacteraceae (Rarobacter)* |
| Family | | | | *Sanguibacteraceae (Sanguibacter)* |
| Family | | | | *Brevibacteriaceae (Brevibacterium)* |
| Family | | | | *Cellulomonadaceae (e.g., Cellulomonas, Oerskovia)* |
| Family | | | | *Dermabacteraceae (Dermabacter, Brachybacterium)* |
| Family | | | | *Dermatophilaceae (e.g. Dermatophilus, Kineosphaera)* |
| Family | | | | *Dermacoccaceae (Dermococcus, Demetria, Kytococcus)* |
| Family | | | | *Intrasporangiaceae (e.g., Janibacter, Tetrasphaera)* |
| Family | | | | *Jonesiaceae (Jonesia)* |
| Family | | | | *Microbacteriaceae (e.g., Microbacterium, Agromyces)* |
| Family | | | | *Beutenbergiaceae (Beutenbergia, Georgenia, Salana)* |
| Family | | | | *Promicromonosporaceae (e.g., Promicromonospora)* |
| Family | | | | *Catenulisporaceae (Actinospica, Catenulispora)* |
| Family | | | | *Corynebacteriaceae (e.g., Corynebacterium)* |
| Family | | | | *Dietziaceae (Dietzia)* |
| Family | | | | *Gordoniaceae (Gordonia, Skermania, Millisia)* |
| Family | | | | *Mycobacteriaceae (e.g., Mycobacterium)* |
| Family | | | | *Nocardiaceae (e.g., Nocardia, Rhodococcus)* |
| Family | | | | *Tsukamurellaceae (Tsukamurella)* |
| Family | | | | *Williamsiaceae (Williamsia)* |
| Family | | | | *Segniliparaceae (Segniliparus)* |
| Family | | | | *Micromonosporaceae (e.g., Micromonospora)* |

| Family | | | | Propionibacteriaceae (e.g., Propionibacterium) |
|---|---|---|---|---|
| Family | | | | Nocardioidaceae (e.g., Nocardoides, Aeromicrobium) |
| Family | | | | Pseudonocardiaceae (e.g., Pseudonocardia) |
| Family | | | | Actinosynnemataceae (e.g., Lentzea, Saccharothrix) |
| Family | | | | Streptomycetaceae (e.g., Streptomyces, Kitasatospora) |
| Family | | | | Streptosporangiaceae (e.g., Streptosporangium) |
| Family | | | | Nocardiopsaceae (e.g., Nocardopsis, Thermobifida) |
| Family | | | | Thermomonosporaceae (e.g., Thermomonospora) |
| Family | | | | Frankiaceae (Frankia) |
| Family | | | | Geodermatophilaceae (e.g., Blastococcus) |
| Family | | | | Sporichthyaceae (Sporichthya) |
| Family | | | | Acidothermaceae (Acidothermus) |
| Family | | | | Kineosporiaceae (e.g., Kineosporia, Cryptosporangium) |
| Family | | | | Nakamurellaceae (Quadrisphaera, Nakamurella) |
| Family | | | | Glycomycetaceae (Glycomyces, Stackebrandtia) |
| Order | | | Bifidobacteriales | |
| Family | | | | Bifidobacteriaceae (e.g., Bifidobacterium) |

## About this phylum

### Diversity

Familiar actinobacteria, such as *Mycobacterium, Corynebacterium, Micrococcus,* and *Streptomyces,* are members of a single order, the *Actinobacteriales,* which spans a relatively small phylogenetic range but a large number of families, genera, and species. The outlying branch containing *Sphaerobacter, Thermoleophilum, Acidimicrobium,* and relatives is generally considered part of the actinobacteria (as shown here), but there are contrary data suggesting instead that these organisms might be members of the *Thermomicrobium* branch of the *Chloroflexi.*

**Figure 11.8** Phylogenetic tree of representative actinobacteria.
doi:10.1128/9781555818517.ch11.f11.8

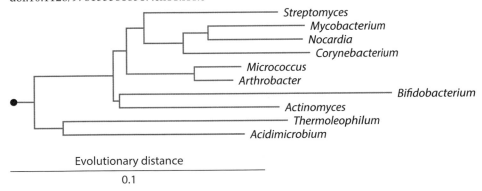

## Metabolism

The actinobacteria are generally aerobic respirers. There are a few exceptions, such as *Propionobacterium* and *Bifidobacterium*, which are anaerobic or aero-tolerant. A few are moderately thermophilic (up to ca. 60°C), but most are mesophilic. They are heterotrophic, but growth substrates vary widely. Members of this group (most notably *Streptomyces* and *Actinomyces*) are well known for their ability to produce antibiotics; these and *Bacillus* are probably the most common bacterial sources of antibiotics.

## Morphology

These organisms are typically nonmotile rods, filaments, or sometimes cocci. Rod-shaped cells are usually uneven, irregular, or club shaped (coryneform). Endospores are not produced, but filamentous species often form spores (sometimes called arthrospores to distinguish them from endospores) that are not particularly resistant but rather are reproductive and important in dispersal.

Many actinobacteria, including the mycobacteria, corynebacteria, nocardiae, and rhodococci, have a mycolic acid outer membrane. This is not the typical outer membrane seen in gram-negative bacteria; typical membrane lipids and lipopolysaccharides are not present. Instead, the typically thick gram-positive-type cell wall is covered by an arabinogalactan polysaccharide layer, which in turn is covered by a mycolic acid bilayer. This is a true lipid outer membrane, not a "waxy coating," as it is often described. As in gram-negative bacteria, this outer membrane incorporates a variety of proteins, including porins. The mycolic acid outer membrane is a potent permeability barrier, even more so than the outer phospholipid membrane of gram-negative bacteria. It is not known whether this mycolic acid outer membrane is a highly altered descendant of an ancestral gram-negative outer membrane or an independently derived addition to an ancestral gram-positive envelope.

## Habitat

Most actinobacteria are soil organisms; one genus, *Streptomyces*, forms a white growth commonly seen in decaying wood (and easily mistaken for fungus) and gives good soil its rich earthy odor. A few are symbionts or pathogens of plants and animals, including some notorious examples: *Mycobacterium tuberculosis* and *M. leprae*, and *Corynebacterium diphtheriae*. More common than pathogens are the commensals, such as *Bifidobacterium*, which is common in the gut and beneficial, and is often used in probiotics.

## *Coryneform actinobacteria*

Coryneform actinobacteria are typified by their club-like or irregular rod-shaped cells. Pairs of cells after division are angled or V-shaped; this is referred to as either "snapping" or "Chinese letter" division. This is caused by asymmetric fracturing of an outer layer of the cell wall after cytokinesis. These organisms are common symbionts of the skin and mucous membranes of animals and the surfaces of plants, as well as being abundant in the soil. The best-known genera in this group are *Corynebacterium* and *Arthrobacter*. Arthrobacteria are very

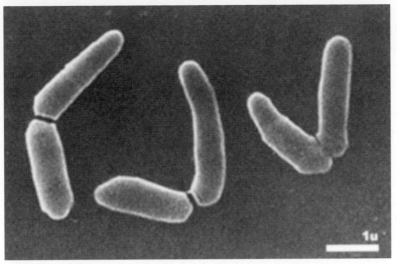

**Figure 11.9** Scanning electron micrograph of snapping division in *Arthrobacter globiformis*. (Courtesy of T. Tamura, The Society for Actinomycetes Japan, http://www.actino.jp/.) doi:10.1128/9781555818517.ch11.f11.9

common soil and root surface inhabitants that are typically pleomorphic: small cocci in stationary phase, and irregular rods with jointed or V-shaped pairs during rapid growth. Corynebacteria are common symbionts of animals; a few species are pathogens, including of course *C. diphtheriae*, the causative agent of diphtheria. Members of this genus are typically club-shaped and also often pleomorphic.

### EXAMPLE *Arthrobacter globiformis*

*A. globiformis* (Fig. 11.9) is one of the most numerically abundant, easily cultivated inhabitants of neutral-pH or alkaline soils and is also abundant on the aerial surfaces of plants. This genus is easily recognized morphologically, but distinguishing among the species requires analysis of the cell wall sugars and amino acids, or small-subunit ribosomal RNA (rRNA) sequence analysis. Stationary-phase cells are small cocci; upon transfer to typical rich media, these cocci swell and then produce outgrowths to generate irregular rod-shaped cells, which divide by snapping division, producing V-shaped pairs of cells or sometime longer pseudohyphae. As cells enter stationary phase, divisions continue as growth slows, resulting in the formation of cocci. Colonies on plates contain both coryneform/rod-shaped cells and cocci, and when examined microscopically they are often mistakenly thought to be impure or contaminated. These organisms can use a remarkably wide range of organic substrates for growth and energy, including nicotine, the antibiotic puromycin, and a range of herbicides. Most are also nitrogen fixers.

### *Filamentous actinobacteria*

*Streptomyces* and related genera form branched filamentous hyphae and, although usually much thinner, otherwise resemble the filamentous fungi. This is no coincidence; it represents an evolutionary convergence because of their common habitat and lifestyle rather than any specific evolutionary relationship.

The filamentous actinobacteria have a complex life cycle that includes programmed cell death and cellular differentiation; these are truly multicellular bacteria. Initial growth from spores on solid media occurs in the form of branching vegetative hyphae. These hyphae are mostly nonseptated; DNA replication produces new nucleoids, but no cytokinesis occurs, and so the filaments share a common syncytial cytoplasm. Filament growth occurs only at the tips; branching is required to allow logarithmic growth because, of course, individual hyphal tips have a limit to their growth rate. Vegetative hyphae give rise to waxy aerial hyphae that grow upward away from the growth substrate. This growth is at the expense of the underlying vegetative hyphae, which undergo programmed cell death (although their cell walls remain largely intact and serve as a supporting structure for the aerial hyphae). The growth tips of the aerial hyphae then undergo cytokinesis to create a series of individual cells, which develop into dormant spores. Again, the growth and development of spores are at the expense of the aerial hyphae.

It is important to remember that the "arthrospores" of these actinobacteria are distinct from the endospores of the firmicutes. Both are metabolically inactive resting stages of the life cycle, but arthrospores are produced in great numbers from each aerial hyphum (they are reproductive) and are readily dispersed by the air or water, although they are not particularly resistant to harsh treatment. Endospores, on the other hand, are extremely resistant to heat and chemical assault but are not reproductive (a mother cell produces a single spore) and are not readily dispersible.

Species of filamentous actinobacteria are distinguished morphologically, mostly on the basis of the structure and morphology of their spore-bearing hyphae (sporangia). These organisms are metabolically diverse; most can use a very wide range of growth substrates. They produce a wide range of antibiotics, including aminoglycosides (e.g., streptomycin), macrolides (e.g., erythromycin), tetracycline, and chloramphenicol, just to name a few. Interestingly, the filamentous actinobacteria have linear rather than circular chromosomes, with unique telomeres.

### EXAMPLE *Streptomyces antibioticus*

*Streptomyces antibioticus*, like *S. coelicolor* (Fig. 11.10), is a typical member of the genus *Streptomyces*; *S. antibioticus* has long been used in the industrial production of the important antibiotic actinomycin. More recently, *S. antibioticus* is being used as a model system for more detailed examination of the life cycle of filamentous actinobacteria. Although still poorly understood, this life cycle is turning out to be far more complex than previously imagined, particularly in terms of the horizontal (as opposed to vertical) spatial organization and waves of growth in specific spatial arrangements. For example, the initial germination hypha is cellular rather than syncytial, and upon reaching a specified density, alternating cells in these filaments die. Vegetative, syncytial hyphae are the outgrowths of the surviving cells from these initial filaments.

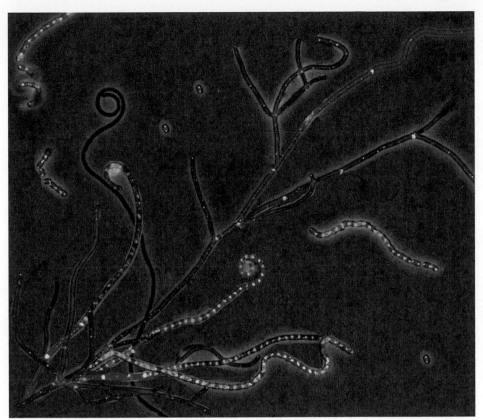

**Figure 11.10** Overlay of phase-contrast and red and green fluorescent images of sporulating *Streptomyces coelicolor* grown on SFM agar plates for 3 days. Red fluorescence results from the DNA stain 7-AAD; chromosomes are stained unevenly because the image was made by using live cells. Green is the fluorescence of SsfA-GFP. SsfA, a protein that is upregulated during sporulation and localizes to sporulation septa, has been fused to the green fluorescent protein GFP. Images were taken with a Zeiss fluorescence microscope and further artistically rendered with Adobe Photoshop. (Courtesy of Nora Ausmees.)
doi:10.1128/9781555818517.ch1.f1.7

## Acid-fast bacteria

The mycolic acids of the outer membrane of the genus *Mycobacterium* and relatives are much longer than those of other actinobacteria that contain mycolic acids. As a result, these species can be specifically stained using the "acid-fast" stain first developed by Robert Koch during his work to identify the cause of tuberculosis. In culture, the mycobacteria typically grow as branched or unbranched filaments, but these filaments are chains of individual cells rather than syncytial hyphae. Most of the mycobacteria are soil inhabitants, and some are important in the bioremediation of pollutants that are otherwise recalcitrant. A few are human and animal pathogens, most notably *Mycobacterium tuberculosis* and *M. leprae*.

EXAMPLE *Mycobacterium ulcerans*

*Mycobacterium ulcerans* (Fig. 11.11) is the causative agent of Buruli (or Bairnsdale) ulcer, a necrotic disease of the skin and surrounding soft tissue and bone. Lesions are most common on the arms and legs, primarily in children, and

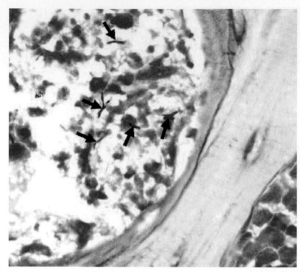

**Figure 11.11** Stained sample of human tissue (blue) infected with *Mycobacterium ulcerans* (red). [Reprinted from Portaels F, Meyers WM, Ablordey A, Castro AG, Chemlal K, et al., *PLoS Negl Trop Dis* **2**(3): e178, 2008. doi:10.1371/journal.pntd.0000178, with permission. Courtesy of the U.S. National Library of Medicine.] doi:10.1128/9781555818517.ch11.f11.11

although not generally fatal they typically result in permanent disfigurement. Although not as well known (or understood) as leprosy or tuberculosis, which are caused by related species of *Mycobacterium*, Buruli ulcer has surpassed these diseases in frequency in some parts of impoverished central and western Africa. Unlike *M. leprae* or *M. tuberculosis*, *M. ulcerans* is an environmental pathogen, probably transmitted to humans from aquatic insects.

## Deeply branching questionable members

There are some species that, although currently classified as members of the *Actinobacteria*, are on such deep branches that their placement among the *Actinobacteria* is uncertain. One of these, *Sphaerobacter*, is almost certainly really a deep branch of the *Chloroflexi* rather than the *Actinobacteria*. The opposite may be the case for the genus *Thermoleophilum*; originally classified a member of the green nonsulfur bacteria (*Chloroflexi*), there is some evidence that it may instead be an actinobacterium, where it is currently classified as a relative of *Rubrobacter*. Alternatively, it may represent an independent phylum of the *Bacteria*.

### EXAMPLE *Thermoleophilum album*

*T. album* (Fig. 11.12) is one of only two species of the genus *Thermoleophilum* (the other is *T. minutum*). These organisms are very small (ca. 0.5 by 1 μm), obligately aerobic gram-negative (in terms of staining) rods. All isolates of these species are thermophilic, growing optimally at about 60°C, and have been isolated from hot-spring sediments and dark muds exposed to solar heating from a wide range of sites scattered around the United States. The unique feature of these organisms is that they can grow only on long-chain *n*-alkanes, e.g., wax. No other substrates, not even the alcohol derivatives of growth substrates, can be used for either carbon or energy.

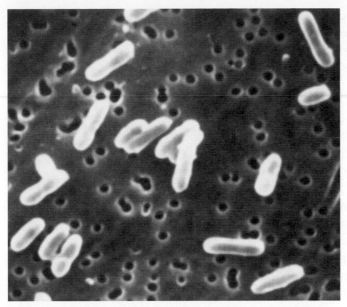

**Figure 11.12** Scanning electron micrograph of *Thermoleophilum album* resting on a traditional 0.22-μm sterilizing filter. (Courtesy of Jerome Perry.) doi:10.1128/9781555818517.ch11.f11.12

## Bacterial development

Log-phase cells are very different from stationary-phase cells: log-phase cells are adapted for maximal growth rates in nutrient-saturated conditions, where growth rate is often limited only by the physical ability of the organism to import nutrients, process them into cell material, and replicate. They are usually large cells with the ability to replicate DNA and make RNA and proteins rapidly. Stationary-phase cells are usually smaller and are adapted for maximum competitiveness. In many species, such as *Arthrobacter* (described above), the morphologies of these cells types are strikingly different. Keep in mind, as well, that microbes generally spend most of their time in stationary phase in the environment.

The shift to stationary phase (and later back to log phase) is a complex developmental process, controlled by sigma-factor cascades (like phage infection or sporulation in *Bacillus*). In log-phase bacteria, the "vegetative" sigma subunit ($\sigma^{70}$) is the predominant sigma factor directing RNA polymerase promoter recognition. In late log phase, expression of the stationary-phase sigma ($\sigma^{S}$) is turned on. Because of its higher affinity for the core RNA polymerase, $\sigma^{S}$ progressively replaces $\sigma^{70}$. Genes needed during log-phase growth require $\sigma^{70}$-containing RNA polymerase for expression, and so these genes are progressively turned off as the concentration of $\sigma^{70}$ declines. Genes for stationary-phase growth (including the gene encoding $\sigma^{S}$) are expressed by $\sigma^{S}$, and so the expression of these genes increases until they reach normal levels for stationary phase. In more complex developmental pathways, such as sporulation, heterocyst formation, or phage infection, many sigma factor transitions occur sequentially, driven by the fact that each sigma factor initiated expression of the sigma factor to follow. Each sigma factor directs expression of the genes required at that stage of the developmental pathway. These sigma factor-directed developmental pathways are in many ways analogous to the homeo-box-directed developmental pathways of animals.

### Secondary metabolites

Production of antibiotics and other microbials occurs in many bacteria, but it is very common in *Streptomyces* and *Bacillus*. Antibiotics are not produced during rapid growth but in stationary phase; i.e., they are secondary metabolites. Secondary metabolites are compounds produced (and typically secreted into the environment) only during stationary phase, and so are not required for growth. Other secondary metabolites include iron-binding compounds (siderophores) and other compounds that help the organism compete for limited resources.

There are two ways an organism can increase its competitive fitness in a tight environment: increasing its own fitness (self-improvement) or decreasing the fitness of its competitors. Siderophores and high-affinity uptake mechanisms are examples of secondary metabolites that directly increase an organism's supply of nutrients; this is self-improvement. Antibiotics and bacteriocins are examples of secondary metabolites that provide a competitive advantage to an organism by crippling the competition, allowing the producing organism access to resources that otherwise would have gone to the competition or even providing the producer with the nutrients released by the extinguished competitor.

## Bacterial multicellularity

We generally think of bacteria (and archaea and protists, for that matter) as being unicellular—every bug for itself. But this is not a very sophisticated viewpoint. It is true that bacteria do not develop into large interdependent clusters that walk around talking to one another on cell phones, but multicellularity is a spectrum, not just an "is" or "is not." Most bacteria communicate by secreting small compounds called pheromones; cell-to-cell communication is a kind of multicellularity, in the sense that the cells act and react as a population rather than individually. Many bacteria, during their life cycle, form aggregates of various kinds in which cells differentiate. In some cases, there is no doubt that the bacterium is multicellular, for example, the filamentous cyanobacteria in which some cells undergo terminal differentiation in nitrogen-fixing heterocysts, or the colonies of *Streptomyces* described above that undergo complex morphological development and programmed cell death. Specialization of a group of cells into different forms, especially if that differentiation is terminal (i.e., those cells will not contribute to future generations), is the hallmark of multicellularity.

One characteristic of multicellular behavior in bacteria (and archaea and unicellular eukaryotes), in contrast to plants, animals, and fungi, is that the groups of cells are often composed of more than one species; an example is the photosynthetic mat. Here we have complex layers of different kinds of cells, working together to make a living: a tissue, of a sort, made up of cells of various kinds specialized for different functions. This is not very different from the tissue of a plant or animal, except that the constituent cells are of different species.

## Questions for thought

1. Given that gram-positive bacteria evolved from a gram-negative ancestor, what might be the advantages, or disadvantages, of this change in cell envelope structure?

2. The complete genome sequence of *Mycoplasma genitalium* is 0.6 Mbp in length. All of the genes that it does not absolutely need for its parasitic lifestyle have been lost. What genes do you think remain? What genes would a minimal free-living heterotrophic species need?

3. Endospores of *Bacillus* or *Clostridium* seem to be able to wait more or less indefinitely for favorable conditions to germinate. However, how can they monitor the environmental conditions without metabolic activity (and therefore a continuous energy cost)? What do you think might limit the life span of an endospore? Why not longer?

4. What do you think are some of the differences in metabolism and gene expression between log-phase and stationary-phase cells? How might an organism evolve to grow faster, or subsist better without growth?

5. How would you go about seeing if the mechanism underlying programmed cell death in *Streptomyces* is in any way related mechanistically to programmed cell death (apoptosis) in animals?

6. Can you give any other examples of multicellularity in microbes?

# 12 Spirochetes and Bacteroids

The phyla *Spirochaetae* and *Bacteroidetes* contain many common but generally unfamiliar organisms. Although the *Spirochaetae* bring to mind some important pathogens, both groups are common in the environment and as components of the normal flora of animals. Although not specifically related (Fig. 12.1) or alike morphologically, these two groups share a common metabolic theme: most are saccharolytic and are often involved in the decomposition of long-chain polysaccharides such as starch, cellulose, and chitin. In addition, each group is motile by a mechanism different from the common flagellar propeller most commonly found in bacteria.

**Figure 12.1** Phylogenetic tree of the bacterial phyla, with the bacteroids and spirochetes highlighted. doi:10.1128/9781555818517.ch12.f12.1

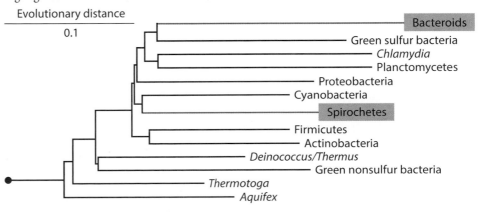

# Phylum *Spirochaetae*

Figure 12.2 is a phylogenetic tree of representative spirochetes.

| Phylum | *Spirochaetae* | | | | |
|---|---|---|---|---|---|
| Class | | *Spirochaetes* | | | |
| Order | | | *Spirochaetales* | | |
| Family | | | | *Spirochaetaceae* | |
| Genus | | | | | *Spirochaeta* |
| Genus | | | | | *Borrelia* |
| Genus | | | | | *Brevinema* |
| Genus | | | | | *Clevelandina* |
| Genus | | | | | *Cristispira* |
| Genus | | | | | *Diplocalyx* |
| Genus | | | | | *Hollandina* |
| Genus | | | | | *Pillotina* |
| Genus | | | | | *Treponema* |
| Family | | | | *Brachyspiraceae* | |
| Genus | | | | | *Serpulina* |
| Genus | | | | | *Brachyspira* |
| Family | | | | *Leptospiraceae* | |
| Genus | | | | | *Leptospira* |
| Genus | | | | | *Leptonema* |
| Genus | | | | | *Turneriella* |

## *About this phylum*

### Diversity

Most familiar organisms in this group (the pathogens and their relatives) are closely related members of the family *Spirochaetaceae*. Many species that were originally thought to be members of this group, even species of the genera *Spirochaeta* and *Treponema*, have more recently been shown by phylogenetic analysis to constitute a separate group, the family *Serpulinaceae*. This family is

**Figure 12.2** Phylogenetic tree of representative members of the *Spirochaetae*. doi:10.1128/9781555818517.ch12.f12.2

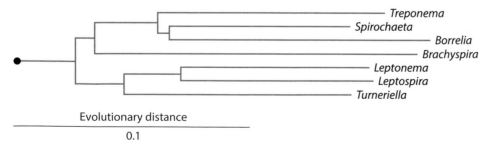

known primarily for animal and human intestinal parasites. The environmental species are more diverse, and because they are typically difficult to grow, most of the species, even the conspicuous ones, have not been grown or studied in culture, and so their phylogenetic affiliations are generally unknown.

## Metabolism

The spirochetes are uniformly heterotrophic and generally microaerophilic or anaerobic and saccharolytic, although some (e.g., *Treponema denticola*, a member of the normal flora of human teeth and gingiva) ferment amino acids. $H_2$ and $CO_2$ are the main products of this fermentation, although some species can convert these two waste products to acetate. In contrast, the leptospiras are generally aerobic degraders of fatty acids.

## Morphology

The spirochetes share a common body plan (Fig. 12.3). The long, thin body of the cell is generally helical or a two-dimensional wave. Polar flagella (typically one at each end, but more are present in the larger species) are anchored subterminally in the cytoplasmic membrane but do not emerge from the outer membrane; they are contained in the periplasm. These flagella wind their way along the curved or helical shape of the cell body, usually overlapping in the medial part of the cell. Flagellar rotation causes the body of the cell to rotate within the outer membrane. In flat, wavy species, this causes the cell to move across a surface or through a liquid environment similar to the way a snake moves along the ground or in the water. This also works for helical species because the viscosity of their surroundings retards rotation of the outer membrane surface; the shape of the cell therefore rotates like a corkscrew, providing propulsive force.

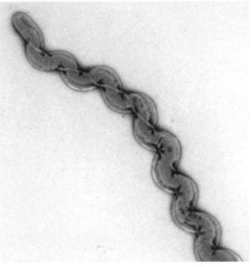

**Figure 12.3** Shadow-cast electron micrograph of a typical spirochete, showing the periplasmic filaments. (Source: Institute Pasteur/E. Couture-Tosi and M. Picardeau.) doi:10.1128/9781555818517.ch12.f12.3

## Habitat

Spirochetes are common inhabitants of sediments, especially sediments rich in decomposing plant material, in which they are involved in the decomposition of cellulose and other polysaccharides. The best-studied spirochetes, however, are symbionts or parasites of the gastrointestinal tract of animals, including humans. A few parasites (e.g., *Treponema pallidum*) invade the tissues of the host, but most inhabit the surface of the mucosa. The richest, most readily available source of large numbers of diverse spirochetes is the hindgut (homologous to the colon of vertebrates) of wood-eating insects, especially termites.

## Are spirochetes the progenitors of eukaryotic flagella?

Termites and other wood-eating insects subsist primarily on a diet of cellulose. Degradation of cellulose and generation of the nutrients required by the insect constitute an involved process carried out by a complex population of symbiotic bacteria (including spirochetes), archaea (methanogens), and eukaryotes (protists) (Fig. 12.4). Some of the protists involved harbor symbiotic spirochetes, which are attached to the surface of the protists at one end of their wavy or

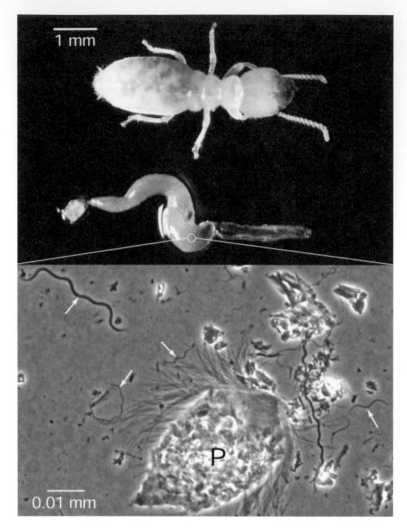

**Figure 12.4** (Top) A typical termite and a gastrointestinal tract dissected from a sibling. The expanded portion found in the abdomen is the hindgut. (Bottom) A phase-contrast image of the contents of the hindgut, including a large protist (P). Various spirochetes are indicated by arrows. (Courtesy of John Breznak, Michigan State University.) doi:10.1128/9781555818517.ch12.f12.4

helical cells. Movement of this "collective" is driven by movement by the spirochetes but apparently is directed by the protist. The spirochetes physically resemble flagella or long cilia, to the extent that in many cases it is difficult to distinguish normal flagella borne by the protist from symbiotic spirochetes attached to the same creature. This similarity in form and function led to the suggestion that eukaryotic flagella, and the associated structures, including cilia, basal bodies, and the spindle apparatus, all of which are hallmarks of the eukaryotic cell structure, may have originated by symbiosis of a progenitor eukaryote with a spirochete. We know this is the case for the origin of mitochondria (by symbiosis with an α-proteobacterium) and plastids (by symbioses with cyanobacteria). However, unlike mitochondria or plastids, neither flagella nor their associated cellular structures contain DNA, nor is there evidence for transfer of significant numbers of spirochete genes into the nuclear genome. Neither do the cytoskeletal structures of spirochetes or their periplasmic flagella resemble in structure, mechanism, or molecular sequence the tubulin/microtubule structures of eukaryotic flagella or their associated structures.

## Family Spirochaetaceae

The familiar genera in this family are *Spirochaeta* (free-living species), *Treponema* (animal symbionts and pathogens), and *Borrelia* (human pathogens with arthropod vectors). These organisms are anaerobic, microaerophilic, or facultatively aerobic. Cultivated *Spirochaeta* and *Borrelia* species are saccharolytic. Cultivated *Treponema* species are also generally saccharolytic, but a few can grow on fatty acids or amino acids; most have not been grown in pure culture, and so their growth substrates are not known. Linear (rather than circular) genomes and plasmids are common in this group.

### EXAMPLE SPECIES

### *Treponema denticola*

Although the genus *Treponema* is best known for the important pathogen *Treponema pallidum*, treponemes are common in the mouth and gastrointestinal tract of healthy humans and other animals. The pathogenicity of *Treponema denticola* (Fig. 12.5) is not clear; it is found in dental plaque and gingiva of healthy humans (and other primates) but is more abundant in people with gingivitis. Although usually described as helical, this species is probably a two-dimensional wave like its close relative *T. pallidum*. Unlike most spirochetes and treponemes, *T. denticola* is strongly proteolytic and so presumably is capable of fermentation of amino acids in its natural environment.

### *Borrelia recurrentis*

Although the best known *Borrelia* species is *B. burgdorferi*, the causative agent of Lyme disease, a wide variety of *Borrelia* species cause similar zoonotic infections. *B. recurrentis* (Fig. 12.6) is the cause of louse-borne relapsing fever. This

**Figure 12.5** *Treponema denticola*. (A and B) Fluorescent micrographs and (C) electron micrograph showing the central filament (CF) and some of the flagella that comprise it released from the outer membrane (PFF). (Izard J. 2010. *Treponema denticola* cell observed by dark-field microscopy. Visual Resources. American Society for Microbiology, Washington, DC. www.microbelibrary.org. Accessed 3 November 2014.) doi:10.1128/9781555818517.ch12.f12.5

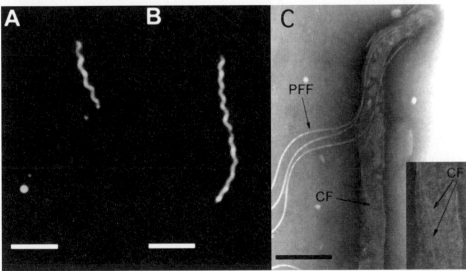

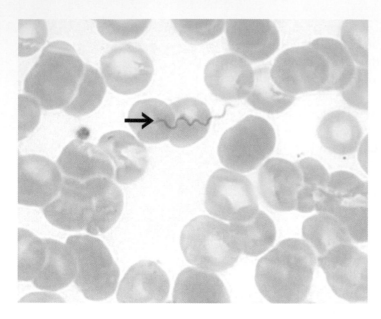

**Figure 12.6** Electron micrograph of *Borrelia recurrentis*. This spirochete is not helical but is a flattened wave. (Source: Heinz Diem.)
doi:10.1128/9781555818517.ch12.f12.6

relapsing fever is caused by the organisms' use of antigenic variation to evade elimination by the host immune response. Like other *Borrelia* species, the ends of the cells are tapered to fine points. Cells are relatively open helices and contain 8 to 10 periplasmic flagella. Although it has not been tested in the case of this species, other species of *Borellia* have linear genomes.

## *Family* Leptospiraceae
In contrast to the other spirochetes, the leptospiras are obligately aerobic and use fatty acids as their growth substrate. Leptospiras are common in aquatic environments and in association with mammals, including humans. Human pathogenesis is zoonotic; humans are apparently not effective hosts or carriers. Animal carriers harbor the parasites in their kidneys and transmit the infection via urine. Human infections occur primarily in those with direct contact with wild or domestic mammals. Leptospirosis is easily confused with other fevers, especially yellow fever. Only two periplasmic flagella are present, and the cells are very thin and tightly coiled, usually with bent or curved ends.

### EXAMPLE *Leptospira biflexa*
*Leptospira biflexa* (Fig. 12.7) is a nonpathogenic aquatic species that is very closely related to several pathogenic and opportunistically pathogenic species and so serves as a model system for spirochete virulence. It is also a model system for examining motility in spirochetes. During movement, the trailing end of the cell is bent in a hook shape and the leading end forms an open left-handed spiral (over and above the usual right-handed helical cell shape). Both of these rotate as the cell progresses forward. Although rotation of the helical cell body is sufficient to propel the organism through low-viscosity environments, the rotation of the hook and spiral ends allows the organism to bore through viscous environments efficiently.

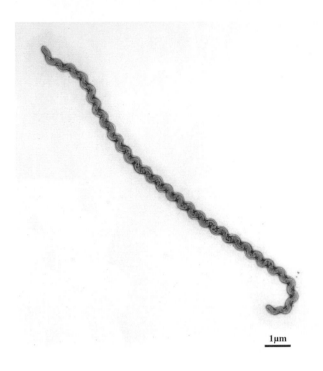

**Figure 12.7** Electron micrograph of *Leptospira biflexa*. The cell is wrapped in a helix around the axial filament, and there is a long helical wave at the leading end of the cell (upper left) and a hook at the trailing end (lower right). (Source: Institute Pasteur/E. Couture-Tosi and M. Picardeau.) doi:10.1128/9781555818517.ch12.f12.7

1µm

## Phylum *Bacteroidetes* (sphingobacteria or *Bacteroides*/*Flavobacterium*/*Cytophaga* group)

Figure 12.8 is a phylogenetic tree of representative members of the *Bacteroidetes*.

| Phylum | *Bacteroidetes* | | | |
|---|---|---|---|---|
| Class | | *Bacteroidia* | | |
| Order | | | *Bacteroidales* | |
| Family | | | | *Bacteroidaceae* (e.g., *Bacteroides*, *Pontibacter*) |
| Family | | | | *Rikenellaceae* (e.g., *Rikenella*, *Alistipes*) |
| Family | | | | *Porphyromonadaceae* (e.g., *Porphyromonas*) |
| Family | | | | *Prevotellaceae* (e.g., *Prevotella*, *Xylanibacter*) |
| Class | | *Flavobacteria* | | |
| Order | | | *Flavobacteriales* | |
| Family | | | | *Flavobacteriaceae* (e.g., *Flavobacterium*, *Polaribacter*) |
| Family | | | | *Blattabacteriaceae* (*Blattabacterium*) |
| Family | | | | *Cryomorphaceae* (e.g., *Algoriphagus*, *Cryomorpha*) |
| Class | | *Sphingobacteria* | | |
| Order | | | *Sphingobacteriales* | |
| Family | | | | *Sphingobacteriaceae* (*Sphingobacterium*, *Pedobacter*) |
| Family | | | | *Saprospira* (e.g., *Saprospira*, *Haliscomenobacter*) |
| Family | | | | *Flexibacteraceae* (e.g., *Flexibacter*, *Cytophaga*) |
| Family | | | | *Flammeovirgaceae* (e.g., *Flexithrix*, *Persicobacter*) |
| Family | | | | *Crenotrichaceae* (e.g., *Chitinophaga*, *Rhodothermus*) |

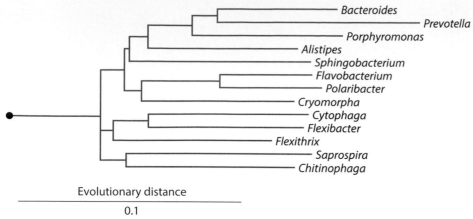

Evolutionary distance

0.1

**Figure 12.8** Phylogenetic tree of representative bacteroids.
doi:10.1128/9781555818517.ch12.f12.8

## *About this phylum*

### Diversity

*Bacteroidetes* is a very large and diverse phylum, encompassing at least 140 genera. It is composed of three major groups, the bacteroids (*Bacteroidales*), flavobacteria (*Flavobacteriales*), and sphingobacteria (*Sphingobacteriales*, sometimes known as "*Cytophaga* and relatives"). Although this phylum is very common in the environment, it contains no spectacular human pathogens or industrial organisms, and so its members are generally unfamiliar and not well studied. The sphingobacteria are specifically related to the *Chlorobi* (chapter 9), and these are sometimes grouped into a single phylum.

### Metabolism

Members of the *Bacteroidetes* are generally saccharolytic, either obtaining sugars directly from the environment or releasing them from long-chain polysaccharides such as starch, cellulose, or chitin. The bacteroids are generally anaerobes, while the flavobacteria and sphingobacteria are aerobes. Most are naturally resistant to aminoglycoside antibiotics. Members of this group synthesize and incorporate sphingolipids into their membranes. Sphingolipids are otherwise found in the cytoplasmic and vesicle membranes of eukaryotes, but especially those of the nervous system of animals.

### Morphology

Members of the *Bacteroidetes* are rod-shaped cells, typically long and thin and often with slightly tapered ends. The most conspicuous exceptions to this are *Bacteroides*, which are typical rod-shaped cells, and *Sporocytophaga*, which are typically thin, tapered rods during log-phase growth but form spherical resting spores in stationary phase. Cells are motile by gliding or nonmotile.

### Habitat

These bacteria are common in a wide range of environments, especially in soils, sediments, and the gut contents of animals: all environments in which long-chain polysaccharides and their decomposition products are degraded. These organisms are found in both moderate temperatures (mesophiles) and cold environments (cryophiles); thermophiles are not known.

## *Class* Bacteroidia

Members of the class *Bacteroidia* are obligately anaerobic, non-spore-forming rods. The most familiar species are *Bacteroides* (a member of the family *Bacteroidaceae*) and *Porphyromonas* (a member of the family *Porphyromonadaceae*). *Bacteroides* species and similar relatives are nonpigmented and saccharolytic or peptidolytic, preferring complex polysaccharides and producing primarily acetate and succinate as their fermentation products. Unlike most polysaccharide degraders, which secrete hydrolytic enzymes into the environment or onto the outer surface of their cells, *Bacteroides* species transport insoluble particles of polysaccharide into the periplasm for degradation; the sugars released by hydrolysis are transported directly into the cell. *Porphyromonas* species are similar morphologically but are heavily pigmented with protoheme and protoporphyrin, producing very dark brown or black colonies. Their preferred growth substrates are peptides, which are converted into a wide range of organic acids. Both genera are abundant symbionts of anaerobic mucous membranes of humans and other animals. Although they are considered members of the normal flora, they can also be opportunistically pathogenic. *Porphyromonas* in large numbers is associated with gum disease such as gingivitis.

EXAMPLE  *Bacteroides thetaiotaomicron*

*Bacteroides thetaiotaomicron* (Fig. 12.9) and related species (e.g., *B. fragilis*) are a major component of the human colonic flora, comprising about one-third of the bacteria in feces (most of the rest are a variety of *Clostridium* species). Gastrointestinal infections caused by *Bacteroides* are usually caused by a close relative, *B. fragilis*. *B. thetaiotaomicron* can degrade and grow on starch but not on other polysaccharides.

## *Class* Flavobacteria

Members of the class *Flavobacteria* are obligately aerobic, yellow-pigmented, usually long rods with slightly tapered ends. They are common in marine and freshwater environments. Most are nonmotile, but some can move rapidly by gliding. Some species can hydrolyze chitin, gelatin, or starch if allowed to attach directly to the surface of these polymers. Many are also proteolytic. They are rarely involved in human disease (usually meningitis in infants), but their innate high-level resistance to many antibiotics often results in a dangerous delay in effective treatment. However, they are very important pathogens of fish, especially in trout and salmon in aquaculture.

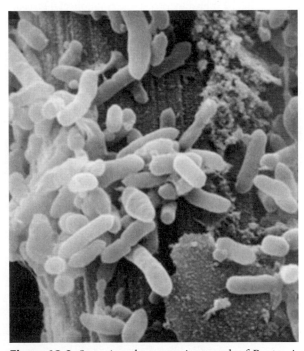

**Figure 12.9** Scanning electron micrograph of *Bacteroides thetaiotaomicron*. (Courtesy of Z. He and L. T. Angenent, Washington University in St. Louis.) doi:10.1128/9781555818517.ch12.f12.9

*Flavobacterium johnsoniae*

*Flavobacterium johnsoniae* (Fig. 12.10) is a common aquatic (freshwater and marine) chitin-degrading organism. It is a rapid glider (up to 10 μm/second), a mode of transport which is often mistaken for flagellar motility. However, this gliding motility requires contact with the substrate; the organisms are incapable of swimming. It is a model system for the study of the unusual mechanism of gliding motility in bacteroids. Interestingly, nonmotile mutants of *F. johnsoniae* are all also incapable of chitin degradation, implying a link between these processes; the nature of this link is not understood.

## *Class* Sphingobacteria

*Cytophaga* and relatives are typically long, thin, flexible rods, but they vary from short rods to filaments or open spirals. These organisms are motile by gliding and are capable of degrading a wide range of biopolymers: chitin, agar, starch, cellulose, pectin, nucleic acids, and proteins. These organisms are abundant in nutrient-rich aquatic environments, sediments, and soils. Some are pleomorphic, forming rod-shaped or filamentous cells during log-phase growth and spherical or short rod-shaped spores in stationary phase. Cultures of these pleomorphs are often thought to be contaminated because of the different distinct cell morphologies. Colonies are usually feathery as a result of cells streaming on the agar surface. Cells placed on glass slides (e.g., in a wet mount) are often arranged in side-by-side monolayers.

**Figure 12.10** Phase-contrast micrograph of *Flavobacterium johnsoniae*. (Reprinted from Nelson SS, Bollampalli S, McBride MJ, *J Bacteriol* **190**:2851–2857, 2008, with permission.) doi:10.1128/9781555818517.ch12.f12.10

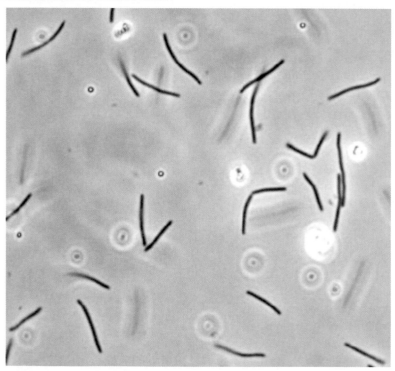

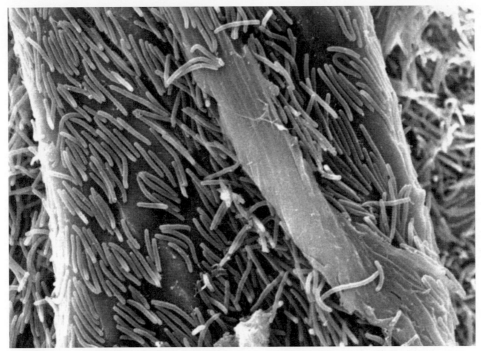

**Figure 12.11** Scanning electron micrograph of *Cytophaga hutchinsonii* cells digesting cellulose fibers. (Reprinted from Xie G, Bruce DC, Challacombe JF, Chertkov O, Detter JC, Gilna P, Han CS, Lucas S, Misra M, Myers GL, Richardson P, Tapia R, Thayer N, Thompson LS, Brettin TS, Henrissat B, Wilson DB, McBride MJ, *Appl Environ Microbiol* **73**:3536–3546, 2007, with permission.) doi:10.1128/9781555818517.ch12.f12.11

EXAMPLE *Cytophaga hutchinsonii*

*Cytophaga hutchinsonii* (Fig. 12.11) is a long, flexible, rod-shaped cellulose degrader isolated from soil. Cellulose degradation requires the organisms to be in direct contact with the substrate; extracellular cellulases are not produced. The cells adhere to cellulose and orient themselves and glide along the axis of the cellulose fibers. Only cellulose and its decomposition products cellobiose and glucose can sustain growth; other simple monosaccharides (except gluconate) cannot be used.

## Bacterial motility

Most bacteria have the ability to move from one place to another in at least some part of their life cycle. The most common and best-understood form of motility is that driven by flagella, but many mechanisms of motility are used by bacteria, and in some cases a species may be capable of more than one type of motility.

Although well studied in only a few flagellated organisms, the *che* gene signaling pathway for chemotaxis are widespread in bacteria, and the same process probably regulates chemotaxis in other organisms regardless of their method of motility.

### Flagella

A wide range of bacteria are motile via flagella. Flagella are proton-gradient-driven helical propellers, allowing the organism to "swim" through an aqueous environment (Fig. 12.12). Cells can have one to many flagella, located at one or both ends or distributed all over. A rare arrangement is to have flagella only on one side of a rod-shaped cell; these cells swim sideways, counter-rotating like a propeller.

Some flagellated organisms are curved (vibrios) or spiral shaped (spirilla). This increases the efficiency of flagellated motility, despite the increase in surface area. Viscous resistance on the rotating flagella in any organism causes a counter-rotation of the cell body; if the cell is properly curved, this lost energy is recaptured by turning the cell into a screw (Fig. 12.13).

Cells "run" (move continuously in a more or less straight line) when the flagella turn in one direction and "twiddle" (tumble) when they turn in the other direction. The length of a run is dependent on whether the cell is moving in the desired direction: long runs if so, short runs if not. Twiddling reorients the cell randomly between runs. The result is a directed random walk, a fairly efficient way to get where you want to go.

Spirilla are little different; they typically have polar flagella, and "twiddle and run" like other flagellated organisms but usually switching from running the flagella at one end of the cell to the other end after each twiddle. They can also switch back and forth between running the flagella at either end of the cell without twiddling, resulting in the cell running directly back and forth. They use this ability to quickly reverse directions when they collide with something or come into abrupt contact with a repellent.

**Figure 12.12** Flagellar motility. Flagella are helical fibers, analogous to propellers, whose rotation is achieved by a proton-driven rotor. doi:10.1128/9781555818517.ch12.f12.12

**Figure 12.13** Vibrios or spirilla are flagellated organisms that recapture some of the energy that is otherwise lost to counter-rotation of the cell. doi:10.1128/9781555818517.ch12.f12.13

## Gliding

Gliding motility is accomplished by at least three fundamentally different mechanisms.

Cyanobacteria, chloroflexi, thiotrichs, probably myxobacteria, and many eukaryotic algae glide using a mechanism that involves the secretion of polysaccharides from pores on the cell surface (Fig. 12.14). Hydration of the polysaccharide as it emerges into the aqueous environment causes it to expand dramatically and provides a reactive force much like a rocket engine. Unicellular gliders usually have the pores at each end of the cells (which are typically rod shaped). Filamentous gliders have these pores along the leading and trailing edges of individual cells, oriented fore and aft. How cells control which pores to activate, so that the cells move in one direction or the other, is not known. Nor

**Figure 12.14** Gliding motility in some organisms is driven by secretion of polysaccharides from pores at the trailing end or edges of cells. Hydration of the polysaccharide upon exposure to the external environment causes it to expand and provides a sort of rocket propulsive force. doi:10.1128/9781555818517.ch12.f12.14

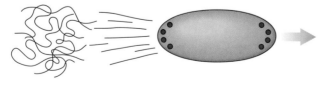

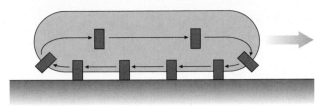

**Figure 12.15** Gliding motility in some organisms is driven by adhesins that move from one end of the cell to the other, analogous to the treads of a tank or bulldozer. doi:10.1128/9781555818517.ch12.f12.15

is the coordination of motility between cells of a filament understood. Gliding leaves a polysaccharide "slime" trail stuck to a substrate or hanging in solution and is an efficient a way to move over the surface of a solid material as well as through liquid media.

The bacteroids glide using a very different and poorly understood mechanism (Fig. 12.15). Adhesins on the surface of the cells seem to move uniformly from one end of the cell to the other. Presumably the adhesins are internalized upon reaching the trailing end of the cell and reemerge at the leading edge. Think of the tracks on a bulldozer, a conveyor belt, or an escalator. These organisms can glide only if in contact with a surface. It has been proposed that gliding in myxobacteria may be similar, except that the surface adhesins follow a helical path along the surface of the cell, following the helical structure of the cytoskeleton by the cytoskeleton.

Gliding in *Mycoplasma* species is poorly understood, and may actually be based on two different unique mechanisms. Rapid gliding in some *Mycoplasma* species is the result of protein "legs" on the leading "neck" of these organisms. Slow gliding in other *Mycoplasma* species may involve surface adhesins and "inchworm" extension and contraction of the leading appendage.

## Twitching

A few organisms, such as myxobacteria (which also glide) and some species of *Pseudomonas* (which can also produce flagella), *Neisseria*, *Nostoc*, and *Clostridium*, can move across surfaces by using retractable pili (Fig. 12.16). Think of this as grappling-hook motility; the cell extends a pilus (type IV, where it has been determined) in the direction it wants to move, the end of the pilus attaches to the substrate, and then retraction of the pilus pulls the cell forward, generally a few cell lengths at a time. Each pull looks like a "twitch": hence the name. Some types of cells can produce many pili simultaneously, so that the cell can move forward more or less smoothly, looking very much as though it is gliding, for

**Figure 12.16** Twitching motility is directed by the extension of pili, which adhere to surfaces and are retracted. Where only a few pili are used, the resulting movements are jerky, from which the name "twitching" motility is derived. Numerous pili produced continuously can result in smooth motility that is often mistaken for gliding. doi:10.1128/9781555818517.ch12.f12.16

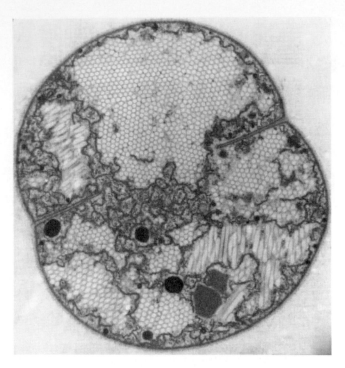

**Figure 12.17** The honeycomb arrays inside the cytoplasm of this dividing *Microcystis* cell are the gas vesicles in cross section. On the lower right and upper left of the cell are gas vesicles, viewed horizontally in this section. (Reprinted from Walsby AE, *Microbiol Rev* **58**:94–144, 1994, with permission. Micrograph by H.S. Pankratz.) doi:10.1128/9781555818517.ch12.f12.17

which this form of motility is often mistaken. To change direction, the cell has to disassemble the pilus apparatus from one end and reassemble it at the other; again, the mechanism for this is not understood.

### Gas vacuoles

Many aquatic phototrophs move themselves up and down in the water column by fine-tuning their buoyancy via gas vacuoles (Fig. 12.17). These vacuoles can be bound by internal lipid membranes or, more commonly, by protein coats. Gas vacuoles usually contain $CO_2$ generated by metabolism. They are very rigid structures and do not compress or expand significantly over a wide range of pressures; this helps simplify the maintenance of constant buoyancy. Otherwise, the gas expansion during ascent or compression during descent would require the organisms to constantly adjust their buoyancy, a problem well known to SCUBA divers.

### Spirochete motility

Spirochetes move by rotation of periplasmic flagella (Fig. 12.18). This method of motility is therefore structurally related to flagellar motility but is

**Figure 12.18** Motility by spirochetes. Rotation of the rigid helical cell body within the outer membrane sheath is driven by rotation of the axial fiber (periplasmic flagella). doi:10.1128/9781555818517.ch12.f12.18

mechanistically very different. Rotation of the periplasmic flagella causes the rigid helical cell body to rotate within the outer membrane. Viscous drag against the surrounding medium prevents the outer membrane from spinning freely. As a result, the shape of the cell relative to the surrounding medium forms a rotation corkscrew, thus driving the cell forward.

Not all spirochetes are helical; many are flattened waves. The same periplasmic flagellar rotation causes the shape of the cell to wave and progress forward in much the same way that a snake moves forward on flat ground. In some cases, spirochetes are bent or curved at the ends (see "Family Leptospiraceae," above) to improve motility through semisolid environments, such as the interstitial spaces between animal host cells.

Spirochetes "run" when the two terminal flagella rotate in opposite directions (i.e., together, since they are from opposite ends of the cell). The cell can switch the direction of motion by switching the rotation of both flagella simultaneously. If the switch is not simultaneous, the cell flexes while the flagella rotate out of sync, analogous to the twiddles of regularly flagellated bacteria.

## Spiroplasma *motility*

*Spiroplasma* is a relative of *Mycoplasma* (see chapter 11). *Spiroplasma* cells are helical and move by a novel mechanism based on changes in the shape of their internal cytoskeleton. This cytoskeleton is a flat ribbon composed of 14 fibrils (seven pairs) that is fixed to the inside surface of the cells along the midline of the cell spiral. Independent contraction of the fibrils in this cytoskeletal ribbon can be used to contract or expand the helical shape of the cell, or even reverse the handedness of the helix, at any point along the length of the cell. Motility is driven by moving a stretched or contracted region of the cell, or a kink produced by a region of reversed handedness, in a wave from one end of the cell to the other; viscous drag on this irregularity in the helix results in rotation of the cell, which drives it forward (Fig. 12.19).

**Figure 12.19** Model of *Spiroplasma* motility. Kinks, where the helix switches from right handed to left handed or vice versa, start at one end and are propagated to the other end. Continuous waves of these kinks propel the organism forward.
doi:10.1128/9781555818517.ch12.f12.19

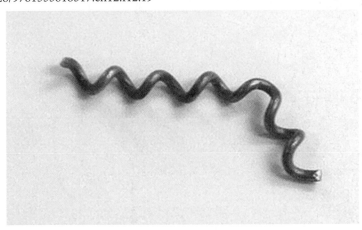

## Questions for thought

**1.** At the size scale of bacteria, momentum is trivial but viscosity is very high; i.e., the Reynolds number is low. From the perspective of a bacterium, it and everything around it is essentially massless but water has the consistency of cold molasses. How does this affect the way you might think about the different forms of motility described above?

**2.** What do you think of the hypothesis that eukaryotic flagella might be derived from spirochetes? How would you test this hypothesis? What observations would confirm or refute this hypothesis in your mind?

**3.** What do you see the relative advantages and disadvantages of each type of motility to be?

**4.** Can you think of any other mechanism that bacteria might be able to exploit for motility?

# 13 Deinococci, Chlamydiae, and Planctomycetes

With this chapter, we round out the traditional 13 "main" phyla of the domain *Bacteria*. Two of these phyla, the *Chlamydiae* and *Planctomycetes*, are probably specifically related (Fig. 13.1). *Deinococcus-Thermus* forms an independent branch and represents another relatively primitive and perhaps deeply branching lineage of the *Bacteria*.

**Figure 13.1** Phylogenetic tree of the bacterial phyla, with the phyla *Chlamydia*, *Planctomycetes*, and *Deinococcus-Thermus* highlighted. doi:10.1128/9781555818517.ch13.f13.1

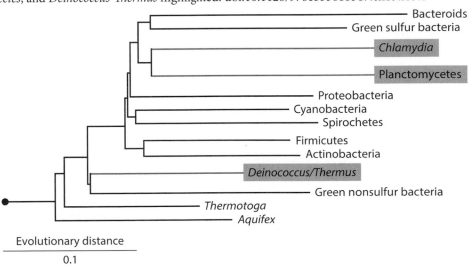

# Phylum *Deinococcus-Thermus*

| Phylum | Deinococcus-Thermus | | | | |
|---|---|---|---|---|---|
| Class | | Deinococci | | | |
| Order | | | Deinococcales | | |
| Family | | | | Deinococcaceae | |
| Genus | | | | | Deinococcus |
| Family | | | | Trueperaceae | |
| Genus | | | | | Truepera |
| Order | | | Thermales | | |
| Family | | | | Thermaceae | |
| Genus | | | | | Thermus |
| Genus | | | | | Marinithermus |
| Genus | | | | | Meiothermus |
| Genus | | | | | Oceanithermus |
| Genus | | | | | Vulcanithermus |

## *About this phylum*

This phylum contains only two well-known genera: *Deinococcus* and *Thermus*. These organisms are quite different phenotypically and phylogenetically, and each represents a small collection of closely related, very similar species.

## Deinococcus *and relatives*

### Diversity

*Deinococcus*, the only genus in the order *Deincoccales*, consists of 18 closely related species and a collection of other partially characterized isolates. A second genus, *Deinobacter*, was previously represented by a single species, *D. grandis*, which has been reclassified as a member of the genus *Deinococcus*. The only exception is an additional single species, *Truepera radiovictrix*, which is more distantly related to *Deinococcus* and shares both the thermophilic phenotype of *Thermus* and the radiation-resistant phenotype of *Deinococcus*.

Figure 13.2 is a phylogenetic tree of the genera of the phylum *Deinococcus-Thermus*.

**Figure 13.2** Phylogenetic tree of the genera of *Deinococcus-Thermus*.
doi:10.1128/9781555818517.ch13.f13.2

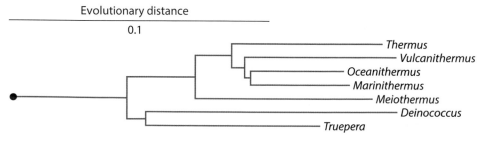

## Metabolism

The deinococci are aerobic heterotrophs, and most are mesophilic. The most striking feature of these organisms is their extreme resistance to ionizing (gamma) radiation (Fig. 13.3), but they are also extremely resistant to UV radiation, desiccation, oxidizing agents, and mutagens. The common thread is that these all cause damage to DNA, and in the most extreme cases result in double-strand breaks. *Deinococcus* has very active DNA repair systems, and by keeping between 4 and 10 copies of the genome in each cell, it can even use homologous recombination to correctly reassemble the DNA after wholesale fragmentation by high-energy gamma-irradiation.

## Morphology

The deinococci are nonmotile cocci, except for the rod-shaped *D. grandis*. Although they have a gram-negative-type envelope, the peptidoglycan layer is very thick and the outer membrane is covered in an S-layer, resulting in typically gram positive staining. Along with the fact that they are commonly pigmented (pink or red to purple or even black), this means that they can easily be mistaken for *Micrococcus* organisms. Binary fission is unusual in deinococci; instead of a cell pinching off into two daughter cells, cells divide by forming a "septal curtain," which closes inward like the shutter on a camera without changing the shape of the original cell. The division plane alternates by ca. 90° in the $x$ and $y$ (but not $z$) axes, resulting in tetrads or larger arrangements of cells in some species.

## Habitat

Most species were isolated from irradiated samples, including foods supposedly sterilized by irradiation, clean rooms (which use UV lights for "sterilization"), and nuclear reactor cooling pools, but the natural environment of these

**Figure 13.3** Kill curves of *Escherichia coli* versus *Deinococcus radiodurans* upon exposure to gamma-irradiation, expressed as percent survival (notice that this is a logarithmic axis) versus Grays (Joules/kilogram) of gamma irradiation. At 500 Gy of exposure, *E. coli* survival is less than 0.1%. At 10 times this exposure (5,000 Gy), there is no detectable killing of *D. radiodurans*. For comparison, 10 Gy is lethal to humans. (Reprinted from Sale JE, *Curr Biol* **17**:R12–R14, Copyright 2007, with permission from Elsevier.) doi:10.1128/9781555818517.ch13.f13.3

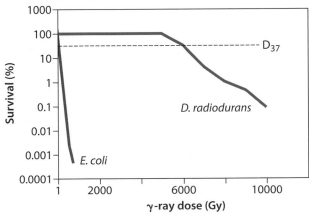

organisms is not known. They have been isolated sporadically from soils, sediments, sewage, rainwater, and many dust-covered surfaces. This suggests that their resistance to radiation might be a by-product of their evolved resistance to desiccation, which also induces double-strand breaks in DNA. It has been suggested that the natural habitat of this organism might be the water droplets that make up clouds.

**EXAMPLE** *Deinococcus radiodurans*

*Deinococcus radiodurans* (Fig. 13.4) is by far the best-studied species of this family. It was originally isolated over 50 years ago from cans of meat treated with large doses of gamma-irradiation during the development of this preservation process. Some cans nevertheless spoiled, and the organism responsible was isolated. Irradiation is now a common method for preservation of packaged food, and the dosage used is based on the need to kill this organism, just as autoclaving time and temperature are based on the need to kill endospores. Although involved in food spoilage, *D. radiodurans* is not pathogenic. *D. radiodurans* has been a model system for the study of the biochemistry of DNA repair. *D. radiodurans* cells contain several copies of each of the two chromosomes in a condensed toroidal nucleoid.

## Thermus *and relatives*

### Diversity

This group is more diverse than are the deinococci, with 17 species in four genera. There are also a large number of partially characterized isolates. *Thermus aquaticus* is the best-known member of this group, and *T. thermophilus*, because of its very high growth temperature (up to 85°C compared with 79°C for *T. aquaticus*), has also been well studied.

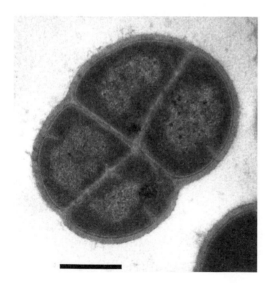

**Figure 13.4** Electron micrograph (thin section) of dividing *Deinococcus radiodurans*, showing perpendicular (in two planes) division, "septal curtains," the dense toroid nucleoid, and the thick cell wall. (Courtesy of M. J. Daly.) doi:10.1128/9781555818517.ch13.f13.4

## Metabolism

Members of the order *Thermales* are all thermophilic heterotrophs, capable of utilizing a wide range of carbon and energy sources but growing best in media with low concentrations of these organic substrates. These organisms are either obligate aerobes or facultative anaerobes, growing anaerobically by nitrate reduction.

## Enzymes

Enzymes from these organisms have proven very useful because of their thermostability. This was demonstrated dramatically in the development of the use of *T. aquaticus* (*Taq*) DNA polymerase in the automated polymerase chain reaction (PCR). Before this, PCR required the user to manually add DNA polymerase (typically from *Escherichia coli*) to each sample during each cycle of the reaction, and so it remained a tedious and obscure method. The use of *Taq* polymerase, because it is not inactivated during the DNA heat denaturation step in each cycle, allowed the automated cycling of PCR amplifications, and now PCR is one of the mainstays of molecular biology. The DNA polymerase from *T. thermophilus* is also now widely used, because of its greater thermostability and reverse transcriptase activity. These enzymes have largely been replaced in PCR by DNA polymerases from thermophilic archaea, which are more processive and accurate (because of their 3′-5′ exonuclease "proofreading" activity), and more thermostable.

These organisms are also the sources of other important thermostable enzymes used in biotechnology and industry. Industrially important enzymes are primarily carbohydrate hydrolases and are useful because their long life span (stability) makes them useful in immobilized enzyme systems.

Enzymes from *Thermus* have been studied extensively by structural biochemists because their thermostability (and therefore their rigidity at moderate temperatures) often results in the ability to easily grow very uniform crystals for X-ray diffraction analysis and determination of three-dimensional structure. In addition, these enzymes are generally readily overexpressed in *E. coli* and easily purified from *E. coli* extracts; a quick heat treatment curdles all but the smallest of the *E. coli* proteins, leaving the protein of interest as the predominant remainder in solution.

## Morphology

Most members of the *Thermales* are filamentous in nature and initially upon isolation, but upon domestication they become pleomorphic rods and short filaments. The outer membrane of their gram-negative envelope is loosely attached to the cell wall, appearing corrugated in electron micrographs. In captivity, some cells produce vesicular blebs and rotund bodies, aggregates of cells bound inside a common outer membrane. *T. filiformis* is a filamentous sheathed species. Most species produce carotinoid pigments and so form yellow, orange, or red-pink colonies.

**Figure 13.5** Octopus Spring, Yellowstone National Park, from which *Thermus aquaticus* is readily isolated. doi:10.1128/9781555818517.ch13.f13.5

## Habitat

*Thermus* and related genera are readily isolated from neutral-pH to slightly alkaline hot springs, at temperatures between 55° and 80°C (Fig. 13.5). Halophilic species have been isolated from submarine vents. Hot artificial environments can also harbor *Thermus*, including thermally polluted water outflows, soil heated by steam pipes, and household hot water heaters.

EXAMPLE *Thermus aquaticus*

The original isolates of *T. aquaticus* (Fig. 13.6) were obtained from Mushroom Spring, Octopus Spring, and other alkaline hot springs in the White Creek area of Yellowstone National Park during attempts to cultivate the pink filamentous growth that is common in these springs (see the discussion of *Thermocrinis ruber* in chapter 8). It forms pale yellow colonies, growing between 40 and 79°C (optimally at 70°C). *T. aquaticus* is an obligate aerobe; it cannot reduce nitrate.

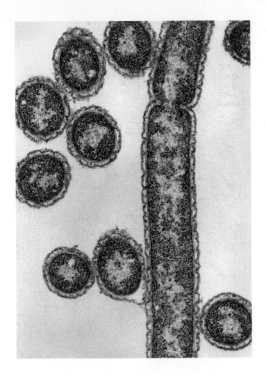

**Figure 13.6** *Thermus aquaticus*, thin-section electron micrograph. Oval cells are cross-sections. Notice the loose and ruffled outer membrane. (Reprinted from Edwards M, Brock T, *J Bacteriol* **104**:509–517, 1970, with permission.) doi:10.1128/9781555818517.ch13.f13.6

## Phylum *Chlamydiae* (*Chlamydia* and relatives)

| Phylum | Chlamydiae | | | | |
|---|---|---|---|---|---|
| Class | | Chlamydiae | | | |
| Order | | | Chlamydiales | | |
| Family | | | | Chlamydiaceae | |
| Genus | | | | | Chlamydia |
| Genus | | | | | Chlamydophila |
| Genus | | | | | Clavochlamydia |
| Family | | | | Parachlamydiaceae | |
| Genus | | | | | Parachlamydia |
| Genus | | | | | Neochlamydia |
| Genus | | | | | Protochlamydia |
| Family | | | | Simkaniaceae | |
| Genus | | | | | Simkania |
| Genus | | | | | Fritschea |
| Family | | | | Criblamydiaceae | |
| Genus | | | | | Criblamydia |
| Genus | | | | | Rhabdochlamydia |
| Genus | | | | | Piscichlamydia |
| Family | | | | Waddliaceae | |
| Genus | | | | | Waddlia |

Figure 13.7 is a phylogenetic tree of representative members of the *Chlamydiae*.

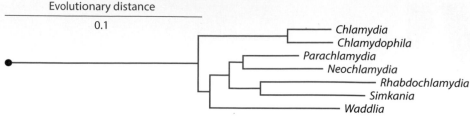

**Figure 13.7** Phylogenetic tree of representative members of the *Chlamydiae*.
doi:10.1128/9781555818517.ch13.f13.7

## About this phylum

### Diversity

Historically, there were only three species in this phylum, all of the genus *Chlamydia*: the human pathogens *C. trachomatis*, *C. psittaci*, and *C. pneumoniae*. (The last two have been moved to the genus *Chlamydophila*.) Although many new species are now known, this phylum remains a collection of very closely related, phenotypically similar organisms. However, environmental surveys using the small-subunit ribosomal RNA (SSU rRNA) PCR approach (see chapter 18) suggest that the diversity of this phylum is very much broader than is represented by known species. Although chlamydiae are sometimes classified with viruses, because they are obligate intracellular parasites transmitted via small, metabolically inert particles, they are bacteria phylogenetically and in every other meaningful way.

Chlamydiae are distantly related to *Verrucomicrobium*, and probably the *Planctomycetes* as well, all of which have little or no peptidoglycan in their cell walls.

### Life cycle

The chlamydiae are obligate intracellular parasites of eukaryotes with a biphasic life cycle (Fig. 13.8). The elementary body (EB) is the infectious phase found in interstitial fluids, secretions, and the environment. EBs are small, only 0.2 to 0.3 μm in diameter, and metabolically inert. Although chlamydiae lack detectable amounts of peptidoglycan, the envelope of EBs is rigid due to heavy disulfide cross-linking of the major outer membrane protein (MOMP). EBs attach to the host cell surface, probably nonspecifically rather than via any specific receptors or adhesins, and are endocytosed. The EBs then develop into vegetative reticulate bodies (RBs), which metabolize, grow, and divide within the endocytic vesicle. RBs are larger (ca. 1-μm) cells and are noninfectious and osmotically fragile, apparently lacking MOMP cross-linking. When the resources of the host cell become limited, most of the RBs differentiate into EBs, which are then released into the surroundings by either host cell lysis or exocytosis.

### Metabolism

Although the genomes of chlamydiae are reduced in animal pathogens to about 1,000 genes, they retain the genes required for information processing (transcription, translation, replication), the cell envelope (including peptidoglycan synthesis, even though this is undetectable in practice), and much of the central

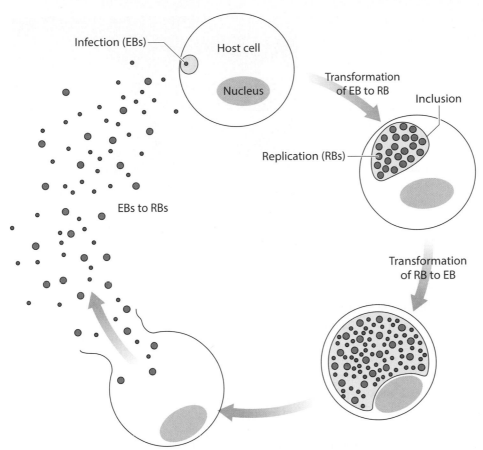

**Figure 13.8** The chlamydial developmental cycle. The small, infectious elementary bodies (EBs) are represented by small red dots; the larger, vegetative reticulate bodies (RBs) are shown in green. Not all infectious cycles end in host cell lysis; some species are released by exocytosis. (Courtesy of Karin D. E. Everett, Ph.D., Editorial Advisory Board, Veterinary Microbiology, www.chlamydiae.com.) doi:10.1128/9781555818517.ch13.f13.8

metabolism. Genes for glycolytic enzymes are present, as are genes for an incomplete tricarboxylic acid (TCA) cycle. Some species have genes for purine and pyrimidine biosynthesis, but they seem to rely on the host for most amino acids and cofactors. Adenosine triphosphatase (ATPase), run in "reverse" at the expense of adenosine triphosphate (ATP), is probably used to generate the proton gradient required to drive active transport of nutrients from the host cell; no electron transport chain is present.

The chlamydiae are largely energy parasites. An ATP/ADP antiport is used to acquire ATP from the host and recycle adenosine diphosphate (ADP). The ability to synthesize ATP may supplement energy parasitism or may be required only to generate ADP to supply the ATP/ADP antiport as the cells grow.

## Habitat

The chlamydiae are all obligate intracellular parasites, predominantly of animals (as far as we know). The "environmental" chlamydiae, despite the name, are also obligate intracellular parasites, but they infect protists, especially amoebas.

In fact, it may be that amoebas can act as intermediate hosts for the more traditional *Chlamydia* species as well. These environmental species have the same life cycle as the other chlamydiae, but the EBs float around in the environment instead of the body fluids of the host.

## EXAMPLE SPECIES

### *Chlamydia trachomatis*

*C. trachomatis* is a human pathogen that causes the most common sexually transmitted disease in the United States, ca. 4 million cases/year. It is easily spread since most infections in females are asymptomatic and untreated. Infection can lead to pelvic inflammatory disease in women and, eventually, sterility and urethritis in men. Repeated ocular infection (trachoma [Fig. 13.9]), usually in children, leads to blindness, primarily in the third world, and is the leading cause of childhood blindness worldwide. Blindness is an indirect result of infection of the inner eyelid; scarring causes the eyelids to curl inward such that the eyelashes rub painfully across the surface of the eye with every blink. This constant irritation clouds the cornea, obscuring vision. (Patients often pull out their eyelashes to help alleviate the pain.) This species also infects koalas, resulting in infertility that, along with habitat loss, is a serious threat to the survival of the species. In fact, the four major infectious diseases of koalas are all chlamydial.

### *Protochlamydia amoebophila*

*Protochlamydia amoebophila* grows symbiotically in amoeba of the genus *Acanthamoeba* (Fig. 13.10). These amoebas are common environmental organisms, although some are opportunistically pathogenic to humans; *A. castellanii* is commonly found in the tear fluid of the eyes, where it can cause infections, especially of contact lens wearers. *P. amoebophila* is a model system for investigation of the evolution of the human-pathogenic chlamydiae. The genome of *P. amoebophila* is less reduced than that of the pathogenic chlamydiae. At about 2.4 MBb in length, with over 2,000 protein-encoding genes, it is twice the size of the genomes of the pathogenic chlamydiae and as large as those of many free-living bacteria. It has a complete TCA cycle, from which it can synthesize glycine, serine, glutamine, and proline. Unlike pathogenic chlamydiae, it cannot synthesize tryptophan; this ability is a virulence factor in the

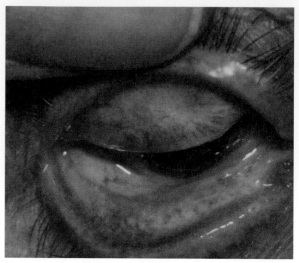

**Figure 13.9** Trachoma. Notice the granular, everted eyelids. This patient is unusual in that he/she has not pulled out his/her eyelashes, which patients often resort to for pain relief. (Source: World Health Organization, http://www.who.int/blindness/causes/priority/en/index2.html.) doi:10.1128/9781555818517.ch13.f13.9

**Figure 13.10** *Protochlamydia amoebophila* (pink) in two cells of its host, *Acanthamoeba* (green). (Reprinted from Poppert S, Essig A, Marre R, Wagner M, Horn M, *Appl Environ Microbiol* **68**:4081–4089, 2002, with permission.) doi:10.1128/9781555818517.ch13.f13.10

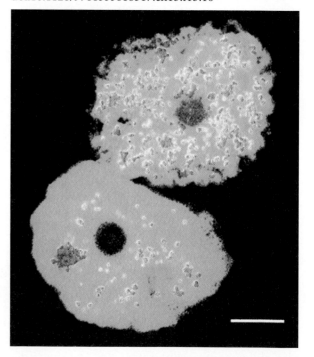

pathogens. *P. amoebophila* has an abbreviated electron transport chain, which is probably used to generate a proton gradient for active transport, but may also be used to generate ATP by oxidative phosphorylation to supplement that acquired from the host.

## Phylum *Planctomycetes* (*Planctomyces* and relatives)

| Phylum | *Planctomycetes* | | | | |
|---|---|---|---|---|---|
| Class | | *Planctomycea* | | | |
| Order | | | *Planctomycetales* | | |
| Family | | | | *Planctomycetaceae* | |
| Genus | | | | | *Planctomyces* |
| Genus | | | | | *Gemmata* |
| Genus | | | | | *Isosphaera* |
| Genus | | | | | *Pirellula* |
| Genus | | | | | *Blastopirellula* |
| Genus | | | | | *Rhodopirellula* |
| Genus | | | | | *Pirella* |
| Genus | | | | | *Singulisphaera* |
| Genus | | | | | *Nostocoida* |
| Incertae sedis (uncertain affiliation) | | | | | |
| Genus | | | | | *Brocadia* |
| Genus | | | | | *Kuenenia* |

### *About this phylum*

#### Diversity

The diversity of the *Planctomycetes* is unclear; although they are morphologically conspicuous because they divide by budding, are stalked, and often form spectacular rosettes, they are rarely isolated in pure culture for study. The phylum is distantly but specifically related to *Chlamydiae* and *Verrucomicrobia*.

Figure 13.11 is a phylogenetic tree of the phylum *Planctomycetes*.

**Figure 13.11** Phylogenetic tree of the genera of *Planctomycetes*.
doi:10.1128/9781555818517.ch13.f13.11

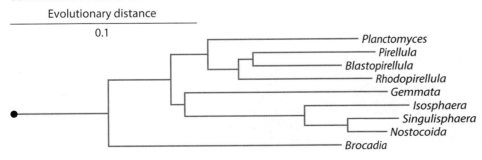

## Metabolism

Cultivated members of the *Planctomycetes* are all aerobic mesophilic oligotrophic heterotrophs, except *Isosphaera pallida*, which is moderately thermophilic (up to 55°C), and *Brocadia anammoxidans* and *Kuenenia stuttgartiensis*, which carry out anaerobic ammonia oxidation. The heterotrophs are capable of growth on a wide range of sugars and sugar derivatives, including polysaccharides. All lack peptidoglycan and are resistant to β-lactam antibiotics such as ampicillin; this trait can be useful in enrichment cultures and isolations.

## Morphology

Typically these organisms are cocci or ovals. Most have stalks, but these can be too thin or short to be apparent by light microscopy. The stalks are external fibrous structures, unlike the cytoplasmic extensions of appendaged bacteria. Stalked forms often form planktonic rosettes (Fig. 13.12). Fimbriae are common, and newly released buds are often flagellated and highly motile. They lack a peptidoglycan cell wall but have an external pitted wall of unknown composition (Fig. 13.13). The nucleoids are quite distinct and condensed.

Members of the *Planctomycetes* contain internal membrane-defined compartmentalization. The cytoplasm is divided into the riboplasm, containing ribosomes and DNA, and the paryphoplasm, containing RNA (of unknown type or function) but not ribosomes. These compartments are separated by a

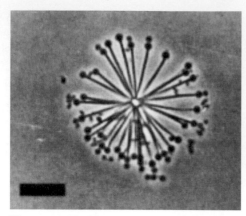

**Figure 13.12** Phase-contrast micrograph of a rosette of *Planctomyces bekefii*. (From Balows A, Truper HG, Dworkin M, Harder W, Schleifer K-H [ed]. 1992. *The Prokaryotes*, 2nd ed. Springer-Verlag, New York, NY, with kind permission from Springer Science and Business Media.)
doi:10.1128/9781555818517.ch13.f13.12

**Figure 13.13** Electron micrograph of *Planctomyces bekefii* showing the external fibrous stalk, fimbriae (close-up on the right), pits, and plate-like cell wall. (From Balows A, Truper HG, Dworkin M, Harder W, Schleifer K-H [ed]. 1992. *The Prokaryotes*, 2nd ed. Springer-Verlag, New York, NY, with kind permission from Springer Science and Business Media.) doi:10.1128/9781555818517.ch13.f13.13

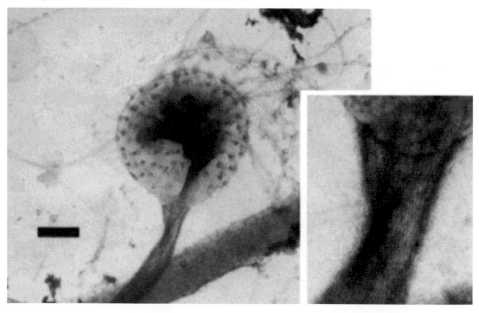

membrane; the central internal compartment containing the riboplasm and nucleoid is termed the pirellulosome. In *Gemmata*, the nucleoid is separated from most of the riboplasm by an additional double-membrane "nuclear" envelope. This "nucleus" is very different from those of eukaryotes, however, in that it contains apparently functional ribosomes. In *Brocadia* (and presumably *Kuenena*) species, there is an additional membrane separating the "anammoxisomes" from the rest of the cell. This membrane has an unusual lipid composition specially designed to keep the toxic intermediates of this reaction contained.

## Habitat

Most members of the *Planctomycetes* are aquatic, and they appear most commonly in eutrophic environments. However, *Isosphaera* is found in the phototrophic mats of hot springs (35 to 55°C), *Brocadia* and *Kuenenia* are found in anaerobic waste digesters, and SSU rRNA sequences from this group are isolated from a wide range of environments.

### EXAMPLE SPECIES

### *Blastopirellula marina*

*Blastopirellula* (previously *Pirellula*) *marina* (Fig. 13.14 and 13.15) is a relatively common freshwater species, is oval with short stalks, and forms small, flower-like rosettes (in contrast to the spectacular "fireworks" rosettes of

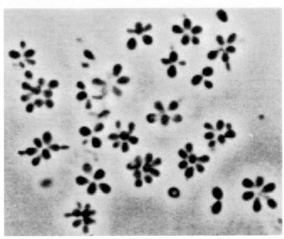

**Figure 13.14** Phase-contrast micrograph of *Blastopirellula marina* rosettes. (From Balows A, Truper HG, Dworkin M, Harder W, Schleifer K-H [ed]. 1992. *The Prokaryotes*, 2nd ed. Springer-Verlag, New York, NY, with kind permission from Springer Science and Business Media.)
doi:10.1128/9781555818517.ch13.f13.14

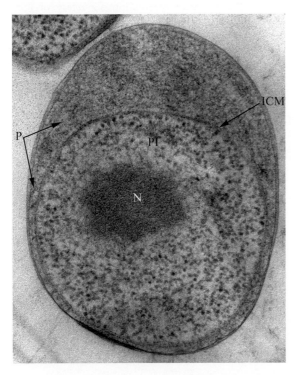

**Figure 13.15** Thin-section electron micrograph of *Blastopirellula marina*. The darkly stained nucleoid (N), intracytoplasmic membrane (ICM), and paryphoplasma (P) are labeled. (Reprinted from Lindsay MR, Webb RI, Fuerst JA, *Microbiology* **143**:739–748, 1997, with permission. doi: 10.1099/00221287-143-3-739)
doi:10.1128/9781555818517.ch13.f13.15

*Planctomyces* species). *Blastopirellula* has the simplest form of compartmentalization of the planctomycetes. The intracytoplasmic membrane (ICM) is a simple ovoid separating the riboplasm, containing both the ribosomes and the nucleoid, from the paryphoplasma. The paryphoplasma is primarily at one end of the cell; i.e., it is polar.

## Isosphaera pallida

*Isosphaera pallida* (Fig. 13.16 and 13.17) is an unusual member of this group phenotypically; individual cells are cocci, and they form filamentous chains that contain gas vacuoles and are motile by gliding. Budding occurs along the axis of the chain, and so daughter cells are formed interstitially. Isolated from hot spring phototrophic mats, this species was originally mistaken for a cyanobacterium. *I. pallida* has a cell structure very similar to that of *Planctomyces*, with a single membrane (ICM) separating the cytoplasm into paryphoplasm and riboplasm. The paryphoplasm is highly polar, located mostly at one end of the cell as a sort of vesicle, but it still forms a thin layer between the cytoplasmic membrane and the ICM all the way around the cell.

## Brocadia anammoxidans

*Brocadia anammoxidans* (Fig. 13.18) is phenotypically very different from the other planctomycetes; it is an anaerobic autotroph, gaining energy by the production of dinitrogen gas by the reaction of ammonia and nitrite. This is known as anaerobic ammonia oxidation, or the

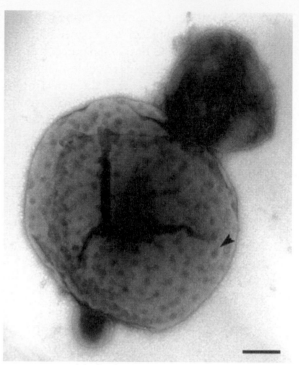

**Figure 13.16** Negatively stained electron micrograph of *Isosphaera*, with buds forming at each end. Additional budding would produce a filament, the commonly observed form of this organism. Notice the surface pits (arrow) common in this phylum. (Reprinted from Wang J, Jenkins C, Webb RI, Fuerst JA, *Appl Environ Microbiol* **68:**417–422, 2002, with permission. Courtesy of John Fuerst.) doi:10.1128/9781555818517.ch13.f13.16

**Figure 13.17** Thin-section electron micrograph of *Isosphaera*. The intracellular membrane (M) and the nucleoid (N) are labeled. (Reprinted from Wang J, Jenkins C, Webb RI, Fuerst JA, *Appl Environ Microbiol* **68:**417–422, 2002, with permission. Courtesy of John Fuerst.) doi:10.1128/9781555818517.ch13.f13.17

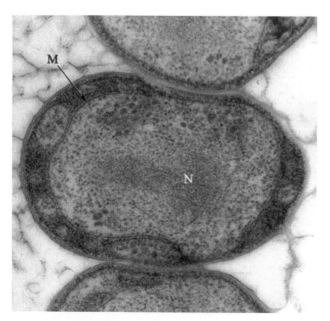

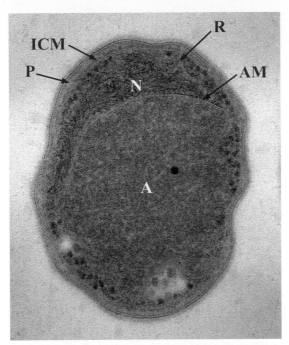

**Figure 13.18** Thin-section electron micrograph of *Brocadia anammoxidans*. The "riboplasm" (cytoplasm) is peripheral to the large internal anammoxisome (A), contains the nucleoid (N), and is surrounded in turn by the intracytoplasmic membrane (ICM)-bound paryphoplasm (P). (Reprinted from Fuerst JA, *Annu Rev Microbiol* **59**:299–328, 2005, with permission from Annual Reviews. doi: 10.1146/annurev.micro.59.030804.121258) doi:10.1128/9781555818517.ch13.f13.18

"anammox" reaction. Carbon fixation is by the acetyl coenzyme A pathway. *B. anammoxidans* has a bit more internal complexity than most planctomycetes but is based on the same cell structure as the previous examples. The paryphoplasm is a relatively thin layer all around the cell (not polarized), and there is an additional membrane-bound structure, the anammoxisome. This is a critical aspect of the anammox reaction, which includes a very highly reactive intermediate, hydrazine (a.k.a. rocket fuel). Note that the anammoxisome membrane is not attached to or part of the ICM, nor is the ICM attached to or part of the cytoplasmic membrane (this is true of all of members of the *Planctomycetes*).

### Gemmata obscuriglobus

*Gemmata obscuriglobus* (Fig. 13.19 and 13.20) is a spherical or ovoid nonstalked species. Although it was previously thought to have a large indentation in the surface of the cell, as seen in many electron microscopic images, this seems to be an artifact of dehydration in preparation for microscopy. *Gemmata* is the most complex planctomycete in its internal structure. As in *Brocadia anammoxidans* (above), the paryphoplasm forms a relatively thin layer all around the cell, between the cytoplasmic membrane and the ICM. In *Gemmata*, however, there is an additional double-layer membrane within the riboplasm surrounding the nucleoid. This "nuclear envelope" is studded on both sides with ribosomes and continuous with the cell membrane. Openings in this membrane allow movement of riboplasm contents between two compartments.

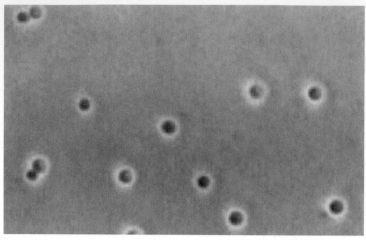

**Figure 13.19** Phase-contrast micrograph of *Gemmata obscuriglobus*. Notice the budding cell on the left. (From Balows A, Trüper HG, Dworkin M, Harder W, Schleifer K-H [ed]. 2006. *The Prokaryotes*, 2nd ed. Springer-Verlag, New York, NY, with kind permission from Springer Science and Business Media.) doi:10.1128/9781555818517.ch13.f13.19

**Figure 13.20** Thin-section electron micrograph of *Gemmata obscuriglobus*. Ribosomes (R), paryphoplasm (P), and internal membrane surrounding the nucleoid (E) are labeled. (From Lindsay MR, Webb RI, Strous M, Jetten MSM, Butler MK, Forde RJ, Fuerst JA. 2001. *Arch Microbiol* **175**:413–429, with kind permission from Springer Science and Business Media.) doi:10.1128/9781555818517.ch13.f13.20

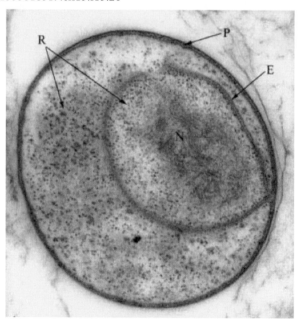

## What is the difference between paryphoplasm and periplasm?

A substantial issue with these cellular structures in members of the *Planctomycetes* is distinguishing the paryphoplasm from typical gram-negative periplasm. These organisms have a cell wall outside the cell membrane, but it is not a peptidoglycan (its actual composition is not known). They do not have a membrane outside the cell wall; they lack the hallmark gram-negative outer membrane. Or do they? What if the ICM is really the cytoplasmic membrane and the cytoplasmic membrane is really the outer membrane? In this case, the riboplasm becomes traditional cytoplasm and the paryphoplasm becomes periplasm (sometimes pretty substantial, as in *Thermotoga*). The difference, then, with the *Planctomycetes* would be that they lost the peptidoglycan cell wall and reinvented it outside of the outer membrane, much like the S-layer of many members of the *Bacteria* and *Archaea*.

However, even if this is the case (and the authorities on the *Planctomycetes* argue that it is not), it does not change the most interesting observation, i.e., that *Gemmata* has a sort of nucleus with a nuclear envelope. Yes, it contains ribosomes, but maybe eukaryotic nuclei do too (involved in nonsense-mediated decay), and surely there is some sort of functional differentiation between riboplasm inside and outside the nuclear envelope. Just for example, the ribosomes translating outside of the "nucleus" cannot be translating messenger RNAs (mRNAs) that are still being transcribed; this lack of linkage between transcription and translation is usually cited as an important distinction between "prokaryotes" (this misguided term is discussed in detail in chapter 2) and eukaryotes.

## Reductive evolution in parasites

It is common for parasites, especially obligate endoparasites, to evolve by simplification, and the more the parasite relies on its host for the things it needs, the more it can simplify. This "reductive evolution" (which used to be called "degeneration," or sometimes "devolution": hence the name of the inexplicably popular band of the early 1980s, Devo) allows the parasite to focus its resources on reproduction. For example, many parasitic worms lack digestive systems (they absorb their nutrients directly through their cuticle), a circulatory system, and so forth, and in extreme cases are little more than genitals that hook onto the gastrointestinal tract (or elsewhere) of their host. Even ectoparasites often become simplified; there are a slew of ectoparasites that are little more than stomachs, sucking mouths, and reproductive organs (i.e., leeches).

Even bacterial parasites can evolve by simplification; good examples are the chlamydias, rickettsias, and mycoplasmas. Perhaps more extreme examples would be plastids and mitochondria; some viruses may also have originated by simplification of cellular intracellular parasites (see chapter 17). In bacteria, this simplification is most easily seen in their genomes; the sizes of the genomes of these obligate intracellular parasites are drastically smaller than those of their free-living relatives. Any gene that can be done without is eliminated. The genomes of the human-pathogenic chlamydias are only 1 Mbp in length—about 1,000 genes, only one-fourth as big as wild *E. coli* and about one-half the size of the smallest genomes of free-living bacteria. The reasons for simplification are many: in addition to the usual reasons for simplification in parasites generally, the smaller the genome, the faster it can replicate, and the simpler the organism, the faster it can evolve. The latter is especially important for a bacterial parasite, which is in a continuous arms race with its host.

## *Questions for thought*

1.  If you were interested in getting some novel isolates of *Deinococcus* or *Thermus*, where would you look and how would you set up the enrichment cultures?

2.  Can you think of any enzymes other than DNA polymerases from thermophilic organisms that might be useful? What about enzymes from organisms that live in extremely cold environments? Saline environments?

3.  How far do you think a bacterial intracellular parasite could minimize itself? Why not further? What could it do without? What could it *not* do without?

4.  What is the difference between a minimized, obligately intracellular bacterial parasite like *Chlamydia* and a virus?

5.  The chlamydiae we know about infect animals (mammals, birds, reptiles, and insects) and amoebas. Where would you look for whole new kinds of chlamydiae? How would you go about this?

6.  *Gemmata* has a double membrane around its nucleoid. How would you determine how closely this membrane resembles the nuclear envelope of eukaryotes?

7.  How would you go about determining whether the paryphoplasm was similar to the periplasm of gram-negative bacteria?

8.  Many bacteria divide, like the planctomycetes, by budding rather than by binary fission. What are the ramifications of this for both the mother and daughter cells, compared to binary fission?

9.  Members of the *Planctomycetes* divide by budding, and *Gemmata* has a double-membrane-enclosed nucleoid. Draw a picture of how you think cell division works in this organism, given that even early buds have a "nucleus." How is your scheme similar to and different from the eukaryotic cell cycle?

# 14
# Bacterial Phyla with Few or No Cultivated Species

The 13 phyla of bacteria described in chapters 8 to 13 constitute the "standard" main phylogenetic groups as defined by Carl Woese in a classic 1987 review of bacterial diversity, with the more recent addition of *Aquifex*. As more sequences become available from cultivated species and from surveys of natural populations, it has become increasingly clear that the main radiation of bacterial phyla contains many more branches than originally thought (Fig. 14.1); over 200 contain at least two established sequences. Several of these groups contain only a few cultivated species, but in most cases they contain none at all. These groups are known mainly or exclusively from small-subunit ribosomal RNA (SSU rRNA) sequences cloned from DNA extracted directly from environmental samples (see below).

Nevertheless, it is important to remember that most of the cultivated, characterized bacteria and most SSU rRNA sequences from environmental samples fall into only five bacterial phyla: the *Proteobacteria*, the *Firmicutes*, the *Actinobacteria*, the *Bacteroidetes*, and the *Cyanobacteria*. The other groups, the *Aquificae*, *Thermotogae*, *Chlorobi*, *Chloroflexi*, *Planctomycetes*, *Chlamydiae*, *Deinococcus-Thermus*, and even the *Spirochaetae*, qualify in some sense as phyla with few cultivated species.

doi:10.1128/9781555818517.ch14

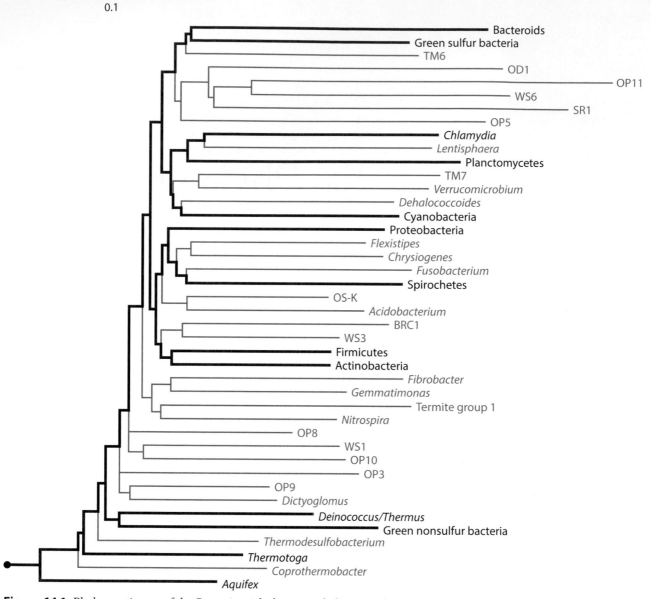

**Figure 14.1** Phylogenetic tree of the *Bacteria*, with the main phylogenetic branches (phyla) discussed in prior chapters highlighted in black. Some examples of less well-known phyla, phyla with few or even no cultivated species, are shown in red. doi:10.1128/9781555818517.ch14.f14.1

The observation that most cultivated species of bacteria come from only a small number of phyla is similar to the situation in animals; the vast majority of animal species belong to only 9 of the ca. 35 animal phyla: mollusks, sponges, cnidarians (coelenterates), flatworms, nematodes, annelids, arthropods, echinoderms, and chordates contain approximately 95% of animal species. The remaining 5%, most of animal *diversity*, are mostly obscure.

## How do we know about these organisms?

The knowledge that these scattered species belong to novel bacterial phyla comes from their SSU rRNA sequence, as described in chapters 3 through 5. Some of these are species isolated long ago that have only recently had their SSU rRNA sequences determined; both the American Type Culture Collection (ATCC) and Deutsche Sammlung von Mikroorganismen und Zellkulturen GmbH (DSMZ) have sequenced the SSU rRNAs from nearly the entirety of their culture collections. Others are newly discovered organisms; obtaining SSU rRNA sequence information is now one of the first steps in the characterization of new isolates.

However, most of these odd phyla are best, or even entirely, represented by SSU rRNA sequences extracted directly from environmental samples: so-called molecular phylogenetic surveys. In a typical molecular phylogenetic analysis, the SSU rRNA sequence is obtained by polymerase chain reaction (PCR) amplification from DNA isolated from a pure culture of the organisms of interest. The resulting sequence is used to determine the place of that organism in the "big tree." It is also possible, however, to start a molecular phylogenetic analysis with DNA extracted directly from an environmental sample instead of a pure culture. The collection of sequences obtained from such PCR amplification products (hopefully) represents the population of organisms in the original sample. This gives us phylogenetic information about organisms in an environment regardless of whether they can be cultivated. This approach is far superior to the traditional cultivation-based approaches and is described in detail (including its weaknesses) in chapter 19.

### *Summary of molecular phylogenetic surveys*

In a 1998 review (*J Bacteriol* **180**:4765–4774), Philip Hugenholtz, Brett Goebel, and Norman Pace summarized the results of 86 SSU rRNA molecular phylogenetic surveys of microbial populations from a wide range of environments: geothermal sites, soils, freshwater and saltwater environments, wastewater, and many others. The final distillation of these surveys is summarized in Table 14.1.

They found that nearly 3,000 bacterial sequences were reported, 90% of which fell into the proteobacteria, gram-positive bacteria (*Firmicutes* and *Actinobacteria*), or *Cytophagales* (*Bacteroidetes*). The remaining were widely scattered among the other groups, and many were not related to any known cultivated species. There were nine groups of such sequences that did not fall into any of the standard bacterial groups. One of these groups (called OP11, from the first such sequence reported, originating from Obsidian Pool) seems to be a major constituent of subsurface environments and is common in most other places too.

Unfortunately, no recent such compilation has been published, but this early work is probably representative of what would be found today in a compilation of the millions of environmental SSU rRNA sequences now available from environmental surveys. However, Table 14.2 tabulates the numbers of SSU rRNA sequences in release 10 of the Ribosomal Database Project, divided into type strains, isolates with various degrees of characterization, and sequences from uncultivated organisms. These numbers are largely consistent with those in Table 14.1.

**Table 14.1** Summary of 16S rRNA-based clonal analyses of diversity of uncultivated bacteria[a]

| Habitat type | No. of studies | No. of sequences[c] | Proteobacteria[d] α[c] | β | γ | δ[d] | ε[e] | Cytophagales | Actinobacteria | Low-G+C gram positive[e] | Acidobacterium | Verrucomicrobia | Spirochetes | Nitrospira | GNS | OP11 | Planctomycetes | Green sulfur | TM7 | TM6 | Thermus/Deinococcus | Cyanobacteria[c] | Synergistes | OP8 | Termite group I | OS-K | Chlamydia | OP3 | OP10 | WS1 | OP5 | Marine group A | Fibrobacter | Flexistipes | Dictyoglomus | Thermotogales | Thermodesulfobacterium | Aquificales |
|---|---|---|---|---|---|---|---|---|---|---|---|---|---|---|---|---|---|---|---|---|---|---|---|---|---|---|---|---|---|---|---|---|---|---|---|---|---|---|
| Geothermal | 10 | 212 | ○ | ○ | ○ | | ✓ | ○ | ✓ | | ○ | ✓ | ○ | ○ | ○ | ✓ | ○ | ○ | | | ○ | ✓ | ✓ | ✓ | | ○ | | ✓ | ✓ | ✓ | ✓ | | | | ✓ | ○ | ✓ | ○ |
| Soil | 16 | 743 | ● | ○ | ✓ | ○ | ✓ | ○ | ● | ○ | ● | ● | ✓ | ✓ | ○ | ✓ | ○ | ✓ | ✓ | ✓ | ✓ | ✓ | ✓ | ✓ | ✓ | ✓ | ✓ | | ✓ | ✓ | ✓ | | | | | | ✓ | |
| Marine | 23 | 687 | ● | ○ | ○ | ○ | ○ | ○ | ○ | ○ | ✓ | ○ | ○ | | ✓ | ✓ | ○ | | ✓ | ✓ | | ○ | | | | ✓ | | | | | | ✓ | | | | | | |
| Freshwater | 4 | 107 | ● | ○ | ✓ | | ✓ | ✓ | ○ | ○ | ○ | ✓ | ✓ | ✓ | | ✓ | | | | | | | | | | | | | | | | | | | | | | | |
| Wastewater | 5 | 430 | ● | ● | ● | ○ | ✓ | ○ | ● | ○ | ○ | ✓ | ✓ | ✓ | ○ | | | ✓ | | ✓ | | ✓ | ✓ | ✓ | ✓ | ✓ | ✓ | ✓ | | ✓ | | | | ✓ | | | | |
| Pollutant associated | 7 | 202 | ○ | ○ | ● | ○ | ○ | ○ | ○ | ○ | ○ | ✓ | | ✓ | ✓ | ✓ | | | | | | | | ✓ | ✓ | | ✓ | | ✓ | ✓ | ✓ | | | | | | | |
| Acid metal leaching | 2 | 21 | ● | ○ | ● | | | ○ | ● | ○ | ○ | | | ● | | | | | | | | | | | | | | | | | | | | | | | | |
| Subsurface | 6 | 229 | ○ | ● | ● | ○ | ○ | ✓ | ○ | ● | | | ✓ | ○ | ○ | ○ | ✓ | ✓ | | ✓ | ✓ | | ✓ | ✓ | ✓ | ✓ | | ○ | | | ✓ | | | | | | | |
| Symbionts and commensals | 10 | 280 | ✓ | ✓ | ✓ | ✓ | ○ | ✓ | ✓ | ✓ | | | ○ | | | ○ | ✓ | | | | | | ✓ | | ✓ | | | | | | | | ✓ | | | | | |
| Disease associated | 3 | 7 | | ○ | | | | ○ | ○ | ○ | | | ○ | | | ○ | | | | | | | | | | | | | | | | | | | | | | |
| **Total** | **86** | **2,918** | | | | | | | | | | | | | | | | | | | | | | | | | | | | | | | | | | | | |

[a] Reprinted with permission from the erratum (*J Bacteriol* **180**:6793) to Hugenholtz P, Goebel BM, Pace NR, *J Bacteriol* **180**:4765–4774, 1998. Note that in this table, the proteobacteria are so numerous that they are divided into their five sub-branches.

[b] Incidence of division-level representatives in studies of particular habitat types ranked from most represented to least represented divisions: >75% (●), 25 to 75% (○), or <25% (✓) of studies have representatives of division. The absence of a symbol indicates that the division was not detected.

[c] Excluding organelles.

[d] Proteobacteria are presented at the subdivision level due to the extensive sequence representation of this division.

[e] Cannot be established as a monophyletic group in all analyses.

**Table 14.2** Phylogenetic tabulation of sequences in the Ribosomal Database Project release 10 (2010)

| Phylum | Type strains | Other isolates | Uncultivated |
|--------|-------------:|---------------:|-------------:|
| *Aquificae* | 18 | 121 | 668 |
| *Thermotogae* | 27 | 74 | 37 |
| *Thermodesulfobacteria* | 4 | 8 | 67 |
| *Deinococcus-Thermus* | 40 | 400 | 189 |
| *Chrisiogenetes* | 1 | 3 | 0 |
| Green nonsulfur bacteria | 9 | 62 | 1,603 |
| *Nitrospira* | 5 | 89 | 461 |
| *Deferribacteres* | 10 | 32 | 193 |
| *Cyanobacteria* | 14 | 4,353 | 1,492 |
| Green sulfur bacteria | 9 | 154 | 58 |
| *Proteobacteria* | 1,927 | 59,028 | 38,046 |
| *Firmicutes* | 2,151 | 27,635 | 54,075 |
| *Actinobacteria* | 1,286 | 22,306 | 5,310 |
| *Planctomycetes* | 7 | 140 | 1,674 |
| *Chlamydiae* | 4 | 192 | 43 |
| *Spirochaetae* | 54 | 1,397 | 1,367 |
| *Fibrobacteres* | 2 | 58 | 135 |
| *Acidobacteria* | 0 | 185 | 2,566 |
| *Bacteroidetes* | 355 | 5,592 | 22,959 |
| *Fusobacteria* | 34 | 172 | 513 |
| *Verrucomicrobia* | 10 | 96 | 2,203 |
| *Dictyoglomi* | 1 | 5 | 3 |
| *Gemmatimonadetes* | 1 | 4 | 323 |
| *Lentisphaerae* | 2 | 4 | 66 |
| BRC1 | 0 | 0 | 27 |
| OP10 | 0 | 0 | 98 |
| OP11 | 0 | 0 | 43 |
| TM7 | 0 | 0 | 253 |
| WS3 | 0 | 0 | 50 |
| Dehalococcoides | 0 | 19 | 77 |
| SR1 | 0 | 0 | 2 |
| OD1 | 0 | 0 | 32 |
| Unclassified | 1 | 371 | 4,985 |
| Total | 5,972 | 122,500 | 139,618 |

## Phyla with few cultivated species

There are several bacterial phyla that have only a few cultivated, well-known species. Several, such as the *Thermotogae*, *Aquificae*, *Chloroflexi*, and *Chlorobi*, are described in previous chapters. The dividing lines between the classical 13 bacterial phyla and the phyla described below are, of course, arbitrary and historical. In reality, some of these groups are apparently far more abundant in the environment than are the groups already discussed and now have many more known species (although not as well known).

## EXAMPLE PHYLUM *Verrucomicrobia*

A phylogenetic tree of the *Verrucomicrobia* is shown in Fig. 14.2. This phylum contains a couple of appendaged bacteria, *Verrucomicrobium* and *Prosthecobacter*; these were previously thought to be related to the appendaged α-proteobacteria. The remaining cultivated members of this phylum are poorly characterized. The uncultivated species in this group are apparently abundant in the soil and subsurface, making up significant fractions of the total number of sequences isolated. Fluorescent probes targeting EA25 (which appeared frequently in an analysis of soil) indicate that the organism it comes from can make up 1 to 10% of soil microbes: $10^7$ to $10^8$ per gram of soil. This phylum is probably related to the *Chlamydiae*.

**Figure 14.2** Phylogenetic tree of the phylum *Verrucomicrobia*, including cultivated species (thick lines and bold fonts) and organisms known only from environmental sequences (thin lines and light fonts). doi:10.1128/9781555818517.ch14.f14.2

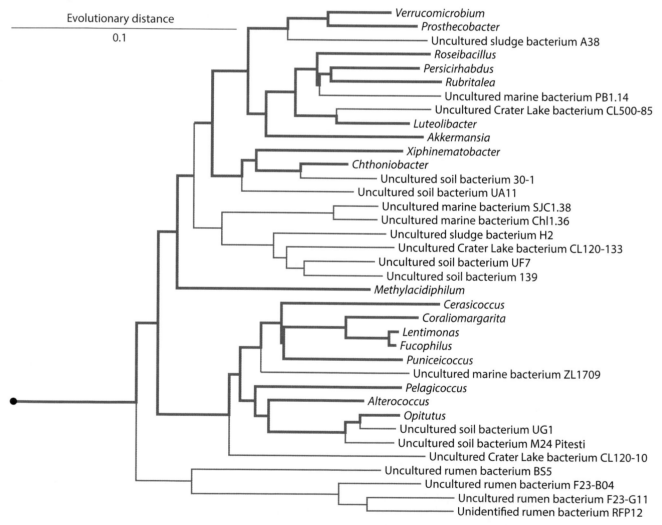

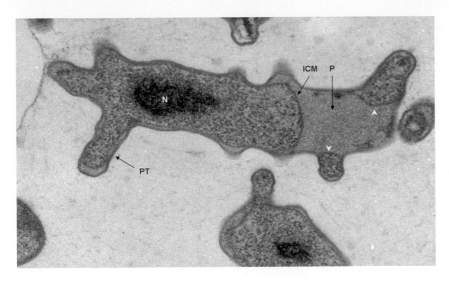

**Figure 14.3** Scanning electron micrograph of *Verrucomicrobium spinosum*. (Reprinted from Lee et al., *BMC Microbiology* **9**:5, 2009. http://creativecommons.org/licenses/by/4.0/n.) doi:10.1128/9781555818517.ch14.f14.3

## EXAMPLE SPECIES

*Verrucomicrobium spinosum* (Fig. 14.3) is a mesophilic nonmotile heterotrophic aerobe isolated from a small eutrophic lake (Lake Plußsee). Unlike the appendaged α-proteobacteria, the appendages of *V. spinosum* have bundles of fimbriae at their tips (a bit like the stereotypical hairs from the nose warts of a witch). It is covered in many short (ca. 0.5-μm) appendages that appear conical under phase-contrast microscopy and often also has a single large polar appendage.

*Prosthecobacter fusiformis* (Fig. 14.4) is a heterotrophic, mesophilic, obligate aerobe isolated from freshwater contact slides. The organisms generally resemble *Caulobacter* but are more fusiform, and dividing cells have a stalk at both cells. They do not have a dimorphic life cycle and are nonmotile throughout their life.

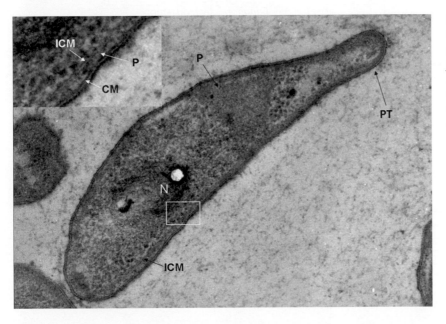

**Figure 14.4** Thin-section electron micrograph of *Prosthecobacter dejongeii*, a close relative of *P. fusiformis*. (Reprinted from Lee et al., *BMC Microbiology* **9**:5, 2009. http://creativecommons.org/licenses/by/4.0/n.) doi:10.1128/9781555818517.ch14.f14.4

***Opitutus terrae*** is a heterotrophic, motile (monotrichous), obligately anaerobic coccus isolated from rice paddy soil. Cells are very small, only ca. 0.5 μm in diameter; this organism and its relatives were previously informally known as *Ultramicrobium*. Pairs or longer chains of cells can be mistaken for rod-shaped cells.

## EXAMPLE PHYLUM *Acidobacteria*

This group contains only six cultivated species (Fig. 14.5), none of them well known; however, it also contains a large number of SSU rRNA sequences from environmental samples, including mostly soils but also a wide variety of other habitats, including a jet-fuel-contaminated aquifer, a hot spring, a sponge symbiont, and freshwater and marine environments. Fluorescently labeled oligonucleotide probes specific for one subgroup of this phylum (subgroup 6) hybridize to cells of all shapes and sizes, suggesting a broad phenotypic range to match the broad phylogenetic diversity of this group.

## EXAMPLE SPECIES

***Acidobacterium capsulatum*** (Fig. 14.6) is an acidophilic aerobic heterotroph isolated from acid mine drainage. It is heavily encapsulated, saccharolytic, rod shaped, and nonmotile. *Acidobacterium* is related to a number of SSU rRNA sequences isolated from acidic environments (e.g., peat bog, acid mine drainage), consistent with its acidophilic phenotype.

**Figure 14.5** Phylogenetic tree of the phylum *Acidobacteria*, including cultivated species (thick lines and bold fonts) and organisms known only from environmental sequences (thin lines and light fonts). doi:10.1128/9781555818517.ch14.f14.5

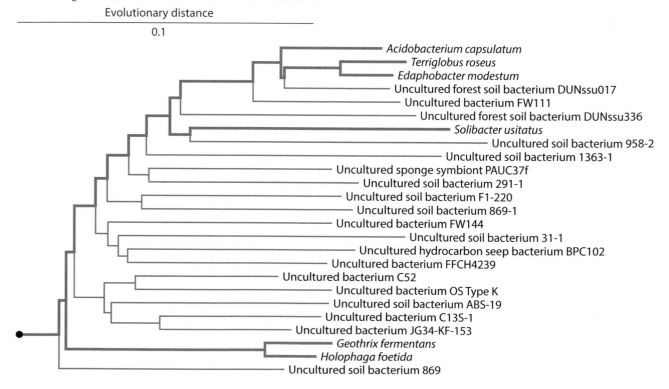

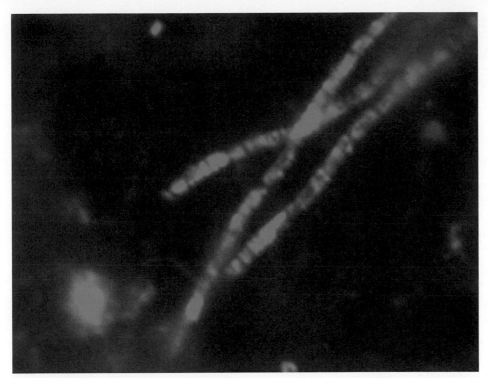

**Figure 14.6** Fluorescence micrograph of *Acidobacterium capsulatum*. (Source: U.S. Department of Energy [http://genomics.energy.gov] and Wikimedia Commons.) doi:10.1128/9781555818517.ch14.f14.6

**Figure 14.7** Phylogenetic tree of the phylum *Nitrospira*, including cultivated species (thick lines and bold fonts) and organisms known only from environmental sequences (thin lines and light fonts). doi:10.1128/9781555818517.ch14.f14.7

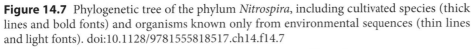

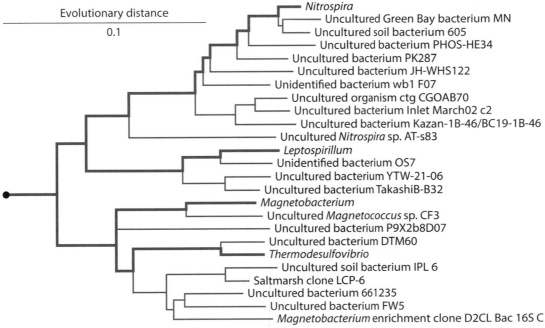

## EXAMPLE PHYLUM *Nitrospira*

A phylogenetic tree of the members of the *Nitrospira* is shown in Fig. 14.7. This phylum contains only five cultivated species in four genera, all very different phenotypically and not well characterized. Nevertheless, its members are apparently common in acidic and nitrogen-cycling environments and anaerobic marine sediments.

## EXAMPLE SPECIES

***Nitrospira marina*** is, together with other species of this genus, the predominant nitrite oxidizer (to nitrate) in marine environments and aquaria. The species in this genus are poorly distinguished, and many of the original species have been lost. Nevertheless, mixtures of *Nitrospira*, *Nitrobacter* (another nitrite oxidizer), and *Nitrosomonas* (an ammonia oxidizer, which produces nitrite) are sold commercially to help start a productive nitrogen cycle in aquaria.

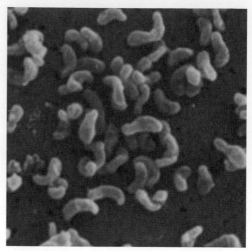

**Figure 14.8** Scanning electron micrograph of *Leptospirillum ferrooxidans*. (Courtesy of D. Barrie Johnson.) doi:10.1128/9781555818517.ch14.f14.8

***Leptospirillum ferrooxidans*** (Fig. 14.8) is also a chemolithoautotroph, oxidizing $Fe^{2+}$ to $Fe^{3+}$ in mine tailings and contributing to acid mine drainage, perhaps the pollution "worst-case scenario." The natural habitat for this organism is presumably the deep aquifers infusing iron-rich ores.

***Magnetobacterium bavaricum*** (Fig. 14.9) is a magnetotactic organism that swims north following Earth's magnetic field, directed by internal magnetite beads. These bacteria are anaerobic heterotrophs that live in aquatic sediments; in the Northern Hemisphere, the magnetic field lines lead both north and down, toward their desired habitat. Other magnetotactic bacteria (e.g., *Magnetospirillum*) are members of the *Proteobacteria*, but few are cultivated.

**Figure 14.9** Electron micrograph of *Magnetobacterium bavaricum*. (From *The Scientist* [http://www.the-scientist.com/?articles.view/articleNo/36722/title/A-Sense-of-Mystery/] courtesy of Marianne Hanzlik.) doi:10.1128/9781555818517.ch14.f14.9

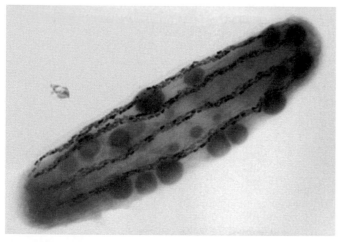

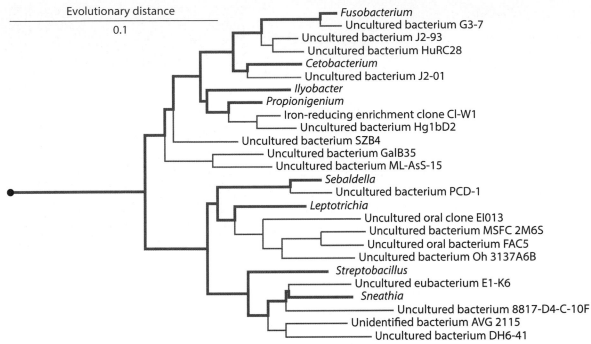

**Figure 14.10** Phylogenetic tree of the phylum *Fusobacteria*, including cultivated species (thick lines and bold fonts) and organisms known only from environmental sequences (thin lines and light fonts). doi:10.1128/9781555818517.ch14.f14.10

EXAMPLE PHYLUM  *Fusobacteria*

A phylogenetic tree of *Fusobacteria* is shown in Fig. 14.10. Most of the few cultivated members of this group are animal symbionts, probably opportunistic rather than serious pathogens. By far the best-studied genus of this phylum is *Fusobacterium*. Some members of the *Fusobacteria* are also soil organisms, and environmental sequences have come primarily from oral samples (where *Fusobacterium* is common), fecal samples, soil, and sediments.

EXAMPLE SPECIES

***Fusobacterium nucleatum*** (Fig. 14.11) is part of the normal flora of the oral cavity and is particularly abundant in dental plaque, where it plays a central role in nucleating the accumulation of various types of bacteria, including organisms such as *Porphyromonas* that can lead to periodontal disease. It is an anaerobic heterotroph, primarily fermenting sugars to butyric acid. *F. nucleatum* is spindle shaped to filamentous with tapered ends and so is easily mistaken for oral bacteroids.

## *Other examples of phyla with few cultivated species*

- *Chrysiogenetes: Chrysiogenes arsenatis* and relatives
- *Deferribacteres: Deferribacter* and relatives
- *Fibrobacteres: Fibrobacter succinogenes, F. intestinalis*, and relatives
- *Dictyoglomi: Dictyoglomus thermophilum* and relatives

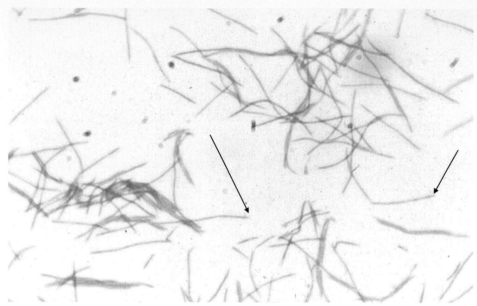

**Figure 14.11** Gram stain of *Fusobacterium nucleatum*. (Gini G. 2006. Gram-stained Fuso-bacterium. Visual Resources. American Society for Microbiology, Washington, DC. www .microbelibrary.org. Accessed 3 November 2014.) doi:10.1128/9781555818517.ch14.f14.11

- *Gemmatimonadetes: Gemmatimonas aurantiaca* and relatives
- *Lentisphaerae*: *Lentisphaera araneosa*, *Victivallis vadensis*, and relatives

## Phyla with no cultivated species

The extreme case of a phylum with few known cultivated species is, of course, a phylum with no known cultivated members. There are many of these; some are large groups that are commonly seen in microbial surveys (e.g., OP10 and TM7), but most are small groups, and many are only one or a few sequences that are not specifically related to any bacterial phylum. Some of these will be found to be deep branches in known phyla once additional related sequences are obtained. Most, however, probably represent the hidden bulk of bacterial diversity.

EXAMPLE PHYLUM  OP11

This is a large group (Fig. 14.12), containing about 100 unique sequences (only 43 are nearly full length) from a very wide range of environments, but there are no known cultivated species in this group, so nothing is known about their phenotype. This phylum is particularly interesting because of its high evolutionary rate, reflected in its long branch length; this is an unusually "advanced" group of bacteria. Years of attempts to cultivate anything from this group have proved unsuccessful.

Like most of these phyla, OP11 gets its name from a sequence designation of the original sequence identified in this group. OP11 was sequence number 11 from a collection of cloned bacterial SSU rRNA sequences from Obsidian Pool (OP [Fig. 14.13]).

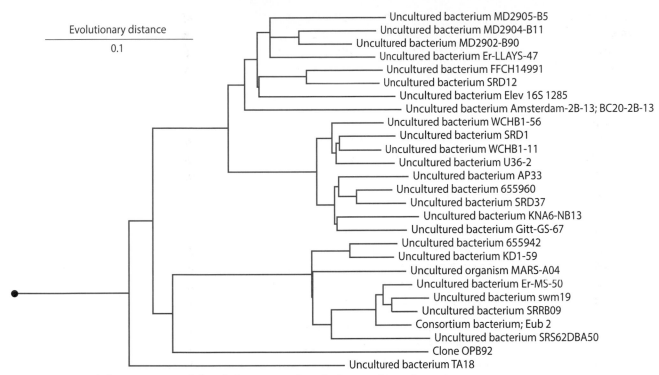

Figure 14.12 Phylogenetic tree of the phylum OP11, for which no cultivated species are known. doi:10.1128/9781555818517.ch14.f14.12

Figure 14.13 The author collecting samples in Obsidian Pool, Yellowstone National Park. doi:10.1128/9781555818517.ch14.f14.13

## EXAMPLE PHYLUM  SR1

A large number of environmental SSU rRNA sequences do not fall into any of the recognized bacterial phyla; conceptually, each group of these is a cryptic phylum. Even if we demand that at least two related, nonidentical, nearly full-length sequences be identified before describing them as a new phylum, there are several hundred such cryptic phyla. The phylum SR1 is an example (Fig. 14.14), being composed of only 2 nearly full-length and 23 shorter sequences (SR is named for Sulfur River).

## Phylogenetic groups at all levels are dominated by uncultivated sequences

Although it is easy to think of sequences from uncultivated organisms only in terms of the unusual phyla described in this chapter, in reality sequences from uncultivated organisms fall into phylogenetic groups at all levels. Even in well-studied phyla, there are orders, classes, families, and genera with large numbers of species represented solely or largely by sequences from uncultivated organisms. The tree in Fig. 14.15 was generated from an arbitrarily selected collection of sequences from cultivated and uncultivated organisms in the family *Enterobacteriaceae*. These organisms are familiar to any microbiologist, but there are many uncultivated species hidden even within this family.

## How much of the microbial world do we know about?

This is a difficult question to answer; in fact, it cannot be realistically answered at this time within even a factor of a thousand. Some people think that there may be only a few thousand species, or perhaps tens of thousands. This is ridiculous. There are over 350,000 described species of beetles, and even if there is only a single specific bacterial symbiont for each of these beetle species, that alone would account for at least 10 times more species of bacteria than these folks would believe exist. If you plate a typical environmental sample out onto rich medium after counting cells microscopically, you typically see that less than one observable bacterium in a thousand grows to produce a colony (averaging about one in a million), and of course these are only from the most abundant species. As poorly characterized microbiologically as the world around us is, we know nothing at all about some very large microbial habitats: the subsurface world, the deep-aquifer world, the hydrothermal-field world, and so on.

A more fundamental problem is that we really do not have a very good idea of what a bacterial "species" is. This is a general problem with asexual organisms; species of plants and animals are defined in terms of breeding populations, and so this only applied to organisms that "breed." The concept of a species is critical to biology; this is part of why *On the Origin of Species* was so important, and most of this book was spent creating a rational description of a species in the plant and animal worlds. (Actually, most of this is spent showing how *indistinct* the divisions really are between species.) A rational "concept of a bacterial species" does not yet exist. This is probably the most important open

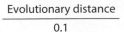

Evolutionary distance

0.1

Uncultured bacterium SRB67
Uncultured bacterium SRB44

**Figure 14.14** Phylogenetic tree of the phylum SR1, known only from two similar (but not identical) environmental sequences. doi:10.1128/9781555818517.ch14.f14.14

**Figure 14.15** Phylogenetic tree of the family *Enterobacteriaceae*, including an arbitrary collection of cultivated species (bold) and organisms known only from environmental sequences (light). doi:10.1128/9781555818517.ch14.f14.15

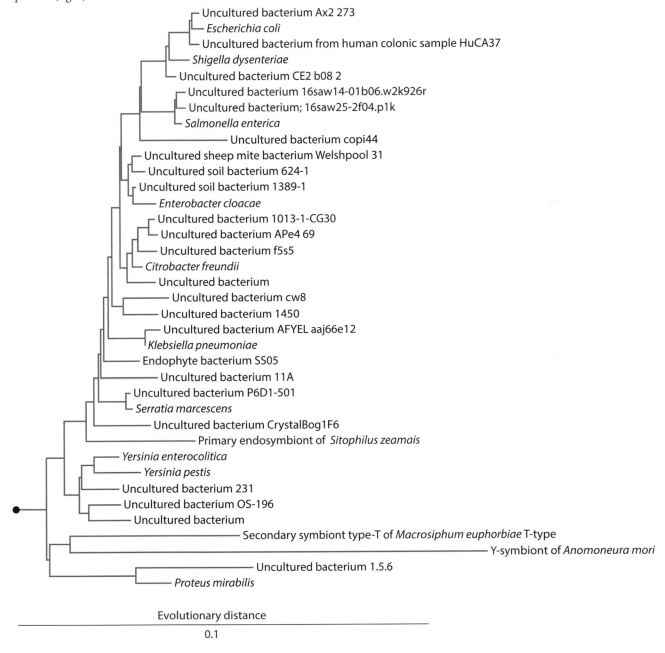

theoretical question in microbiology. Seventy percent DNA:DNA hybridization is sometimes used as an operational definition of a species, but this is an arbitrary definition without a theoretical underpinning. Until we have a meaningful definition of species, how can we count them? In fact, it has been argued that asexual organisms do not even have species, in which case some other term (and definition) might be needed.

## Questions for thought

**1.** How many species of bacteria would you predict there are? Why not more?

**2.** What do you think defines a bacterial phylum? The question is, is a phylum just any coherent deep branch? How deep does it have to be before it is a distinct phylum? If more than just branch depth is required for the definition, then what is it that defines a phylum?

**3.** *Opitutus terrae* has a volume of only about 0.1 $\mu m^3$, less than 5% of that of *E. coli*. What do you think would limit how small a free-living organism could be?

**4.** Kirk is a graduate student in a famous laboratory at a big university, and the project he and his faculty advisor and committee agreed on is to obtain a pure culture of a species in the OP11 group for further microbiological characterization. How might Kirk go about trying to do this?

# 15 Archaea

The *Archaea* represent a monophyletic group distinct from both *Bacteria* and *Eukarya*. The *Archaea* fall into two major phyla: *Euryarchaeota* and *Crenarchaeota* (Fig. 15.1). Phenotypically, cultivated *Crenarchaeota* and some *Euryarchaeota* are sulfur-metabolizing thermophiles (yellow in Fig. 15.1) whereas the *Euryarchaeota* also include the methanogens (green) and extreme halophiles (red). Both major phyla are also home to a large number of small-subunit ribosomal RNA (SSU rRNA) sequences apparently from planktonic marine species; the phenotypes of these organisms are not known. Two potential minor phyla are the *Nanoarchaeota* and *Korarchaeota*; the relationships between these minor phyla and either the *Euryarchaeota* or *Crenarchaeota* remain unclear.

## General properties of the *Archaea*

Although they superficially resemble the *Bacteria*, the *Archaea* are neither bacteria nor eukaryotes. In some respects, they do resemble bacteria. Most metabolic proteins are similar to those of bacteria, as are gene and chromosome structure; the processes of replication, transcription, and translation; and the cytoskeleton. In many other respects, however, archaea are more similar to eukaryotes. Most of the information-processing machinery (replication, transcription, and translation) is like that of eukaryotes. Archaea also have nucleosomes, nucleolar enzymes, and cell cycle proteins not present (or at least very different) in bacteria. And in some ways, archaea are different from both bacteria and eukaryotes, the most obvious being membrane lipids.

doi:10.1128/9781555818517.ch15

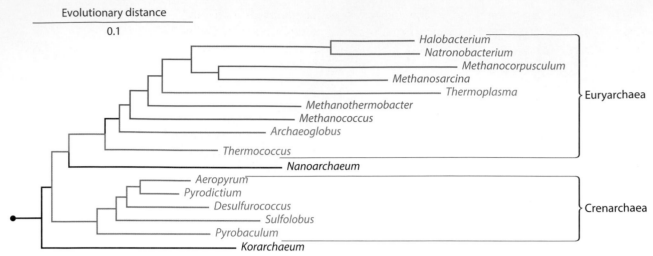

**Figure 15.1** Phylogenetic tree of the *Archaea*. Extremely halophilic members are highlighted in red, methanogenic members are in blue, and sulfur metabolizers are in green. *Nanoarchaeum* is a parasite, and *Korarchaeum* has not been grown in pure culture. doi:10.1128/9781555818517.ch15.f15.1

## Morphology

Phenotypically, archaea are much like bacteria. Most are small (0.5- to 5-μm) rods, cocci, spirilla, and filaments. Archaea most often reproduce by fission, like most bacteria and most unicellular eukaryotes, and reproduction is asexual.

## Cell envelope

One of the unique features of archaea is their membrane lipids, which are quite different from those of either bacteria or eukaryotes. They are ether-linked (not ester-linked) glycerol derivatives of 20- or 40-carbon branched (isoprenyl) lipids rather than fatty acids. Unsaturations in the lipid chain are generally conjugated (those of bacteria and eukaryotes are unconjugated). Forty-carbon lipids are ether-linked to glycerol at both ends, and if these glycerol moieties are on opposite sides of the membrane they form lipid monolayers rather than bilayers. These lipids can be used as chemical signatures for the presence of archaea in a sample.

Archaea have a cytoplasmic membrane only; they do not have an outer membrane. They have a wide variety of cell walls, but none contain peptidoglycan, the signature of bacterial cell walls. Protein or glycoprotein S-layers are common, as are cell walls containing pseudomurein, which is chemically related to peptidoglycan in that it is a fabric of linear disaccharide polymers cross-linked with oligopeptides.

## Flagella

Although the flagella of archaea (those that have them) resemble bacterial flagella superficially, they are fundamentally different structures. Like bacterial flagella, archaeal flagella are composed of a motor embedded in the cell envelope and a long, semi-flexible helical filament that is turned by the motor and drives the cell. The similarity ends there. The archaeal flagellar motor proteins and structure are related to type II secretion systems and type IV pili, whereas bacterial flagella are related to type III secretion systems. The archaeal flagellar motor is driven directly by adenosine triphosphate (ATP) hydrolysis rather than the flow of protons across the cell membrane; it is a chemical motor rather than an electric motor. Archaeal flagellar filaments are much thinner than those of bacteria and are not hollow; protein subunits are added to the base to lengthen the filament, rather than to the tip as in bacteria. Clearly these structures are analogous rather than homologous; they evolved from independent origins and are a classic example of evolutionary convergence.

## Transcription and translation

The transcriptional/translational machinery in archaea is fundamentally like those of both bacteria and eukaryotes. As in bacteria, genes are arranged in cotranscribed clusters (operons). Ribosomes recognize translational start sites and bind to the messenger RNAs (mRNAs) directly at Shine-Dalgarno sequences as in bacteria. Also as in bacteria, transcription and translation are linked; that is, they occur simultaneously.

However, in many ways transcription and translation in archaea are like those in eukaryotes. Although each promoter drives the expression of an entire operon, the promoters are very much like eukaryotic RNA polymerase II promoters and, as in eukaryotes, are binding sites for transcription factors rather than the RNA polymerase itself, as in bacteria. Archaea have a single RNA polymerase (like bacteria), but this RNA polymerase is essentially a eukaryotic RNA polymerase II and requires the same general transcription factors for promoter recognition. Translation is initiated with methionine (as in eukaryotes) rather than formylmethionine (as in bacteria). Translation is inhibited by diphtheria toxin, as are eukaryotic ribosomes, but is not inhibited by most bacterial-translation-inhibiting antibiotics. Chimeric archaeal/eukaryotic ribosomes are functional; neither bacterial/archaeal nor bacterial/eukaryotic ribosomes are functional.

## Genomes

The genomes of archaea are generally ca. 2 to 4 Mbp, usually in one main circular DNA molecule, as in most bacteria. However, as in eukaryotes, archaea have abundant histone-like proteins and the DNA is packaged in the form of nucleosome-like particles. Hyperthermophilic species contain a reverse gyrase that positively supercoils this genomic DNA, rendering it more resistant to thermal denaturation.

In the same way that transcription and translation processes are generally like those of bacteria although the machinery that carries these out is essentially like those of eukaryotes, so too the genome replication process is similar in archaea and bacteria, with one (or a small number of) replication origin and a replication terminus, but this process is carried out by replication machinery that closely resembles the eukaryotic replisome. In addition, chromosome replication and segregation/mitosis seem to be distinct processes in the cell cycle of archaea, as in eukaryotes.

# Phylum *Crenarchaeota*

| Phylum | *Crenarchaeota* | | | |
|---|---|---|---|---|
| Class | | *Thermoprotei* | | |
| Order | | | *Thermoproteales* | |
| Family | | | | *Thermoproteaceae* (e.g., *Thermoproteus*, *Pyrobaculum*) |
| Family | | | | *Thermofilaceae* (*Thermofilum*) |
| Order | | | *Caldisphaerales* | |
| Family | | | | *Caldisphaeraceae* (*Caldisphaera*) |
| Order | | | *Desulfurococcales* | |
| Family | | | | *Desulforococcaceae* (e.g., *Desulfurococcus*, *Aeropyrum*) |
| Family | | | | *Pyrodictiaceae* (*Pyrodictium*, *Pyrolobus*, *Hyperthermus*) |
| Order | | | *Sulfolobales* | |
| Family | | | | *Sulfolobaceae* (e.g., *Sulfolobus*, *Acidianus*, *Desulfurolobus*) |
| Incertae sedis | | | Group I marine *Archaea* (e.g., *Cenarchaeum symbiosum*) | |

A phylogenetic tree of representatives of the phylum *Crenarchaeota* is shown in
Fig. 15.2.

**Figure 15.2** Phylogenetic tree of representatives of the phylum *Crenarchaeota*
doi:10.1128/9781555818517.ch15.f15.2

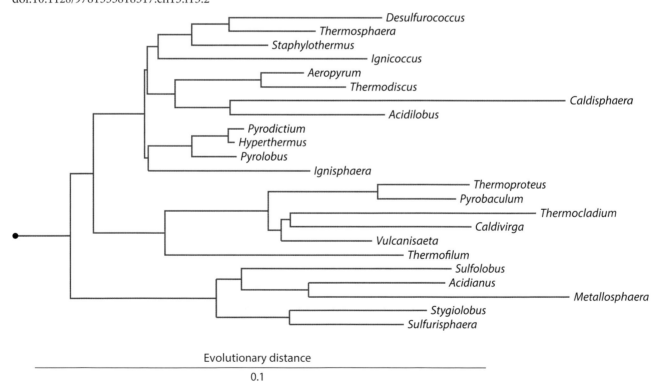

## About this phylum

### Diversity

Cultivated crenarchaea are all relatively closely related and primitive (at least in terms of their SSU rRNA sequences). Crenarchaeal sequences from environmental surveys are much more diverse and are often abundant in non-thermophilic environments from which no cultivated crenarchaea are known. A large phylogenetic group of cultivated crenarchaea known as the "marine group I *Archaea*" was first seen in marine water surveys but has since been found in many other nonthermophilic environments.

### Metabolism

Cultivated crenarchaea generally oxidize and/or reduce sulfur and sulfur compounds (at least facultatively) by one of three biochemical processes:

**1.** Sulfur reduction

$$Sulfur + H_2 \rightarrow H_2S + protons$$

These organisms are autotrophic anaerobes that fix carbon from $CO_2$. Hydrogen is the electron donor for electron transport and elemental sulfur (or sulfur compounds such as thiosulfate) is the terminal electron acceptor.

**2.** Sulfur respiration

$$Sulfur + organics \rightarrow CO_2 + H_2S$$

These organisms are heterotrophic anaerobes. Both carbon and energy are from organic compounds. Organics are the electron donors for electron transport, and sulfur (or sulfur compounds) is the terminal electron acceptor. This process is much like aerobic respiration, except that sulfur compounds take the place of $O_2$. In fact, many organisms that grow by sulfur respiration can also grow by aerobic respiration.

**3.** Sulfur oxidation

$$Sulfur + O_2 \rightarrow H_2SO_4$$

These organisms can usually grow heterotrophically, getting fixed carbon from low concentrations of organics in the medium. Most can also be grown autotrophically, fixing carbon from $CO_2$ via the reverse tricarboxylic acid (TCA) cycle. All are aerobes, of course, since it is the terminal electron acceptor (sulfur is the electron donor) for electron transport.

In most cases, reduced sulfur compounds such as thiosulfate or sulfite are also usable in place of elemental sulfur. Cultivated crenarchaea are all thermophilic, and most are extremely thermophilic, with optimal growth temperatures above 80°C. As a group, these are the most thermophilic organisms known. Many are also acidophilic and autotrophic. Because this phenotype is shared by the most primitive and deepest branches of the euryarchaea, it is probably the primitive phenotype of the archaea.

At least some of the "marine group I" crenarchaea are anaerobic ammonia oxidizers.

## Morphology

Cellular morphology of crenarchaea generally follows the phylogenetic subgroups: most *Thermoproteales* are rod shaped or filamentous, most *Desulfurococcales* are flattened ovoids, and most *Sulfolobales* are irregular cocci.

## Habitat

Cultivated crenarchaea are common in hydrothermal environments, especially acidic hot springs, solfataras, and marine hydrothermal vents (Fig. 15.3). Uncultivated crenarchaea are apparently very common in some non-thermophilic environments, including ocean water, soil, and cave rock surfaces. One uncultivated species, *Cenaerchaeum symbiosum*, is a symbiont of a marine sponge, but no other crenarchaeal symbionts or parasites of plants or animals are known.

**Figure 15.3** "The Pit" (an informal name), an acidic hot spring in the Mud Volcano area of Yellowstone National Park, adjacent to Obsidian Pool. doi:10.1128/9781555818517.ch15.f15.3

### Thermoproteus tenax

*Thermoproteus tenax* (Fig. 15.4), like other *Thermoproteales*, is a strict anaerobe that grows best (at least in cultivation) by sulfur respiration. However, in the wild it probably grows via autotrophic sulfur reduction. Therefore, *T. tenax* can either respire heterotrophically or reduce sulfur autotrophically for a living, and the switch between growth modes is a distinct developmental process.

T. tenax is a long rod-shaped organism that reproduces by branch formation; the branch bud forms near the end of a cell and grows into a new individual cell. It is a common solfatara inhabitant, with an optimal growth temperature of 85°C. *T. tenax* is motile, but this is not usually seen microscopically; after all, a microscope slide at room temperature is 65°C below the optimal growth temperature of the organism!

### Pyrodictium occultum

*Pyrodictium occultum* (Fig. 15.5) is a marine organism common in deep-sea hydrothermal vents. It is a flat, irregular coccus (think of the body of a prickly pear cactus) with a network of tubular fibrils that connect the cells. The optimal growth temperature is 105°C, and cultures grow well at temperatures up to 115°C, making it one of the most thermophilic species known. It is also one of the most primitive organisms known; it is a general rule that thermophiles, and especially extreme thermophiles, are primitive, at least in terms of their SSU rRNA sequences. *P. occultum* can grow by either sulfur respiration or sulfur reduction.

### Sulfolobus solfataricus

*Sulfolobus solfataricus* (Fig. 15.6) is a lobed coccus; the lobes seem to be budding scars from reproduction, and often these buds can be almost like appendages that hold the cell to the sulfur granules on which they grow. *S. solfataricus* and its relatives are common, even predominant, organisms of solfataras and boiling mud pots.

S. solfataricus is an obligate aerobe or microaerophile, capable of autotrophic sulfur oxidation, chemolithoheterotrophy (using sulfur oxidation for energy but requiring organic carbon for growth), or heterotrophy by oxidative respiration. It is an obligate thermoacidophile, requiring a pH of ca. 4.5 and temperature of 87°C for optimal growth. Like most acidophiles, it is sensitive to fatty acid toxicity; fatty acids are protonated and so uncharged at these environmental pHs, diffuse readily through the cytoplasmic membrane, then ionize in the more moderate cytoplasmic pH. This results in an uncoupling of the proton gradient and acidification of the cytoplasm.

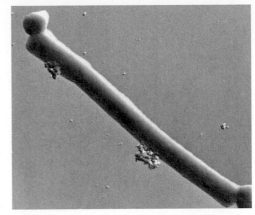

**Figure 15.4** Shadow-cast electron micrograph of *Thermoproteus tenax*. (Courtesy of Reinhard Rachel, Harald Huber, Reinhard Wirth, and Michael Thomm, University of Regensburg.)
doi:10.1128/9781555818517.ch15.f15.4

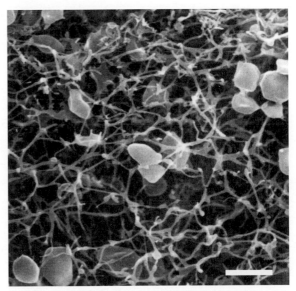

**Figure 15.5** Scanning electron micrograph of *Pyrodictium occultum*. (Courtesy of Gertraud Rieger and Reinhard Rachel, University of Regensburg, and René Herrmann, ETH Zürich.)
doi:10.1128/9781555818517.ch15.f15.5

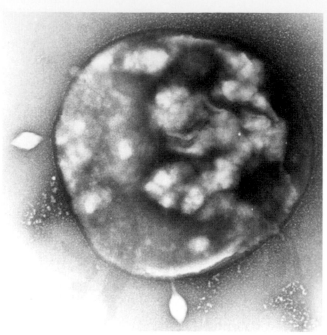

**Figure 15.6** Negative-stain electron micrograph of *Sulfolobus solfataricus* infected with the virus STSV1. (Source: Wikimedia Commons.) doi:10.1128/9781555818517.ch15.f15.6

A phylogenetic tree of representative euryarchaea is shown in Fig. 15.7.

**Figure 15.7** Phylogenetic tree of representatives of the phylum *Euryarchaeota*. Extremely halophilic members are highlighted in red, methanogenic members are in blue, and sulfur metabolizers are in green. doi:10.1128/9781555818517.ch15.f15.7

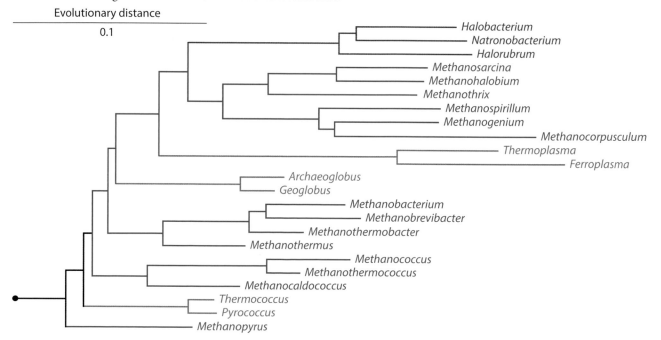

# Phylum *Euryarchaeota*

| Phylum | *Euryarchaeota* | | | |
|---|---|---|---|---|
| Class | | *Methanobacteria* | | |
| Order | | | *Methanobacteriales* | |
| Family | | | | *Methanobacteriaceae* (e.g., *Methanothermobacter*) |
| Family | | | | *Methanothermaceae* (*Methanthermus*) |
| Class | | *Methanococci* | | |
| Order | | | *Methanococcales* | |
| Family | | | | *Methanococcaceae* (e.g., *Methanococcus*) |
| Family | | | | *Methanocaldococcaceae* (e.g., *Methanocaldococcus*) |
| Class | | *Methanomicrobia* | | |
| Order | | | *Methanomicrobiales* | |
| Family | | | | *Methanomicrobiaceae* (e.g., *Methanogenium*) |
| Family | | | | *Methanocorpusculaceae* (e.g., *Methanocorpusculum*) |
| Family | | | | *Methanospirillaceae* (*Methanospirillum*) |
| Order | | | *Methanosarcinales* | |
| Family | | | | *Methanosarcinaceae* (e.g., *Methanosarcina, Methanolobus*) |
| Family | | | | *Methanosaetaceae* (*Methanosaeta, Methanothrix*) |
| Family | | | | *Methanocalculus* (*Methanocalculus*) |
| Class | | *Methanopyri* | | |
| Order | | | *Methanopyrales* | |
| Family | | | | *Methanopyraceae* (*Methanopyrus*) |
| Class | | *Halobacteria* | | |
| Order | | | *Halobacteriales* | |
| Family | | | | *Halobacteriaceae* (e.g., *Halobacterium, Natronobacterium*) |
| Class | | *Thermoplasmata* | | |
| Order | | | *Thermoplasmatales* | |
| Family | | | | *Thermoplasmataceae* (*Thermoplasma*) |
| Family | | | | *Picrophilaceae* (*Picrophilus*) |
| Family | | | | *Ferroplasmataceae* (*Ferroplasma*) |
| Class | | *Thermococci* | | |
| Order | | | *Thermococcales* | |
| Family | | | | *Thermococcaceae* (e.g., *Thermococcus, Pyrococcus*) |
| Class | | *Archaeoglobi* | | |
| Order | | | *Archaeoglobales* | |
| Family | | | | *Archaeoglobaceae* (e.g., *Archaeoglobus, Geoglobus*) |

## About this phylum

### Diversity

The euryarchaea are more diverse both phylogenetically (as measured by SSU rRNA sequence) and phenotypically than are the crenarchaea. The primitive phenotype of the euryarchaea seems to have been sulfur-metabolizing thermophily, and because this is also the general phenotype of the crenarchaea, this is the most likely phenotype of the ancestral archaeon. However, methanogenesis arose early in the evolution of the euryarchaea and is the predominant phenotype of this phylum, at least among cultivated species. Two groups with methanogenic ancestry have reverted to sulfur-metabolizing thermophily (defined broadly): the *Thermoplasmata*, relatives of the *Methanomicrobia* and *Halobacteria*, and the *Archaeoglobi*, perhaps related to the *Methanococci*, although their placement in the tree is not entirely clear. The extreme halophiles are a third group with methanogenic ancestry and are, like the *Thermoplasmata*, specifically related to the *Methanomicrobia*.

The placement of *Methanopyrus* as a very primitive branch at the base of the euryarchaea in SSU rRNA-based trees seems to be a long-branch attraction artifact; most trees based on non-rRNA sequences indicate that *Methanopyrus* is a relative of the *Methanobacteria*.

### Methanogens

Other than the fact that they all make a living in the same general way (by methanogenesis), methanogens are a diverse phenotypic and ecological group. Methanogens fall into three main classes: *Methanococci*, *Methanobacteria*, and *Methanomicrobia*. They are chemolithotrophic, generating energy by the production of methane, and often autotrophic. All are extreme anaerobes, much more sensitive to oxygen or oxidizing environments than are most other anaerobes, and so cultivation requires specialized equipment and techniques. Many methanogens are thermophilic, but this is not a general property of these organisms; mesophily is also common.

### Metabolism

Methanogens obtain energy by the reduction of one- or two-carbon compounds to methane (Fig. 15.8). $C_1$ compounds (formate, $CO_2$, CO) are reduced using hydrogen as the electron donor and covalently attached to methanofuran in the form of a formyl group. This formyl group is transferred to tetrahydromethanopterin ($H_4MPT$) and sequentially reduced through methenyl and methylene to methyl, which is transferred to coenzyme M and finally reduced to free methane. The source of reducing power in these steps is hydrogen; the $F_{420}$ hydrogenase transfers the electrons from hydrogen to cofactor $F_{420}$, which supplies the electrons for $C_1$-$H_4MPT$ reduction. In the final step of methanogenesis, electrons from $H_2$ are transferred to $F_{430}$ rather than $F_{420}$ by a specific $F_{430}$ hydrogenase. The protons released from hydrogen are released externally from the membrane hydrogenase, and protons required (in addition to electrons) during the $C_1$ reduction are taken from the cytoplasm; the result is a proton gradient

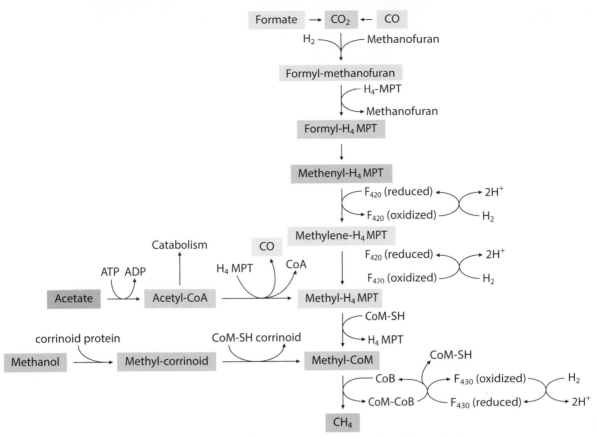

**Figure 15.8** Methanogenesis from $C_1$ compounds (top), or acetate or methanol (lower left). MPT, methanopterin; $F_{420}$, coenzyme $F_{420}$ (8-hydroxy-5-deazaflavin); CoA, coenzyme A; CoM, coenzyme M (2-mercaptoethanesulfonate); $F_{430}$, coenzyme $F_{430}$ (a Ni-tetrapyrrole); CoB, coenzyme B (7-mercaptoheptanoylthreonine phosphate).
doi:10.1128/9781555818517.ch15.f15.8

that can be used to generate ATP via a traditional adenosine triphosphatase (ATPase).

Some methanogens, the *Methanomicrobia*, can make methane from acetate and/or methanol rather than $CO_2$, CO, or formate; in these cases, only the ultimate or penultimate steps of methanogenesis are used. In methanogenesis from acetate, the methyl group of acetate is transferred directly to $H_4MPT$—the carboxy group is released as $CO_2$, which in some methanogens can be fed into the methanogenesis pathway as well. In methanogenesis from methanol, the methyl group of methanol is transferred to a corrinoid protein and then to the final methyl carrier, coenzyme M, before release of methane. Other methyl-containing substrates for methanogenesis (e.g., methylamines) can be utilized by some methanogens by transfer of their methyl groups to the corrinoid protein.

Most methanogens are autotrophs and use the methanogenic pathway for carbon fixation as well as energy production (notice the similarity between methanogenesis and the reductive acetyl coenzyme A (acetyl-CoA) pathway discussed at the end of chapter 9). Methyl-$H_4$MPT is carboxylated and transferred to CoA to produce acetyl-CoA; this is the reverse of the reaction otherwise used by acetoclastic (acetate-utilizing) methanogens to make methane from acetate. Acetyl-CoA can then be fed into catabolism via the "intermediate" reaction or TCA cycle.

## Morphology

Members of the *Methanococci* and *Methanobacteria* are, as might be expected, cocci and rod shaped, respectively. The *Methanomicrobia* are more diverse in morphology; usually the genus name is a good indication of their general morphology.

## Habitat

The enzymes in the methanogenic pathway are extremely oxygen sensitive, and so all methanogens are extreme anaerobes. However, they are common organisms, found in all types of anaerobic environments, and are certainly the most prevalent cultivable archaea in the "moderate" world. For example:

- Sediments and soils—swamp gas *is* methane that, because of its low ignition temperature and threshold concentration, is readily ignited and glows very faintly as will-o'-the-wisps visible at night in swamps. Methanogens are also crucial components of the microbial populations of the rhizosphere (the plant root environment).
- Animal gastrointestinal tracts—especially wood-eating insects and ruminants, but most other animals as well. African termite mounds are scrupulously aerated by the insects, not just for oxygenation but also to keep methane concentrations below ignitable levels. Cows may produce enough methane to be a significant source of this potent greenhouse gas.
- Wastewater and landfills—the wastewater process converts organics in the wastewater into methane and $CO_2$. Landfills must be carefully vented; houses near older unvented landfills have exploded because of the buildup of methane that seeped through the ground into their basements. Alternatively, methane can be collected from wastewater or landfill facilities and used for energy production.
- Oil deposits—natural gas *is* methane, and at least some natural gas is produced not geochemically but by methanogens living in subterranean oil deposits.

Methanogens form a variety of symbioses with plants, animals, and protists, but despite these close associations there are no known pathogenic methanogens. None of the other archaea are pathogens either, but considering the conditions under which most of them grow, this is perhaps not surprising. Methanogens also form close syntrophic associations with heterotrophic bacteria that generate hydrogen (i.e., use protons as the terminal electron acceptor). Hydrogen-generating heterotrophism is only energetically favorable where the ambient concentration of hydrogen is kept extremely low. Methanogens associate with these organisms, utilizing the hydrogen and $CO_2$ they generate for methanogenesis, and keep the hydrogen concentration low enough for the heterotrophs to make a living. Neither of these organisms could persist in the environment alone, but together they are successful.

## Methanococci

The methanococci are typically found in marine and freshwater sediments. Some species are thermophiles, but many mesophilic species have been isolated as well. They are motile via tufted flagella, but they are so sensitive to oxygen that motility can be detected only if samples are taken and observed microscopically in a strictly anaerobic environment. Their cell walls are made up of an extracellular protein S-layer. The methanococci can only grow on $H_2$ + $CO_2$ or formate for energy, although some can use organics to avoid the need for carbon fixation. Some are complete prototrophs: they can make everything they need from inorganic compounds. They can fix their own carbon from $CO_2$, synthesize all their vitamins, fix their own nitrogen from $N_2$, and fix sulfur from $H_2S$. These organisms are more sensitive to oxygen than are any other cultivated species, and they are also very sensitive to ultraviolet (UV) light, because they lack the enzymes required for repair of UV-damaged DNA (photolyase).

### EXAMPLE *Methanocaldococcus jannaschii*

*Methanocaldococcus* (previously *Methanococcus jannaschii* [Fig. 15.9]) is a motile coccus with a single "tuft" of many flagella. It is an obligate autotroph (using the Calvin cycle for carbon fixation), reducing only $CO_2$ or CO with $H_2$ to produce methane. *M. jannaschii* is an extreme thermophile, growing optimally at 80°C; it was isolated from a deep-sea hydrothermal vent environment. The

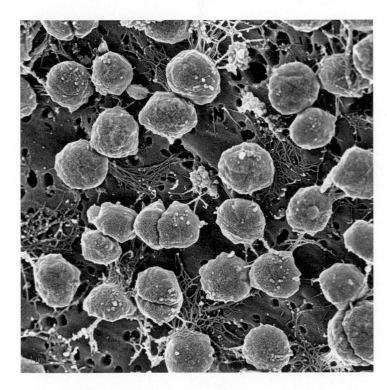

**Figure 15.9** Scanning electron micrograph of *Methanocaldococcus jannaschii*. (Source: Electron Microscope Laboratory, University of California, Berkeley.) doi:10.1128/9781555818517.ch15.f15.9

sequence of the genome of *M. jannaschii* was the first archaeal genome sequence available. It consists of three circular chromosomes: one large (ca. 2-Mbp) chromosome containing all of the identifiable genes, and two small (138- and 38-kbp) chromosomes that could also be considered single-copy plasmids.

## Methanobacteria

The methanobacteria are nonmotile rod-shaped or filamentous organisms with pseudomurein cell walls. They can only use $H_2 + CO_2$ (sometimes CO and/or formate) to make energy, and they fix carbon using the acetyl-CoA pathway (see chapter 9). They are mostly thermophiles and are more easily isolated than other methanogens because they are more resistant to transient exposure to oxygen. Mesophilic or moderately thermophilic species are common colon and rumen inhabitants in animals.

EXAMPLE *Methanothermobacter thermautotrophicus*

*Methanothermobacter thermautotrophicus* (previously *Methanobacterium thermoautotrophicum* strain ΔH [Fig. 15.10]) was isolated from municipal wastewater sludge and is one of the best-studied methanogens. Cells are rod shaped, ca. 0.5 by 3 to 7 μm, that form chains or filaments. *M. thermautotrophicus* is moderately thermophilic, growing optimally at 65°C, and requires no growth factors.

## Methanomicrobia

Members of the methanomicrobia are usually nonmotile and have various shapes: rods, cocci, spirals, filaments, and pleomorphs. Most can use only $H_2 + CO_2$ or formate for energy, but some are capable of methanogenesis using

**Figure 15.10** Scanning electron micrograph of *Methanothermobacter thermautotrophicus*. (Courtesy of Gerhard Wanner, Biocentre, University of Munich, Planegg, Germany.) doi:10.1128/9781555818517.ch15.f15.10

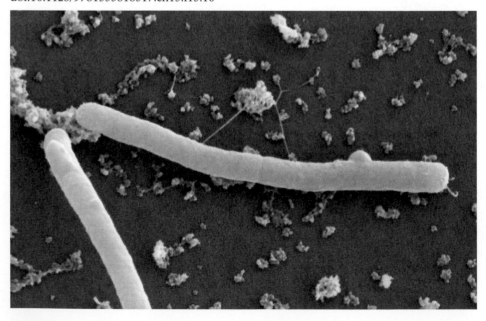

acetate, methanol, or methylamines. Although they are generally autotrophic, some species cannot fix their own carbon; they require acetate for growth. In these instances, they do not make methane from acetate but use it only as a carbon source. Unlike the methanococci and methanobacteria, this group is mostly mesophilic. These organisms are specifically related to both *Thermoplasma* and the extreme halophiles.

**EXAMPLE** *Methanosarcina barkeri*

A particularly important genus in this group is *Methanosarcina* (Fig. 15.11), exemplified by *Methanosarcina barkeri*. These organisms are unique in that they can make methane from hydrogen and $CO_2$ or CO, like other methanogens, or from acetate, methylamines, or methanol alone. *M. barkeri* is common in soils, sediment, swamps, and wastewater treatment sludge. In fact, it is the organism responsible for the success of wastewater treatment. In this process, organic material is concentrated and converted to biomass during aerobic digestion and then converted to acetate (and some $CO_2$) by bacteria during anaerobic digestion. *M. barkeri* converts this acetate to $CO_2$ and methane, which bubbles away. Because *M. barkeri* gets so little energy from this abbreviated form of methanogenesis, it turns vast quantities of acetate into methane, generating only a little biomass in the process. This is, after all, the point of wastewater treatment: to convert as much of the organic carbon into gas as possible.

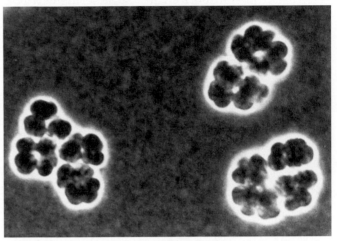

**Figure 15.11** Phase-contrast image of *Methanosarcina barkeri*-like organisms in an enrichment culture from rice paddy sediment/soil. (Reprinted with permission from Großkopf R, Janssen PH, Liesack W, *Appl Environ Microbiol* **64:**960–969, 1998.)
doi:10.1128/9781555818517.ch15.f15.11

## Extreme Halophiles

The extremely halophilic archaea are mesophilic organisms that require at least 2 M NaCl or equivalent ionic strength for growth. Most grow in saturated or near-saturated brines. They are the primary inhabitants of salt and soda lakes. Red pigments make it obvious when large numbers of these organisms are present; blooms often occur after rain carries organic material into a salt lake, and the Red Sea gets its name from such blooms. So does the well-known "red herring," from foul-smelling but hound-diverting salted fish spoiled by *Halobacterium*.

Other halophilic organisms (e.g., some fungi and brine shrimp) have essentially normal cytoplasmic salt concentrations, expend energy to continuously pump salt out of and water into the cell, and contain organic osmolytes such as glycerol or sugars. Halophilic archaea generally grow at even higher salt concentrations but do not fight back at all; the internal salt concentrations are as high as or higher than they are outside! However, $Na^+$ and $Cl^-$ are not particularly biologically friendly, so the extremely halophilic archaea substitute $K^+$ for $Na^+$ (by active transport) and organic acids (e.g., glutamate) for $Cl^-$ in the cytoplasm.

## Metabolism

Halophilic archaea are generally facultative phototrophs. Under aerobic conditions, they are traditional heterotrophs, using organic material from the environment for both carbon and energy. Energy production is respiratory, using oxygen as the terminal electron acceptor. Under anaerobic conditions, they grow photoheterotrophically, using light for energy but requiring organic material for carbon.

Halophilic archaea do not contain the usual photosystems, nor do they use their electron transport chain for gathering energy from light, as do familiar phototrophs. Phototrophy is driven by a single protein, bacteriorhodopsin, which is a light-driven proton pump. This proton pump generates a proton gradient used to make ATP via ATPase, as in other organisms. It is not as efficient as the bacterial photosystems, but light is rarely limiting for growth in the desert salt lakes where these organisms predominate.

Some halophiles grow at high pH (up to pH 10 to 10.5), e.g., *Natronobacterium* in soda lakes. This is a problem for them (or at least for us, trying to understand how they get away with it). At high pHs, protons pumped to the outside by either electron transport or bacteriorhodopsin react with hydroxide in the environment very quickly. Protons are present at too low a concentration outside the cell to drive ATPase. Although the electrical potential is still present, it cannot be harvested by an ATPase unless the enzyme can get protons from the outside, and in any case the cell needs the protons back to maintain the internal pH of the cytoplasm. This problem probably limits the maximum pH compatible with life.

## Morphology

Most halophilic archaea are rod shaped (often irregular) or coccoid. However, there is little net osmotic pressure on the cell wall (high salt concentrations both inside and out), and some species take advantage of this by adopting high-surface-area flattened shapes (disks, squares, or triangles) that are not possible for organisms with "normal" cell turgor. Gas vacuoles are common, and many species are motile.

## Habitat

Extremely halophilic archaea are common in hypersaline seas and lakes, salt evaporation pools, salted meats, salt marshes, and subterranean salt deposits. They can also be found in unexpected low-water-content environments such as soil and sludges.

### EXAMPLE *Halobacterium salinarum*

*Halobacterium salinarum* (this species also includes what was classically known as *H. halobium* [Fig. 15.12]) is a very common extreme halophile, originally isolated from salted cod. The bacteriorhodopsin of this organism, which accumulates to such high concentrations in the cell membrane that it forms a semicrystalline array and is known as a purple membrane, is the model system for the study of these proteins for use as biosensors. *H. salinarum* is motile via a

**Figure 15.12** A bloom of halophilic *Archaea* in a saltern in Namibia. (Courtesy of Alice Lee.) doi:10.1128/9781555818517.ch15.f15.12

single polar flagellum, and in stationary phase it produces gas vacuoles. It is nonfermentative; anaerobic or microaerophilic growth is strictly phototrophic. It requires a minimum of 3 M NaCl for growth, and cultures lyse immediately if exposed to less than 1.5 to 2 M NaCl. Cells are rod shaped, but are often irregular in cultivation or in stationary phase. Most strains of this species contain one large and two small chromosomes, but the smaller "plasmids" are sometimes integrated or rearranged relative to the main chromosome. The genome is extraordinarily rich in transposons and insertion elements, resulting (at least in domesticated strains) in a great deal of genetic instability.

## Sulfur-metabolizing thermophiles

Sulfur-metabolizing members of the *Euryarchaeota* can be separated into two classes: the *Thermococci*, a primitive deep branch in the tree that probably retain this phenotype from the ancestral state shared with the *Crenarchaeota*, and the *Archaeoglobi* and *Thermoplasmata*, which seem to have reverted to sulfur metabolism from a methanogenic ancestry. All are thermophilic, and all are sulfur metabolizers (a *very* general phenotype, to be sure), but they are otherwise not much alike and represent independent branches on the euryarchaeal phylogenetic tree.

## *Pyrococcus furiosus*

*Pyrococcus* and its close relative *Thermococcus* are perhaps the most primitive organisms known; i.e., they are closer to the root of the universal tree than are any other known organisms. *Pyrococcus furiosus* (loosely translated from Latin, this means "great balls of fire") is a neutral-pH heterotroph and is extremely thermophilic, growing optimally at 100°C; as such, it is one of only a few of known organisms that grow at or above the boiling point of water at atmospheric pressure (cultures are kept under more than atmospheric pressure). *P. furiosus* is a heterotroph, growing by anaerobic sulfur respiration using a wide range of peptides, sugars, and polysaccharides. This species is apparently common in deep-sea hydrothermal vent areas (Fig. 15.13) and marine hydrothermal sediments. Cells of all members of this group are motile cocci with a distinct tuft of flagella.

**Figure 15.13** Deep-sea "black smoker" hydrothermal vents, the habitat of *Pyrococcus furiosus*. (Courtesy of Wolfgang Bach.) doi:10.1128/9781555818517.ch15.f15.13

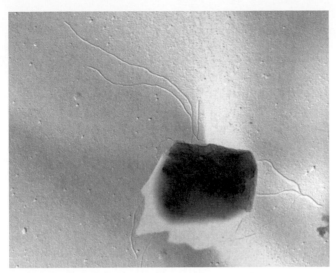

**Figure 15.14** Shadow-cast electron micrograph of *Archaeoglobus fulgidus*. (Courtesy of Reinhard Rachel, Harald Huber, Reinhard Wirth, and Michael Thomm, University of Regensburg.)
doi:10.1128/9781555818517.ch15.f15.14

## *Archaeoglobus fulgidus*

*Archaeoglobus fulgidus* (Fig. 15.14) is an inhabitant of deep-sea hydrothermal vents and heated marine sediments. It is a thermophilic (ca. 85°C) coccus; some species are motile with tufted flagella (much like *Thermococcus* [see above]), and others are nonmotile. *Archaeoglobus* can grow autotrophically by sulfate reduction, using $H_2$ as the electron donor. Carbon fixation is apparently by the reductive (or "reverse") TCA cycle, although the genome contains two very different ribulose bisphosphate carboxylase (RuBPCase) genes (the key enzyme in the Calvin cycle). Alternatively, it can grow heterotrophically from lactate or acetate, plucking the methyl group from these and in essence using the methanogenic pathway in reverse to generate $H_2$ and $CO_2$.

## *Thermoplasma acidophilum*

*Thermoplasma acidophilum* (Fig. 15.15) is a facultatively anaerobic thermoacidophilic heterotroph, using either $O_2$ (aerobically, of course) or sulfur (anaerobically) as the terminal electron acceptor for respiration. *T. acidophilum*, as the name suggests, is also acidophilic, with most isolates growing best at a pH of about 2, but some isolates grow at pHs somewhat below 1. This species is also moderately thermophilic, preferring about 60°C for growth. It is irregular in shape with cytoplasm extensions similar to the pseudopods of amoeboid eukaryotes, but is motile via monotrichous flagella. It lacks a traditional cell wall; cross-linking of the carbohydrate chains of membrane glycoproteins provides cell rigidity and osmotic tolerance. Like *Archaeoglobus*, it reveals

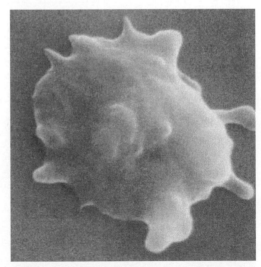

**Figure 15.15** Scanning electron micrograph of *Thermoplasma acidophilum*. (Dennis Searcy)
doi:10.1128/9781555818517.ch15.f15.15

its methanogenic ancestry by containing $F_{420}$ (a major methanogenic hydrogenase cofactor) and other components of the methanogenic pathway, but their function is unknown. *Thermoplasma* has been isolated almost exclusively from smoldering coal refuse piles, and it is presumed that subterranean coal deposits are its natural habitat.

## Phylum *Korarchaeota*

The phylum *Korarchaeota* is known almost exclusively from SSU rRNA sequences from a variety of hydrothermal environments. These organisms are probably best known from the site of their original discovery: Obsidian Pool in Yellowstone National Park (Fig. 15.16). This anaerobic pool varies from 65°C to boiling (94°C at this altitude), has a pH of 6.5, and is a slurry of silica, pyrite, and elemental sulfur. The exact placement of the korarchaeal branch relative to other archaea remains uncertain; it may originate before the split separating *Euryarchaeota* and *Crenarchaeota*, or it may be specifically affiliated with the *Crenarchaeota*. Resolving this issue will probably require additional genome sequences. No members of this group have been grown in pure culture.

**Figure 15.16** Obsidian Pool, Yellowstone National Park, samples of which yielded the first SSU rRNA sequences of *Korarchaea*. doi:10.1128/9781555818517.ch15.f15.16

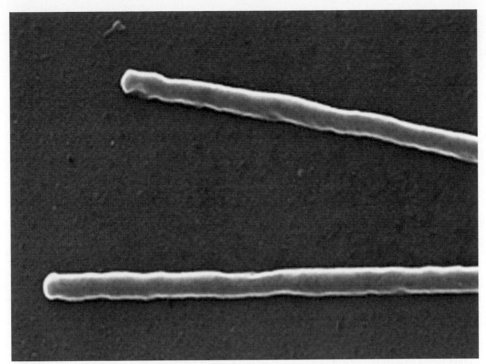

**Figure 15.17** Scanning electron micrograph of purified *Korarchaeum cryptophilum* from enrichment cultures. (Reprinted from Elkins JG, et al, *Proc Natl Acad Sci USA* **105**:8102–8107, 2008, with permission. Copyright 2008 National Academy of Sciences, USA.) doi:10.1128/9781555818517.ch15.f15.17

**EXAMPLE** *Korarchaeum cryptophilum*

*Korarchaeum cryptophilum* (Fig. 15.17) is a very thin filamentous (0.17 by 5 to 100 μm) thermophilic heterotrophic korarchaeote. It has not been grown in pure culture but has been maintained in an 85°C anaerobic community culture originating from a sample from Obsidian Pool. The genome sequence of its single 1.59-Mbp circular chromosome has been determined from cells physically isolated from this culture on the basis of their unusually high resistance to the detergent sodium dodecyl sulfate; this resistance is presumably due to its very dense and orderly S-layer. The composition of its genome suggests that *K. cryptophilum* is a peptidolytic heterotroph, but unlike other peptidolytic hyperthermophilic archaea (e.g., *Pyrococcus*), it seems to use only protons as the terminal electron acceptor (generating $H_2$), lacking the ability to use either oxygen or sulfur (or anything else) for this purpose. The genes required for the biosynthesis of a number of cofactors/vitamins are absent; this may explain the inability of *K. cryptophilum* to grow in pure culture.

## Phylum *Nanoarchaeota*

This phylum, which may be a deep sister-group to the *Euryarchaeota*, is known only from a single cultivated species and a small number of environmental SSU rRNA sequences from thermal and hypersaline environments. Because commonly used primers are not able to amplify SSU rRNA sequences from this group, they may have been missed in surveys of other environments. Therefore, the phylogenetic and phenotypic diversity of this group remains largely unknown.

**EXAMPLE** *Nanoarchaeum equitans*

*Nanoarchaeum equitans* (Fig. 15.18) is an obligate parasite of the crenarchaeote *Ignicoccus*. The cells are very small cocci, only about 400 nm in diameter, that grow attached to the outside of the host cells (which are also cocci). They cannot be cultivated in the absence of the host but have been isolated as a pure coculture from a single *Ignicoccus* cell harboring a single *N. equitans* parasite. *N. equitans* is the only hyperthermophilic symbiont known and is the only archaeal parasite or pathogen. It is also one of the smallest cellular organisms known. The *N. equitans* genome consists of a single circular molecule of slightly less than 0.5 Mbp and lacks almost all biosynthetic genes. Even the ATPase is a minimal version; it may be used in reverse to generate a proton gradient (at the expense of ATP) for use by active-transport pumps. In is probably an energy parasite. Most of the genes that remain are those for information processing: replication, transcription, translation, signal transduction, and the cell cycle.

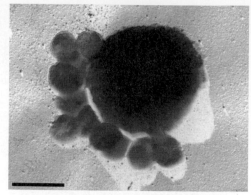

**Figure 15.18** Shadow-cast micrograph of *Nanoarchaeum equitans* (small irregular cocci) attached to its host *Ignicoccus hospitalis* (large irregular coccus). (Courtesy of Reinhard Rachel, Harald Huber, Reinhard Wirth, and Michael Thomm, University of Regensburg.) doi:10.1128/9781555818517.ch15.f15.18

An interesting aspect of the *N. equitans* genome is how disorganized it is. In most bacteria and archaea, genes are organized in operons, with structurally or functionally related proteins generally encoded together. It is not so in *N. equitans*; not even the ribosomal proteins are grouped together in operons (this is otherwise very highly conserved), nor are the rRNAs (which are almost always encoded together in the order 16S-23S-5S in bacteria and archaea, and the homologous 18S-5.8S-28S rRNA genes in eukaryotes). Even the functional domains of some enzymes are encoded separately, and some of the transfer RNAs (tRNAs) are encoded in two pieces that are joined by splicing in *trans*.

# *Archaea* as . . .

## *The "missing link" between* Bacteria *and* Eukarya

Although all species of life on Earth are alike in most ways, bacteria and eukaryotes do differ in significant ways. Traditionally, it is assumed that where bacteria and eukaryotes differ, the bacterial version is primitive, because bacteria are generally simpler than are eukaryotes. But *simple* and *primitive* are not synonymous, and so this is a bad assumption. What is needed is a tie-breaker, an intermediate third distinct phylogenetic group, and the more primitive the better. Such intermediate groups are often called "missing links"; this term is a leftover of the pre-Darwinian "chain of being" view discussed in chapter 2. If a trait is common to two of the three phylogenetic groups, presumably it was also present in the common ancestor of those two groups. Because the last common ancestor is apparently on the branch between the *Bacteria* and *Archaea/Eukarya*, any trait common to bacteria and *either* archaea *or* eukaryotes probably existed in the last common ancestor.

## A deep branch of Eukarya

In order to understand eukaryotic complexity, it would be useful to have a group of primitive organisms that diverged from the rest of the *Eukarya* early in their evolution. Properties shared by such organisms and more complex eukaryotes, but not by bacteria, would presumably represent the unique traits of the deep ancestor of eukaryotes. If these organisms were also relatively simple, they could tell us a lot about the core functionalities of eukaryotic cells, unobscured by all the bells and whistles that were added later in their evolutionary history. Where might we look for such an organism? As you can see from the rooted tree in Figure 7.5 (chapter 7), these organisms are already known: the *Archaea*. For this reason, archaea are often studied for their eukaryote-like processes, which are simpler and easier to understand than the homologous counterparts in plants, animals, and fungi. Examples include RNA polymerase (archaea have a single RNA polymerase homologous to eukaryotic RNA polymerase II), small nucleolar ribonucleoproteins (involved in RNA modifications), and DNA packaging (archaea have relatively simple nucleosomes). In addition, archaeal complexes are often easier systems to study by standard biophysical processes such as X-ray diffraction, because of the extreme stability of thermophilic or halophilic complexes.

## Reflections of early life on Earth

Archaea are, as a group, more primitive than are either bacteria or eukaryotes. Primitive archaea (e.g., *Thermococcus* and *Pyrococcus*) are similar to primitive bacteria (e.g., *Aquifex*) in being thermophilic sulfur/hydrogen metabolizers, and so this is probably the general phenotype of the last common ancestor, and perhaps of primitive life in general. Although modern archaea are not the ancestors of other modern organisms, they probably do resemble in many ways these ancestral life forms. It may be that they have not changed much because of their thermophilic environment; evolutionary drift in genes is much more constrained in thermophiles than in mesophiles, because their macromolecules are less tolerant of minor perturbations that are allowed in molecules that function at lower temperatures. For example, most single changes in an RNA-encoding gene create a mismatch in the secondary structure of the RNA. This is no big deal for a mesophile, which can tolerate the defect with minimal decrease in fitness until a compensatory change occurs. However, the mismatch is a big problem for a thermophile because of the thermal destabilization it causes, making it much less likely that a compensatory change will occur before the decrease in fitness leads to extinction.

## Questions for thought

1. The observations that most of the deepest branches of the "big tree" are thermophiles, and that primitive organisms (also judged by evolutionary distance estimated in SSU rRNA-based trees) are thermophiles, have led to the conclusion that organisms have a thermophilic ancestry. Why do you think thermophilic organisms seem to have slower rates of evolution than mesophiles?

**2.** Conceptually, at least, RNA secondary structures are likely to slow their evolutionary rate as the growth temperature of the organism increases. Are the other components of the cell also likely to slow their evolutionary rates to a similar extent? How would you expect the relative rate of RNA versus protein evolutionary rate to change at high temperatures? What impact does this have on how we interpret "primitive thermophily" in SSU rRNA-based evolutionary trees?

**3.** If archaea are specifically related to eukaryotes to the exclusion of bacteria, why do we not consider archaea to be eukaryotes (even if primitive ones)?

**4.** Acidophiles always have a proton gradient, since the outside is much lower in pH than the cytoplasm. Why, then, can they not make ATP for free? Why do they still have to run electron transport, pumping protons out?

**5.** Given that some species can grow at very low pH, e.g. pH 0, what impact do you think this has on the kinds of oxidation-reduction reactions from which they can derive energy? Remember that the reaction has to drive the electron transport chain to pump protons out against a very steep proton gradient. How might the organism minimize this resistance to outward proton pumping?

**6.** Methanogens and *Archaeoglobus* use the same enzymatic pathway, in opposite directions, to generate energy. Methanogens make methane and fixed carbon from $H_2$ and $CO_2$, whereas *Archaeoglobus* makes $CO_2$ and $H_2$ from organic carbon. How can these enzymes generate energy forwards and backwards? Isn't this like perpetual motion?

**7.** How do you suppose organisms like *Halobacterium* were studied for so long without it being realized that they really were not much like their supposed relative, *Pseudomonas*?

# 16 Eukaryotes

Obviously it is not possible to cover this huge phylogenetic group at an acceptable level of detail in a single chapter in a small book (Fig. 16.1). So this chapter hits only the tips of the icebergs. However, the question "Why cover the eukaryotes at all in a book on *microbial* diversity?" might be asked. The answer is, of course, that most eukaryotes are microbial. By far the greatest diversity of eukaryotes is unicellular; these are usually lumped together as "protists." But like "prokaryotes" or "invertebrates," the term "protist" does not explain what an organism is, only what it is *not*, and so is not a meaningful scientific term. Interestingly, the term "protist" is not usually applied to most phototrophic unicellular eukaryotes, which are typically called algae, nor to unicellular fungi, which are called yeasts.

Nevertheless, most eukaryotes are unicellular microbes, and most green plants, most fungi, and even most animals are microscopic. We refer to the microscopic green plants as green algae, and these are often at least colonial if not clearly multicellular. Fungi are generally microscopic, although the hyphae of an individual "colony" (actually a single organism by most criteria) may span many acres of soil and produce macroscopic fruiting bodies (e.g., mushrooms). Even most animals are microscopic; the world abounds with microscopic nematodes, arthropods, rotifers, tardigrades, sponges, and who knows what else. Many of the cryptic animal phyla you never hear about are microscopic, and even familiar groups are riddled with microscopic members. Our perception of the plant and animal world is skewed by the fact that we largely ignore creatures that exist in a different size range from ourselves.

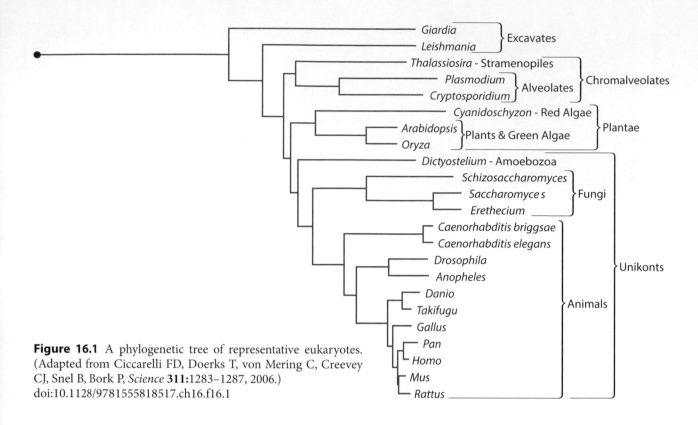

**Figure 16.1** A phylogenetic tree of representative eukaryotes. (Adapted from Ciccarelli FD, Doerks T, von Mering C, Creevey CJ, Snel B, Bork P, *Science* **311**:1283–1287, 2006.) doi:10.1128/9781555818517.ch16.f16.1

Rather than list an unsatisfying taxonomy of eukaryotes, another way to look at the eukaryotic taxonomy is an unrooted tree (Fig. 16.2). Note that this is not a quantitative tree but, rather, just a graph showing general relationships.

# General properties of the eukaryotes

## Diversity

The eukaryotes comprise about five major "superkingdoms": Excavata, Chromalveolata, Plantae, Rhizaria, and Unikonta. How these are related to each other is unclear and is currently the subject of heated debate. Within each of these superkingdoms are groups that are sometimes considered kingdoms, but there is little consistency about how these labels are applied.

As far as the current analysis can tell, early eukaryotic diversification might represent a more or less single radiation. However, rRNA analysis (and other good molecular clocks) suggests that the root of this tree is probably among the Excavata, as shown in Fig. 16.1.

## Metabolism

Eukaryotes are generally heterotrophic or phototrophic; however, in most cases they are supported by organelles derived from endosymbiosis. Phototrophic lifestyles are based on plastids/chloroplasts, which are derived either directly from cyanobacteria or secondarily by endosymbiosis of other eukaryotic phototrophs that, in turn, contain cyanobacterial endosymbionts. Most of the rest

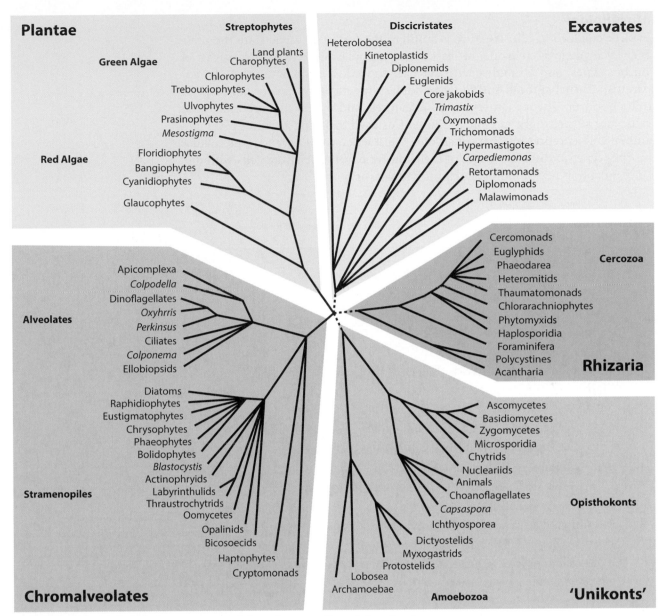

**Figure 16.2** Unrooted phylogenetic tree of the eukaryotes (branch lengths shown are arbitrary), divided into five main "superkingdoms." (Reprinted from Keeling PJ, Burger G, Durnford DG, Lang BF, Lee RW, Pearlman RE, Roger AJ, Gray MW, The Tree of Eukaryotes. *Trends Ecol Evol* **20**:670–676, 2005. Copyright 2005, with permission from Elsevier.) doi:10.1128/9781555818517.ch16.f16.2

of the eukaryotes are aerobic heterotrophs, whose metabolism is based on aerobic respiration in the mitochondria, which are derived from α-proteobacterial endosymbionts. Anaerobic (fermentative) heterotrophs may lack mitochondria (or not use them for oxidative phosphorylation) or contain hydrogenosomes, hydrogen-generating organelles that in some cases are probably derivatives of mitochondria and in others probably represent independent endosymbionts.

## Morphology

Typical eukaryotic cell structure is familiar; the universal aspects are the presence of a nuclear envelope, at least some endoplasmic reticulum, vesicles of various types, and a tubulin/actin cytoskeleton. Most also contain a Golgi apparatus, distinct smooth and rough endoplasmic reticulum, cilia or flagella, and mitochondria. The structures of these features, and others that are more sporadic, are surprisingly variable to those of us used to seeing a typical plant and animal cell as representative eukaryotes. Most eukaryotes have various cellular morphotypes at different stages of their life cycles, or specialized cells with distinct morphotypes in multicellular individuals, or both.

# Unikonta

| Unikonta |
| --- |
| **Amoebozoa** |
| Archamoeba (amitochondrial parasitic amoeba) |
| Lobosea (typical amoeba) |
| Myxogastrids (acellular slime molds) |
| Dictyostelids (cellular slime molds) |
| Protostelids (sometimes called "primitive" slime molds) |
| **Opisthokonta** |
| Fungi |
| Ascomycetes (Sac fungi, mycorrhizal fungi, yeast, fission yeast) |
| Basidiomycetes (mushrooms, rusts, smuts, and *Cryptococcus*) |
| Zygomycetes (common molds) |
| Microsporidia (unicellular intracellular parasites of animals) |
| Chytrids (water molds) |
| Nucleariids (an amoeba with fine, thread-like pseudopods) |
| Mesomycetozoa (nondescript animal parasites) |
| Choanoflagellates (single or colonial flagellates that look like sponge cells) |
| Metazoa (multicellular animals) |

## About the superkingdom Unikonta

The unikonts can be divided into two major subgroups: the amoebozoans (amoebas with large pseudopods, and slime molds) and opisthokonts (fungi, animals, and some unicellular relatives). All have mitochondria with flattened cristae, and they typically have a single flagellum or are amoeboid with no flagella.

## Amoebozoa

The amoeba phenotype has apparently arisen many times in eukaryotic evolution, but this group contains most of those that have robust pseudopods and in which the amoeboid morphotype predominates in their life cycle, i.e., the organisms readily identifiable as "amoebas." Also in this group are the slime molds: both the cellular slime molds (composed of individual amoeboid cells) such as *Dictyostelium* and the acellular slime molds (a syncytial mass; many nuclei in a common cytoplasm) such as *Physarum*.

**EXAMPLE** *Physarum polycephalum*
Also known as the many-headed or yellow slime mold, *Physarum polycephalum* (Fig. 16.3) is a common forest inhabitant that feeds saprophytically on microbes. It is an acellular slime mold with a multinucleate syncytial cytoplasm. Haploid spores germinate to produce swarmer gametes. Two flagellated gametes fuse to form a diploid cell, which transforms into an amoeboid, which grows by enlargement with nuclear division but no cytoplasmic division. Individual plasmodia can become quite large, and because they are brightly colored, they are conspicuous. Upon starvation, the plasmodium produces stalked sporangia, and after meiosis haploid spores are released into the wind. Amazingly, these organisms are capable of learned behavior, including solving simple mazes and anticipating repetitive events.

## Opisthokonts
The opisthokonts include the fungi, animals, and some unicellular relatives of each that are often not considered to be either fungi or animals.

The groups related to fungi include the microsporidia, chytrids, and nucleariids. Microsporidia are spore-forming unicellular intracellular parasites, primarily of insects. Human infections are usually opportunistic. Chytrids are often unicellular, but some are water molds. One species, *Batrachochytrium dendrobatidis*, is responsible for devastating declines in amphibian populations

**Figure 16.3** Photograph of the yellow slime mold *Physarum polycephalum* on the bark of a tree. (Source: Wikimedia Commons. http://commons.wikimedia.org/wiki/File:Slime_Mold_(Fuligo).jpg. http://creativecommons.org/licenses/by/2.0/legalcode.) doi:10.1128/9781555818517.ch16.f16.3

**Figure 16.4** A small (ca. 1-m) great barracuda in Mudjin Harbor, Middle Caicos Island, Turks and Caicos Islands, British West Indies. doi:10.1128/9781555818517.ch16.f16.4

worldwide. Nucleariids are obscure amoeboid soil and aquatic organisms with fine, needle-like pseudopods.

The unicellular relatives of animals are the choanoflagellates and mesomycetozoans. Choanoflagellates are either individual or colonial flagellates, with a single terminal flagellum and a surrounding circle of about three dozen microvilli. Cells are often embedded in an extracellular polysaccharide matrix and/or loose shells of silica. In many ways, the choanoflagellates might resemble primitive sponges. Mesomycetozoans are obscure and nondescriptive fish parasites.

**EXAMPLE SPECIES**

### Sphyraena barracuda

The barracuda (Fig. 16.4) is a large metazoan, reaching a size of up to approximately 2 m. It is heterotrophic, as are all metazoans, feeding carnivorously (mostly as a scavenger) on other teleost fishes. Although it is often considered dangerous, this reputation is undeserved; while barracudas are common and curious, attacks on humans in the water are unsubstantiated. However, humans are often injured by barracudas that are caught by spear or hook and line. Although barracuda is a popular food item in many parts of the world, eating large barracuda can result in ciguatera poisoning.

### Saccharomyces cerevisiae

*Saccharomyces cerevisiae* is a familiar budding yeast (Fig. 16.5), commonly used in baking, brewing, and wine making. Yeasts are saprophytic heterotrophs (as are fungi generally) and can grow either aerobically by respiration or anaerobically by fermentation. Cells in culture can exist as either haploid or diploid or a mixture of the two. Reproduction is by budding, and diploid cells when starved undergo meiosis to produce four haploid spores. Growing haploids can be of either **a** or α mating type, and an **a**/α pair fuses to form a diploid. Although typically unicellular, yeasts can sometimes grow as pseudohyphae.

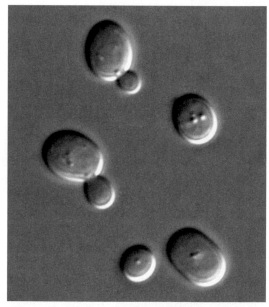

**Figure 16.5** Nomarski interference micrograph of the baker's yeast *Saccharomyces cerevisiae*. (Source: Wikimedia Commons. http://commons .wikimedia.org/wiki/Saccharomyces_cerevisiae.) doi:10.1128/9781555818517.ch16.f16.5

# Plantae

| Archaeoplastida (Plantae) |
| --- |
| Chloroplastida (green algae and plants) |
| Chlorophytes (green algae) |
| Charophytes |
| Plants (streptophytes) |
| Trebouxiophytes |
| Ulvophytes (sea lettuce) |
| Prasinophytes (tiny marine microalgae) |
| *Mesostigma* |
| Rhodophyta (red algae) |
| Floridiophytes |
| Bangiophytes |
| Cyanidiophytes |
| Glaucophyta (with primitive cyanelles) |

## *About the superkingdom Plantae*

This group consists primarily of plants, green algae, and red algae. In this group, but not other groups of eukaryotic phototrophs, the plastids probably originated by direct symbiosis of cyanobacteria; the presence of additional membrane layers around the plastids in other "algae" suggests that these were secondary symbiosis, i.e., symbiosis with a eukaryotic alga rather than a cyanobacterium. The plastids in this group probably have a single common origin; this also is not true of other eukaryotic phototrophs. Most members of this group have cell walls composed of cellulose, store starch, and contain mitochondria with flat cristae but lack centrioles.

## *Chloroplastids*

Most chloroplastids are unicellular, filamentous, or colonial, but a few are multicellular: green plants and sea lettuce. Multicellularity in these two groups arose independently. Chlorophytes (green algae) are mostly aquatic and flagellated, are often colonial, and include the stoneworts and green plants. The other multicellular chloroplastids are the ulvophytes (sea lettuce), common green seaweed. The best known prasinophyte is *Ostreococcus*, an abundant "microalga." *Ostreococcus* is easily mistaken for a bacterium, being only ca. 1 µm in diameter, and contains only a single plastid, a single mitochondrion, and most often a single flagellum. Trebouxiophytes are generally terrestrial algae, often being the algal component of lichens. The most familiar member is *Chlorella*, which is sold as a health food supplement. *Mesostigma* (the scaly green flagellate) is a freshwater genus that was recently shown to be a deep branch within the chloroplastids.

**Figure 16.6** Photograph of common turtle grass in the shallow waters off Bermuda. (Courtesy of Sarah Manuel, Bermuda Department of Conservation Services.)
doi:10.1128/9781555818517.ch16.f16.6

EXAMPLE *Thalassia testinum*

*Thalassia testinum* (turtle grass) (Fig. 16.6) is one of the few truly marine flowering plants (i.e., it is an angiosperm). It is the most abundant sea grass in Florida, the Gulf of Mexico, and the West Indies. It forms dense mats in shallow water, typically from a few feet below low-tide level to 10 m (depending primarily on water clarity and wave surge). It grows very quickly in appropriate conditions, up to an inch per week, and 4 years from seed to seed. It reproduces vegetatively through the rhizomes or sexually via submerged flowers. Turtle grasses are grazed directly by turtles, urchins, and a few fish, but support a complex ecosystem including a wide variety of fish, crustaceans, and molluscs (including the queen conch) that graze the epiphytic film covering the surface of the leaves.

## *Rhodophytes*

Rhodophytes are generally multicellular, abundant, and diverse marine algae found worldwide. Relationships among the red algae are in dispute.

Bangiophytes and floridiophytes probably represent a single group. *Porphyra* is the edible seaweed (nori). Like the chloroplastids, the rhodophytes have an "alternation of generation" reproductive cycle. Cyanidiophytes are obscure unicellular inhabitants of acidic hot springs.

### EXAMPLE *Chondrus crispus*

*Chondrus crispus* (Irish moss, carrageen moss) is a common, small (ca. 2.5-cm) branched red seaweed found in the north Atlantic (Fig. 16.7). It grows attached to rocks in the intertidal zone, and broken fragments commonly wash ashore with the tide. It is a major source of carrageenan (which makes up the bulk of the algal mass), an important food additive. When boiled, Irish moss becomes gelatinous, and when chilled is served as a drink or mixer for "male enhancement." Like other rhodophytes, the life cycle alternates between gametophyte (iridescent in Fig. 16.7) and sporophyte (spotted in Fig. 16.7) stages.

## *Glaucophytes*

Glaucophytes are a small group of freshwater algae that are notable because their plastids are extremely primitive, retaining much more of their cyanobacterial genome and cell structure, including a peptidoglycan cell wall.

**Figure 16.7** Photograph of Irish moss, *Chondrus crispus*, not to be confused with a variety of green terrestrial plants that also grow in Ireland. (Source: Wikimedia Commons. http://commons.wikimedia.org/wiki/File:Chondrus_crispus.jpg. http://creativecommons.org/licenses/by-sa/3.0/legalcode.) doi:10.1128/9781555818517.ch16.f16.7

# Chromalveolata

| Chromalveolata |
| --- |
| **Stramenopiles (heterokonts)** |
| Diatoms |
| Raphidophytes (chloromonads) |
| Eustigmatophytes |
| Chrysophytes (golden algae) |
| Phaeophytes (brown algae and brown seaweeds, kelps) |
| Bolidophytes |
| *Blastocystis* (nonphototrophic parasite commonly mistaken for yeast) |
| Actinophryids (freshwater heliozoans) |
| Labyrinthulids (slime nets) |
| Thraustochytrids (also slime nets) |
| Oomycetes (water molds, animal parasites often mistaken for fungi) |
| Opalinids (obscure animal commensals) |
| Bicosoecids |
| **Alveolata** |
| Apicomplexa (*Plasmodium, Cryptosporidium, Toxoplasma,* and relatives) |
| *Colpodella* |
| Dinoflagellates |
| *Oxyrrhis* |
| *Perkinsus* (agent of Dermo disease of oysters) |
| Ciliates |
| *Colponema* |
| Ellobiopsids |
| **Cryptomonads (extrusome-containing algae)** |
| **Haptophytes (coccolithophore algae and relatives)** |

## About the superkingdom Chromalveolata

Chromalveolates are eukaryotic algae that probably originated by endosymbiosis of a flagellate (a bikont) with a red alga (rhodophyte), although it may also be that different groups within the chromalveolates acquired plastids independently. Regardless, many groups of chromalveolates have lost their plastids and the ability to grow phototrophically. Their mitochondria have tubular cristae.

The stramenopiles (also known as the heterokonts) are an abundant and diverse group of eukaryotes. Most are phototrophic, and those that have lost the ability to grow phototrophically are parasitic.

## Stramenopiles

Most heterokonts are biflagellated at some stage of their life cycles, usually at least as gametes. The two flagella are structurally distinct, with the leading-end flagellum ("tinsel") being branched and the lateral or subapical flagellum being smooth and shorter or even rudimentary. Their plastid envelopes consist of four membrane layers. The innermost two layers are derived from the original cyanobacterial endosymbiont. The next layer is the relic of the cell membrane of the red alga from which the stramenopiles acquired the plastid by secondary endosymbiosis. The outermost layer is actually the host endoplasmic reticulum, inside of which the plastids reside.

Diatoms are the most familiar members of this group and are perhaps the most abundant and diverse as well. By some estimates, they may be responsible for up to half of marine primary production. Most are unicellular. Gametes are flagellated, but diploids are nonmotile or motile by gliding, and are encased in intricate two-part silica (glass) shells. Common in freshwater and marine environments, they produce an exopolysaccharide matrix that, when purified, is used as a nontoxic adhesive (mucilage) for use by children and on old-fashioned lickable postage stamps.

Phaeophytes (brown algae) are common multicellular marine algae. Most are macroscopic; the giant kelp can reach lengths of over 30 m and forms submarine forests. Brown rockweeds are common in the intertidal zone, and sargassum is an abundant free-floating group of species. These "plants" are composed of a root-like holdfast, a stipe (stalk) which may or may not be divided, and fronds (lamina). Gas-filled vesicles in the stipes or fronds are used in many species for buoyancy.

Chrysophytes (golden algae) are generally freshwater unicellular flagellates, with one or two flagella. Some have an amoeboid phase in their life cycle with hairlike pseudopods. Many form a vase-like chitinous shell (lorica), and some of these are colonial. Resting stages have complex silica shells.

Actinophryids are common freshwater heliozoans, amoeboids with many straight, rigid pseudopods supported by internal microtubule structures.

Labyrinthulids and thraustochytrids are primarily marine slime nets. They are unicellular heterotrophs (lacking plastids) that produce a network of polysaccharide filaments, along which they glide. Interestingly, although these filaments are external to the cells, they are nevertheless surrounded by cytoplasmic membrane. They are common symbionts or parasites residing on the surface of marine seagrass.

Oomycetes are heterotrophic (nonphototrophic) and filamentous, and were long thought to be fungi. Most are plant or fish pathogens, but some are water "molds." They cause a variety of wilts and rusts in plants. The best known are the *Phytophthora* species, which cause chestnut ink disease, sudden oak death, and potato blight.

Other stramenopiles are relatively obscure phototrophs or parasites.

## EXAMPLE SPECIES

### *Navicula* spp.

*Navicula* (Fig. 16.8) is a large genus of common diatoms with something like 10,000 phenotypically distinguishable types. Like many diatoms, it has a long slit (raphe) down the long axis of each valve,

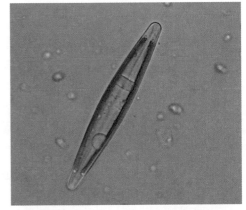

**Figure 16.8** Phase-contrast micrograph of *Navicula* sp. (Source: Texas Tech University Llano River Field Station: http://www.depts.ttu.edu/junction/lrfs/. Copyright 2012 Department of Information Resources, State of Texas. Neither the image nor the information, as it is presented on its website, is endorsed by the State of Texas or any state agency.)
doi:10.1128/9781555818517.ch16.f16.8

from which mucilage is secreted for gliding motility. *Navicula* spp. are bilaterally symmetrical (pennate). Each half of the silica shell has the general shape of a small boat, from which the genus gets its name. The cell contains two large brown-pigmented plastids that lie side by side down the length of the cell.

### *Phytophthora infestans*

*Phytophthora infestans* (Fig. 16.9) is a pathogenic oomycete that causes a range of serious diseases in plants. *P. infestans* infects plants of the family *Solanaceae* (nightshades), including potatoes. It was the cause of the infamous Irish potato blight of the mid-1800s, in which over a million people died of starvation. *Phytophthora* resembles a fungus morphologically, and diseases caused by it resemble fungal wilts and rots. Even today, antifungals are often used, with little effect, to treat *Phytophthora* blights.

## Alveolates

Although the different groups of alveolates are quite distinct, they share a pellicle underlying their cell membrane, composed of many flattened membranous vesicles joined into a tough laminate. Most lack plastids.

Ciliates are the most familiar alveolates. They contain numerous short flagella (cilia) used in locomotion and feeding. Many are quite large, 1 to 2 mm in length, and are anatomically very complex, rivaling in size and mirroring in structure small animals such as rotifers. Ciliates feed by sweeping bacteria and other small organisms into their oral groove into a "gullet" where phagocytosis occurs. When digestion in food vacuoles is complete, the vacuoles fuse with the

**Figure 16.9** Phase-contrast micrograph of *Phytophthora infestans*. The lemon-shaped blebs are zoosporangia. (Courtesy of Kay Yeoman, University of East Anglia.)
doi:10.1128/9781555818517.ch16.f16.9

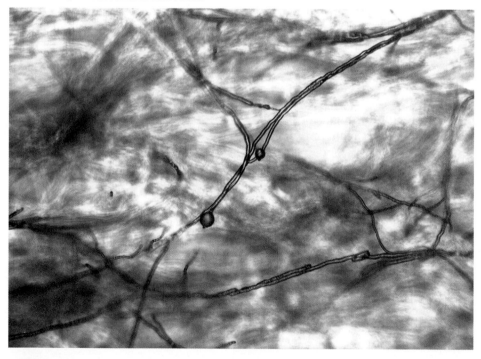

cytoproct (analogous to an anus) to eliminate the waste. Star-shaped contractile vacuoles in freshwater species collect excess water, which is forced out of the cell through pores upon vacuole contraction. Ciliates contain two types of nuclei: a germ line micronucleus containing a pristine diploid copy of the genome, which is apparently transcriptionally inactive, and large active macronuclei. Fission is the most frequent form of reproduction, but cells must periodically undergo sexual reproduction involving meiosis of the micronucleus and exchange of one haploid micronucleus with a partner. The two haploid micronclei fuse in each cell, and new macronuclei are generated from these by mitosis (the old macronuclei are degraded).

Dinoflagellates are common marine flagellates (some are also found in freshwater), often with plastids, although sometimes these are acquired temporarily by partial digestion of food algae (kleptochloroplasts). Blooms of some species along the coast result in red tide. Dinoflagellates have complex life cycles. Alveoli support cellulose plates that cover the cell. Two flagella are present: the coiled posterior flagellum (longitudinal) and the lateral (transverse) flagellum.

Apicomplexa are intracellular parasites of animals with a specialized apical structure (probably a highly modified basal body) used to enter the host cell. They have complex life cycles, often with primary and secondary hosts, and undergo both sexual and asexual reproduction. Although they are not phototrophic, most contain nonpigmented plastids (which are apparently secondary endosymbionts, like those of dinoflagellates) of unknown function. The most familiar (or at least infamous) apicomplexan is *Plasmodium*, the causative agent of malaria, probably the most important infectious disease of humans.

The other alveolates (*Colpodella*, *Oxyrrhis*, *Perkinsus*, *Colponema*, and ellobiopsids) are small, unfamiliar groups.

## EXAMPLE SPECIES

### Vorticella spp.

*Vorticella* spp. (Fig. 16.10) are common freshwater peritrichous ciliates. These are bell shaped, with a stalk (and holdfast) at one end that can contract quickly if disturbed and a ring of cilia and membrane sheets circling the oral cavity. Asexual reproduction occurs by fission; only one cell keeps the stalk, while the other has a ring of cilia at the posterior (stalk) end by which it can swim until attaching to a substrate and developing a stalk. In poor environments, stalked cells can develop posterior cilia, lose their stalk, and swim in search of better conditions.

### Karenia brevis

*Karenia brevis* (formerly *Gymnodinium breve*) is the dinoflagellate that causes Florida red tide (Fig. 16.11). These blooms rarely become dense enough to turn the water red; however, *K. brevis* produces a potent neurotoxin (brevetoxin) that results in serious fish and shellfish kills. Although red tides are a natural phenomenon,

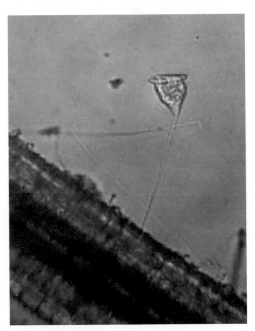

**Figure 16.10** Phase-contrast micrograph of a *Vorticella* sp., the bell-shaped organism attached by a long stalk to what might be a plant root hair at the bottom right. (Courtesy of Simon Andrews, Wikimedia Commons. http://commons.wikimedia.org/wiki/File:Vorticella.JPG. http://creativecommons.org/licenses/by-sa/3.0/legalcode.) doi:10.1128/9781555818517.ch16.f16.10

**Figure 16.11** Aerial photograph of Florida red tide. (Source: P. Schmidt, *Charlotte Sun*, Punta Gorda, FL.) doi:10.1128/9781555818517.ch16.f16.11

they were historically rare and occurred only in the summer; nutrient-rich run-off from human development has resulted in red tides becomes much more frequent and severe, and they now occur year round. Breathing brevetoxin aerosolized in the surf causes "tourist's cough."

## Rhizaria

| Rhizaria |
|---|
| **Cercozoa (amoeba with fine, hairline pseudopods)** |
|     Cercomonads |
|     Euglyphids |
|     Phaeodarea |
|     Heteromitids |
|     Thaumatomonads |
| **Chlorarachniophytes** |
| **Phytomyxids (plasmodiomorphs, plant parasitic slime molds)** |
| **Haplosporidia** |
| **Foraminifera (amoeba with branched fine pseudopods and carbonate shells)** |
| **Radiolaria (amoeba with needle-like pseudopods and silica shells)** |
| **Acantharia (amoeba with needle-like pseudopods and strontium sulfate shells)** |

### About the superkingdom Rhizaria

This superkingdom is a very large, abundant, and diverse group of unicellular eukaryotes. Most are heterotrophic amoeboids, with very thin pseudopods, which are either thread-like (filose) or reticulose (branched or networked) and often with shells. Mitochondria have tubular cristae. Most rhizaria have a distinct division of the cytoplasm into an inner granular "endoplasm" containing the usual organelles (including the nucleus), vesicles, and ribosomes, and a peripheral "ectoplasm" packed with vacuoles and lipid droplets. These apparently function to maintain buoyancy, counteracting the ballast of the mineral shell.

### Foraminifera

Foraminifera (forams) (Fig. 16.12) are common marine amoeboids with carbonate shells punctured by openings for their numerous reticulate pseudopods. The shells (tests) are chambered, with connecting holes. Unlike most rhizaria, pseudopods are used for motility as well as feeding. Most are microscopic (although just barely), but many are several millimeters in diameter. Many contain plastids, which are typically kleptochloroplasts (chloroplasts scavenged from food algae and retained for use in photosynthesis), or endosymbiotic dinoflagellates (zooxanthella).

**Figure 16.12** Ernst Haeckel's classic drawings of foraminiferan tests. (This is one of many pages of such drawings.) doi:10.1128/9781555818517.ch16.f16.12

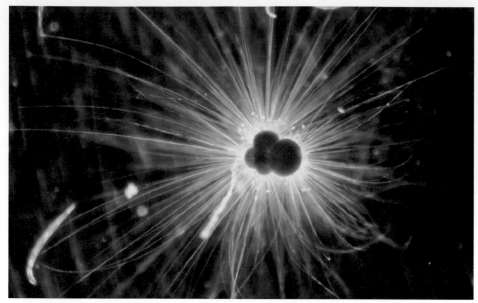

**Figure 16.13** Dark-field micrograph of *Globigerina bulloides*, a living relative of *Rotalipora globotruncanoides*, showing the long pseudopods. (Courtesy of Howard Spero.) doi:10.1128/9781555818517.ch16.f16.13

Because foraminiferan tests fossilize readily and make up a large fraction of ocean sediment and limestones, the fossil record of this group is essentially limitless. Analysis of foraminiferan populations in geologic strata can be used to precisely date limestones and establish detailed histories of ocean depth, temperature, primary production, and climate.

### EXAMPLE *Rotalipora globotruncanoides*

*Rotalipora globotruncanoides* (Fig. 16.13) is an abundant modern planktonic species that first appeared approximately 100 million years ago, and its appearance in the fossil record defines the end of the early Cretaceous and the beginning of the late Cretaceous epoch.

## *Radiolaria*

Radiolarians (Fig. 16.14) are also common marine amoeboids with silica shells (tests). Some are relatively simple spicule-like cones or vases, but most are elaborate, with complex concentric spheres and three-dimensional snowflake-like structures. The test does not usually surround the cell but permeates it; i.e., the inside of the test is embedded in the cytoplasm. Their pseudopods are rigid and needle-like, being supported by microtubule bundles (axopods). The endoplasm is separated from the ectoplasm by a perforated membrane. They usually contain endosymbiotic dinoflagellates (zooxanthella).

Like foraminifera, radiolarian tests fossilize readily and make up a large fraction of marine sediment and limestones, and so the fossil record of this group is essentially limitless. Analysis of radiolarian populations in geologic strata can be used to precisely date limestones and establish detailed histories of ocean depth, temperature, primary production, and climate.

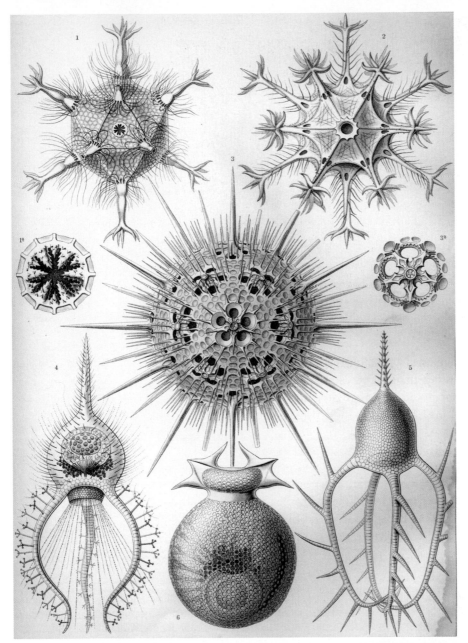

**Figure 16.14** Ernst Haeckel's classic drawings of radiolarian tests. (Again, this is one of many pages of such drawings.) doi:10.1128/9781555818517.ch16.f16.14

**EXAMPLE** *Hexacontium gigantheum*

These are large modern radiolarians of the North Sea, ca. 0.1 mm in diameter (Fig. 16.15). The test is composed of three concentric spheres, connected by 8 to 12 large three-sided spines that originate in the innermost sphere and form radial bars inside the cell. The pores of the outermost (cortical) sphere are surrounded by several small spines (byspines). The thin rigid pseudopods (axiopods) protrude from the pores in the cortical sphere.

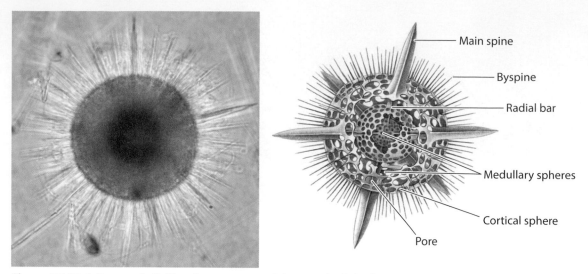

Main spine

Byspine

Radial bar

Medullary spheres

Cortical sphere

Pore

**Figure 16.15** Micrograph (left) and cross-sectional diagram (right) of *Hexacontium gigantheum*. (Courtesy of Jane Dolven.) doi:10.1128/9781555818517.ch16.f16.15

## Cercozoa

Cercozoans are amoebas and flagellates with filose pseudopods that are used for feeding rather than motility. Unflagellated species are motile by gliding. Most have silica or organic plates assembled in a regular pattern to form a simple shell. An opening at the anterior end allows the single thread-like pseudopod to protrude and search for food.

EXAMPLE *Euglypha strigosa*

*Euglypha strigosa* (Fig. 16.16) has flattened oval tests composed of 150 to 300 scale-like plates. Cells are 50 to 100 μm in length. Silica spines project from the test. The oval anterior opening has a ring of specialized plates with tooth-like projections protecting the entrance. *Euglypha* is an inhabitant of rich soil, mosses, and peats.

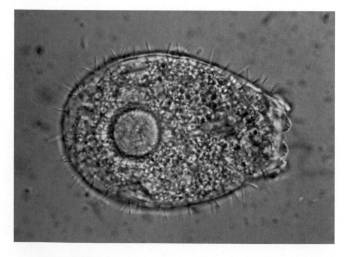

**Figure 16.16** Micrograph of *Euglypha strigosa*. (Courtesy of Eugen Lehle, Wikimedia Commons. http://commons.wikimedia.org/wiki/File:Euglypha.jpg. http://creativecommons.org/licenses/by-sa/2.5/legalcode.) doi:10.1128/9781555818517.ch16.f16.16

# Excavata

| Excavata |
|---|
| **Metamonads** |
| **Fornicata** |
| **Diplomonads (*Giardia* and relatives)** |
| **Retortamonads** |
| *Carpediemonas* |
| **Parabasalians** |
| **Trichomonads** |
| **Hypermastigotes** |
| **Oxymonads (termite gut symbionts)** |
| *Trimastix* |
| **Discobans** |
| **Loukozoans (jakobids, e.g., *Reclinomonas*)** |
| **Heterolobosea (*Naegleria* and relatives)** |
| **Euglenozoa** |
| **Euglenids** |
| **Kinetoplastids (trypanosomes)** |
| **Diplonemids** |
| *Malawimonas* |

## About the superkingdom Excavata

Excavates are flagellates with a well-developed oral groove, the "excavate." They usually lack typical mitochondria. In most cases, they do not entirely lack mitochondria but have organelles (e.g., kinetoplasts, mitosomes, or hydrogenosomes) that probably are very highly specialized derivatives of mitochondria. Those with more or less recognizable mitochondria (which have DNA) are sometimes grouped into the discobans, and those without are grouped into the metamonads. The metamonads were previously thought to lack mitochondria entirely, but mitochondrion-like genes were found in the nuclear genome, and small organelles called mitosomes probably represent the DNA-less relic of mitochondria. On the other hand, the jakobids have primitive mitochondria, retaining many more bacterial features than do other mitochondria.

Excavates may not be a valid phylogenetic group (clade); if the root of the eukaryotic tree were among the excavates, as shown above, then they would be a "grade," like fish or reptiles or "protists." The metamonads are sometimes referred to as the archaeozoans, because they were (or are, depending on the tree) deep branches in the eukaryotic tree, but even if this is the case it is a misnomer, because they are not primitive (short) branches.

## Discobans

Loukozoans (jakobids) are a small group of free-living aquatic biflagellates that feed on bacteria. They have primitive mitochondria (usually only one), which have larger and more complete bacterium-like genomes than do other eukaryotes. Some are free-swimming, while others have an organic (nonmineralized) lorica that covers most of the cell, except for the oral groove, and a stalk for attachment to a surface.

Heterolobosea (percolozoans) have amoeboid, flagellated, and cyst stages in their life cycle and so are also sometimes referred to as schizopyrenids or amoeboflagellates. Common in soil, feces, and freshwater, one species, *Naegleria fowleri*, is an opportunistic pathogen that invades the nervous system via the nasal mucosa and causes necrosis. Patients are usually infected while swimming. Once the central nervous system tissue is infected by these "brain-eating amoebae," mortality is greater than 99%.

Euglenozoans fall mostly into two main phenotypically distinct groups: the euglenids and the kinetoplastids. The free-living euglenids are phagocytic or photosynthetic; chloroplasts, when present, are secondary endosymbionts. They are biflagellated, but often one flagellum is very short or trailing. Often striped, some can move in an "inchworm" fashion in addition to flagellar swimming. Kinetoplastids are generally parasitic and include *Trypanosoma* (the cause of sleeping sickness) and *Leishmania* (the cause of Chagas' disease). Their mitochondrion is a large single organelle containing a compact solid mass of DNA, the kinetoplast. The kinetoplast is located at the anterior end of the cell and is associated with the basal body of the flagella.

## EXAMPLE SPECIES

### *Reclinomonas americana*

*Reclinomonas americana* (Fig. 16.17) is a freshwater jakobid that "reclines" in a wine-glass-shaped lorica (the stem of the lorica points down in Fig. 16.17). One flagellum is free and beats to generate a current, bringing bacteria into the oral

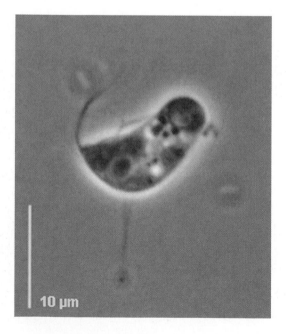

**Figure 16.17** Phase-contrast micrograph of *Reclinomonas americana*. (Image supplied by D. J. Patterson, courtesy of microscope.mbl.edu. http://tolweb.org/Reclinomonas_americana/97410. http://creativecommons.org/licenses/by-sa/2.5/legalcode.) doi:10.1128/9781555818517.ch16.f16.17

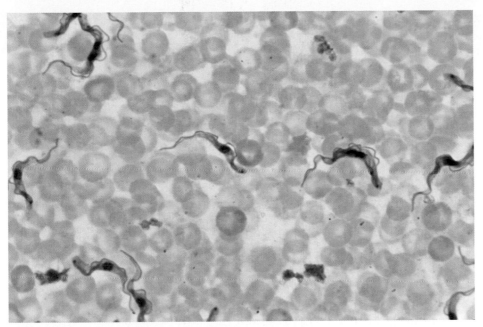

**Figure 16.18** *Trypanosoma brucei* in the blood of a sleeping sickness patient. (Source: Centers for Disease Control and Prevention/Mae Melvin.)
doi:10.1128/9781555818517.ch16.f16.18

groove where the other flagellum is part of an oral vane to trap them. The *R. americana* mitochondrial genome is relatively large (97 kbp), with 97 genes, including many that are not usually seen in mitochondrial genomes, e.g., an RNA polymerase gene (most mitochondria have an RNA polymerase more closely related to phage T3/T3/SP6 RNA polymerases), succinate dehydrogenase genes, *sec* genes, ribosomal protein genes, and a bacterium-type ribonuclease (RNase) P RNA.

### Trypanosoma brucei

*Trypanosoma brucei* (Fig. 16.18) is a euglenozoan and is the cause of African sleeping sickness. It is an obligate parasite with a complex life cycle, including a variety of developmental morphotypes, an insect vector (the tsetse fly, *Glossina* sp.), and a mammalian host (Fig. 16.19). In the insect vector, *T. brucei* is a gut parasite that moves to the salivary gland so that it is injected into the mammalian host when the fly feeds. The parasite lives in the bloodstream of the mammalian host, where it can, in turn, be transmitted to another fly feeding on the infected host.

### Metamonads

Metamonads have a central bundle of microtubules, the axostyle, that run the length of the cell. Undulation of the axostyle seems to be involved in motility, but metamonads are also flagellated.

Fornicata are mostly anaerobic gut symbionts, often parasitic. They lack both typical mitochondria and Golgi complexes, and at least some of them contain

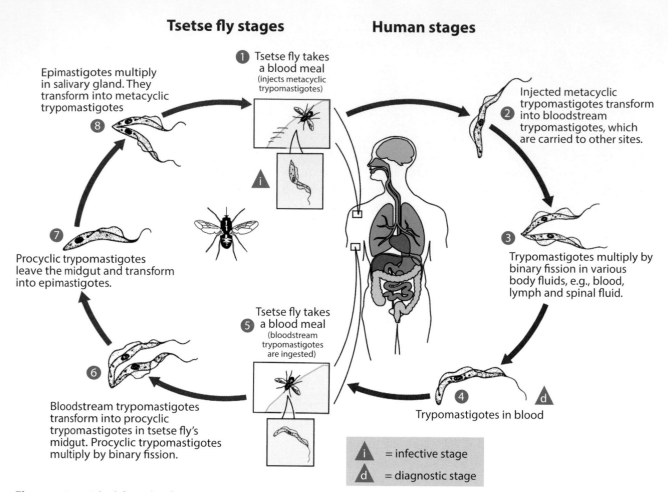

**Tsetse fly stages**

**Human stages**

Epimastigotes multiply in salivary gland. They transform into metacyclic trypomastigotes

8

1 Tsetse fly takes a blood meal (injects metacyclic trypomastigotes)

i

2 Injected metacyclic trypomastigotes transform into bloodstream trypomastigotes, which are carried to other sites.

7

Procyclic trypomastigotes leave the midgut and transform into epimastigotes.

3 Trypomastigotes multiply by binary fission in various body fluids, e.g., blood, lymph and spinal fluid.

5 Tsetse fly takes a blood meal (bloodstream trypomastigotes are ingested)

6

Bloodstream trypomastigotes transform into procyclic trypomastigotes in tsetse fly's midgut. Procyclic trypomastigotes multiply by binary fission.

4 Trypomastigotes in blood

d

i = infective stage

d = diagnostic stage

**Figure 16.19** The life cycle of trypanosomes. (Adapted from the Centers for Disease Control and Prevention.) doi:10.1128/9781555818517.ch16.f16.19

small DNA-less organelles (mitosomes) that may represent the evolutionary relic of mitochondria. The function of mitosomes is unclear. The major groups of fornicates are the diplomonads and the retortamonads; the diplomonads live like "Siamese twins," with two nuclei, each with its associated flagella.

Parabasalians are mostly insect gut symbionts or animal pathogens. Like other metamonads, they lack typical mitochondria, and so are anaerobic. However, they contain hydrogenosomes, small DNA-less organelles that probably represent highly modified mitochondria. They usually have several anterior flagella, the basal bodies of which are connected by fibers (parabasal fibers, hence the name of the group), with obvious Golgi complexes, and the axostyle often protrudes from the posterior end.

Oxymonads are prominent gut symbionts of wood-eating insects that are attached by an anterior stalk to the gut mucosa. They harbor bacterial endo- and ectosymbionts that are involved in lignin and/or cellulose digestion. They lack both mitochondria and Golgi complexes. Their surfaces contain specialized pit structures that serve as receptors for attachment of spirochetes and other bacteria. The relationships between oxymonads and their symbionts and hosts are poorly understood; none have been grown in culture.

## EXAMPLE SPECIES

### *Giardia lamblia*

*Giardia lamblia* (also called *G. intestinalis*) (Fig. 16.20) is a binucleated, flagellated anaerobic gut parasite, the causative agent of beaver fever or traveler's diarrhea. Division occurs by binary fission, creating two uninucleate daughter cells, which quickly undergo mitosis and division of the nucleus to regenerate the normal binucleate structure; mitosis is after cytokinesis, not before. Chromatin is condensed throughout the life cycle. A large ventral disk is used for attachment to the gut mucosa. The function of the paired "medial bodies" (the smile of their "face") is unknown. They contain four pairs of flagella but not mitochondria (although they do have mitosomes), Golgi complexes, endoplasmic reticulum, or lysosomes.

### *Streblomastix strix*

*Streblomastix strix* (Fig. 16.21) is an oxymonad gut symbiont of the damp-wood termite. The anterior end of the cell has four flagella and sometimes an attachment stalk. The remainder of the cell is covered in 100 to 200 very long rod-shaped bacteria, probably of at least three different species. The cell is star shaped in cross section, providing six or seven long lateral vanes, along which the bacteria are attached. Small symbiotic bacteria are found in a central vacuole and in the cytoplasm of the lateral vanes. Mitosomes are apparently absent.

**Figure 16.20** *Giardia lamblia* (a.k.a. *G. intestinalis*) from stool. (Courtesy of Josef Reischig, Wikimedia Commons. http://commons.wikimedia.org/wiki/File:Giardia_intestinalis_ (259_17).jpg. http://creativecommons.org/licenses/by-sa/3.0/legalcode.) doi:10.1128/9781555818517.ch16.f16.20

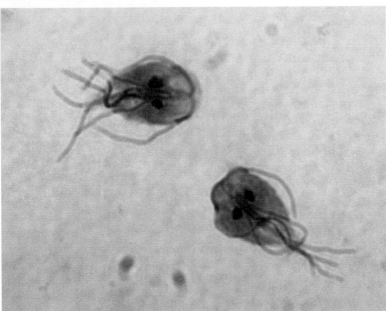

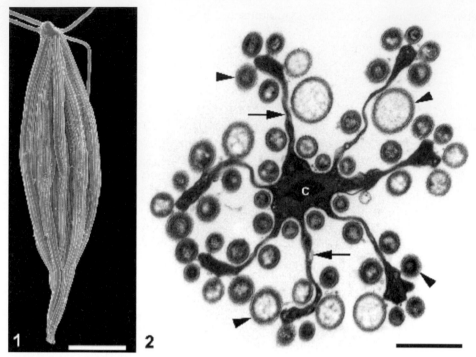

**Figure 16.21** *Streblomastix strix* scanning electron micrograph (1) and cross section (2), showing symbiotic surface bacteria. (Reprinted from Leander BS, Keeling PJ, *J Eukaryot Microbiol* **51:**291–300, 2004, with permission. Copyright 2005, John Wiley & Sons.) doi:10.1128/9781555818517.ch16.f16.21

## Questions for thought

**1.** Think about all the ways in which gene expression is different in bacteria and plants/animals/fungi. How do you think these processes work in other branches of eukaryotes? Why do you think this?

**2.** There has been a lot of uncertainty about whether "algae" represent a single group that acquired chloroplasts (cyanobacteria) once, or whether chloroplasts were acquired by several eukaryotes, each being the origin of one kind of alga. How would you answer this question? What would be the problems with your approach?

**3.** Some unicellular eukaryotes that lack functional mitochondria have other organelles that provide them with better sources of energy metabolism than substrate-level phosphorylation. An example of this is the hydrogenosomes of alveolates, anaerobic ciliates, and even some fungi. These organelles lack DNA. How might you try to determine the origin of these organelles, whether they are related to each other (not just labeled the same), and whether they are descended from mitochondria?

**4.** Given that rRNA trees do not do a reliable job of defining the relationships between these major groups of eukaryotes, or determining where the "root" of the eukaryotic tree is, how might you go about investigating these questions?

# 17

# Viruses and Prions

In this chapter we discuss "acellular life," although some might argue that these things are not really alive. Perhaps this is a case where there is no black-and-white alive-versus-dead distinction but, rather, a gradation where some things might reasonably be considered more alive than others. (The caveat to this would be, How do you measure this? And if you can't, then is it a meaningful scientific distinction?)

## Viruses

Viruses are stretches of DNA or RNA, encapsulated in protein and/or a membrane, that infect a cellular host, merge with its cytoplasm or nucleoplasm, and use the host's cellular machinery for replication (Fig. 17.1). Viruses infect all kinds of living things from all phylogenetic groups.

Rather than attempting to distill an entire virology course into a few pages, this chapter focuses on the evolutionary problem of viruses; in other words, where did they come from?

Since viruses have no ribosomes of their own, they have no ribosomal RNA (rRNA) and cannot be included in the small-subunit-rRNA-based universal tree. However, there are three general views of how viruses may have originated:

1. As genetic offshoots (satellites) of their host's genome
2. As remnants of precellular life
3. As highly reduced (degenerate) remnants of originally cellular parasites

doi:10.1128/9781555818517.ch17

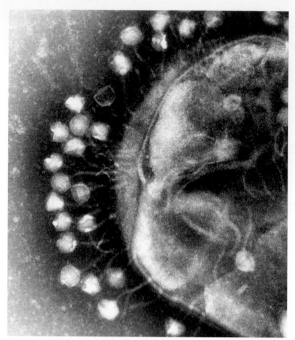

**Figure 17.1** *Escherichia coli* being infected by bacteriophage lambda. (Courtesy of Graham Beards.) doi:10.1128/9781555818517.ch17.f17.1

Keep in mind that viruses are a collection of various very different kinds of "organisms," and there is no reason to believe that viruses are generally related to one another genealogically; different groups of viruses presumably had different origins. Some viruses are probably genetic offshoots of their hosts, some may represent remnants of precellular life, and others may be degenerate cellular parasites. Others are likely to have had other origins that we have not yet considered.

## Viruses as genetic offshoots of their hosts

Although viruses lack rRNAs, it is possible to use other genes present in any particular virus for phylogenetic analysis if they are sufficiently well conserved for good alignment. Many genes are very different in viruses from those in other organisms because of their specialization and rapid evolution, and therefore are not very useful in phylogenetic analysis. Others are acceptable, for example, the tRNA genes found in bacteriophage T4 (and other T-even phages). In most of these cases, the virus is clearly related to the host; i.e., these viruses are almost certainly offshoots of the host's genome. But it is certainly possible that only *some* of the genes are from the host; viruses gather bits of their host DNA all the time. The *core* of the virus may predate the host and may have acquired some genes from the host recently.

Another reason to believe that many (perhaps most) viruses are derived from their host genomes is that they are, without exception, fundamentally dependent on their hosts for cellular processes for replication.

## An evolutionary series between bacterial genomes and bacteriophages

Plasmids are secondary chromosomes. Some "megaplasmids" are not meaningfully distinct from regular chromosomes; the usual distinction is that if a DNA molecule contains essential genes it is considered a chromosome, whereas if it does not it is a plasmid. But as a general rule, plasmids carry important genes, whether they are essential under any specific living conditions or not. Many plasmids (e.g., the "fertility" or "F" plasmid) can be transferred from cell to cell by conjugation, a sort of sexual exchange of DNA not associated directly with reproduction (Fig. 17.2). Some viruses are similar to conjugative plasmids, except that they are transferred via an encapsidated intermediate rather than by direct contact between cells.

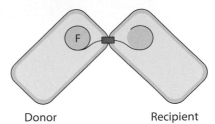

**Figure 17.2** Transfer of F plasmid from donor to recipient. doi:10.1128/9781555818517.ch17.f17.2

These plasmids carry a series of genes that direct replication and conjugative transfer of the plasmid DNA. If the plasmid is replicated to high copy numbers, it can slow the growth of the host; the result is that these viruses/plasmids form cloudy plaques on plates. An example of this is filamentous bacteriophage M13 (and related bacteriophage such as If1 and fd) (Fig. 17.3).

M13, then, is a lot like a conjugative plasmid except that the transfer mechanism does not involve direct donor-recipient contact. This is a fairly clear case of a virus as a genetic offshoot of the host.

Some viruses seem to have derived from transposons (mobile genetic elements). Retroviruses are essentially transmissible retroposons (transposons that encode a reverse transcriptase and transpose via an RNA intermediate, e.g., Ty in yeast, Copia and P-element in *Drosophila*, and human immunodeficiency virus in humans). Bacteriophage Mu (Fig. 17.4) is essentially a transmissible out-of-control transposon (Fig. 17.5).

Some viruses, then, are probably the ultimate "selfish genes."

**Figure 17.3** Transfer of bacteriophage M13 from donor to recipient. doi:10.1128/9781555818517.ch17.f17.3

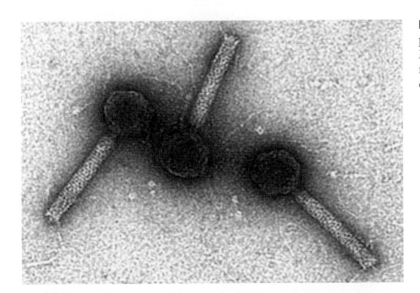

**Figure 17.4** Electron micrograph of bacteriophage Mu. (Reprinted from Inman RB, Schnös M, Howe M, *Virology* **72:**393–401, 1976. Copyright 1976, with permission from Elsevier.) doi:10.1128/9781555818517.ch17.f17.4

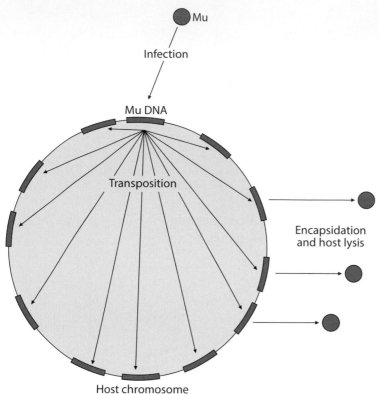

Mu

Infection

Mu DNA

Transposition

Encapsidation
and host lysis

Host chromosome

**Figure 17.5** Very schematic view of the life cycle of bacteriophage/transposon Mu.
doi:10.1128/9781555818517.ch17.f17.5

## *Viruses as remnants of precellular life*

Many believe in an "RNA world" that existed before the evolutionary invention of DNA or protein, or perhaps even lipid membranes. The complexity of this RNA world, presuming it existed at all, is a matter of wildly speculative and largely uninformed debate.

Many RNA viruses (especially virus "satellites," such as plant virusoids, but also hepatitis delta virus) seem similar to what these prebiotic RNAs might have been like, in that they direct their own replication without DNA intermediates and perform at least some of their own replication functions (self-cleavage and ligation of replication forms by ribozymes) (Fig. 17.6 and 17.7). In order to replicate, these viruses simply need their host to provide a primase, an RNA polymerase, nucleoside triphosphates (NTPs), and a decent physical/chemical environment. This is why some people believe that these viruses may have originated deep in time and persist today as remnants of the RNA world.

It is hard to imagine that these RNAs, which are active only in cytoplasm (that of its host), could be directly descended from independent RNAs that lived in the precellular RNA world. If these viruses are remnants of precellular life, they must have originally had mechanisms for replication independent of cytoplasm. It seems unlikely that only parasitic remnants persist, but it is possible that free-living "viruses" have yet to be discovered. After all, how would you look for them or even detect their presence? Could these viruses be remnants of a time before metabolism and the RNA world merged?

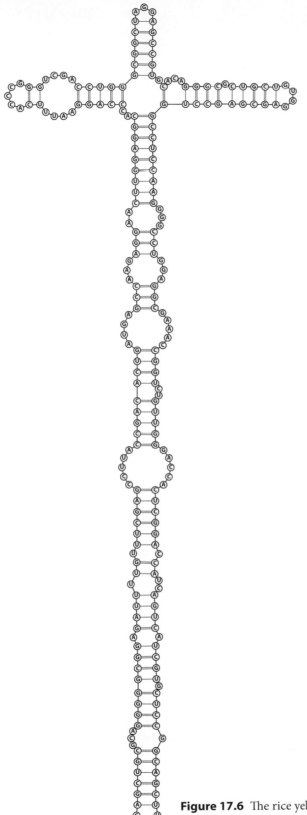

**Figure 17.6** The rice yellow mottle virus-associated viroid: not its genome, the whole thing. doi:10.1128/9781555818517.ch17.f17.6

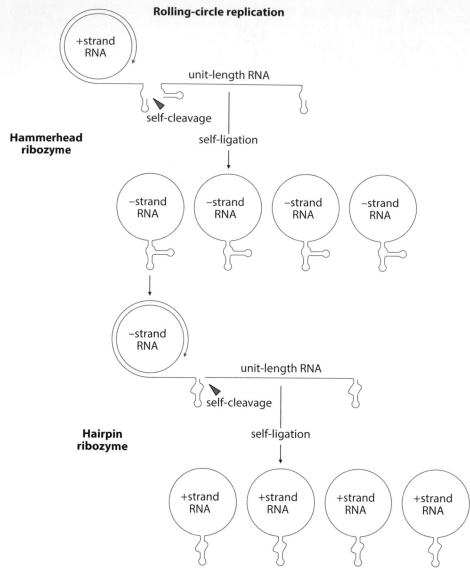

**Figure 17.7** Tobacco ringspot virus satellite RNA-S replication. doi:10.1128/9781555818517.ch17.f17.7

## Viruses as degenerate parasites

Some viruses may be the simplified remnant of microbial intracellular parasites, similar to *Bdellovibrio*, *Chlamydia*, or *Rickettsia*. It is certainly true that parasites often become extremely simple, shedding anything unneeded. Starting with an obligate intracellular energy parasite such as *Chlamydia*, all that conceptually need happen for it to become a virus is the fusion of the cytoplasms of the host and parasite, so that the virus can use the host's translational apparatus and perhaps even the transcription and/or DNA replication machinery. The genome of the parasite could then become even more drastically simplified. Smallpox virus, at 0.2 by 0.3 μm, is as big as a small bacterium, with a genome nearly as large (ca. 0.2 Mbp) as that of the smallest parasitic microbes (*N. equitans* is less than 0.5 Mbp).

The best potential example of this, however, might be mimivirus (Fig. 17.8), a virus of amoebas, which has a capsid 0.4 μm in diameter (as big as *Nanoarchaeum* or *Opitutus*) and a genome of 1.2 Mbp (as big as a great many prokaryotic genomes and bigger than quite a few). The genome encodes a slew of translational proteins, such as EF-Tu, release factor 1, transfer RNA (tRNA) synthetases (tRNA charging enzymes), If1, and topoisomerases. However, this organism lacks genes for rRNAs or proteins, as well as genes for central metabolism.

Interestingly, mimiviruses (there are several known examples) carry out replication and assembly of the core in the nucleus, and then assembly of the large virus particle in specific regions of the host cytoplasm: "virus factories" (Fig. 17.9). Thus, even though the virus fuses with the host and hijacks its ribosomes and other components, it retains some separation between "self" and "host" during replication.

Mimivirus was discovered accidentally during investigation of *Legionella* in amoebas and is large enough that it was first seen in Gram stains and thought to be a gram-positive bacterium! A larger "species," mamavirus, is in turn infected by a smaller satellite virus, the sputnik virophage.

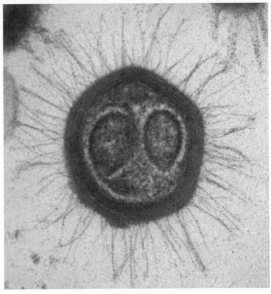

**Figure 17.8** Electron micrograph of mimivirus. (Courtesy of Didier Raoult.) doi:10.1128/9781555818517.ch17.f17.8

**Figure 17.9** Mimivirus (MV)-infected amoeba. Note the "virus factory" (VF), from which assembled virus particles emerge. N, nucleus. (Reprinted by permission from Macmillan Publishers Ltd. [Raoult D, Forterre P, *Nature Rev Microbiol* **6**:315–319, 2008. http://www.nature.com, Copyright 2008.]) doi:10.1128/9781555818517.ch17.f17.9

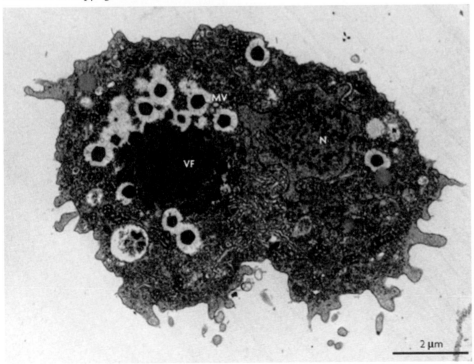

# Prions

## *The prion theory*

Transmissible spongiform encephalopathies (TSEs) are transmissible neurode-generative diseases, leading ultimately to disabling dementia and death. Scrapie is a well-known and historical problem in sheep herding, coming in a variety of versions (strains) and causing symptoms ranging from disabling itching to dementia (Fig. 17.10). Bovine spongiform encephalopathy (mad cow disease; BSE) is a recent problem in cattle, in which cows become infected by scrapie as a result of being fed diseased sheep carcasses as part of their processed feed. The problem, of course, is that mad cow disease can in turn be transmitted to humans who eat meat from infected cows. Creutzfeldt-Jakob disease (CJD) is a human disease, transmitted via infected human immunoglobulins, corneas, and other products, and also found in cannibalistic populations where human brains are ceremonially eaten. TSE-like diseases can also occur spontaneously (very rarely) in individuals, and if present in the germ line and not lethal too young, it can result in familial/inherited forms of the disease.

Brains of TSE-infected individuals have a degenerate spongy appearance (Fig. 17.11), and there is also an accumulation of "plaque," an insoluble, protease-resistant form of a normal brain protein of unclear function. Other than being transmissible, this looks a lot like Alzheimer's disease.

At first, of course, it was assumed that these diseases were caused by a slow-acting virus. But no viral DNA or RNA has ever been found to be associated directly with TSEs, and the fact that infective samples could be heated or

**Figure 17.10** Sheep with an "itchy" form of scrapie. (Courtesy of the University of Idaho.)
doi:10.1128/9781555818517.ch17.f17.10

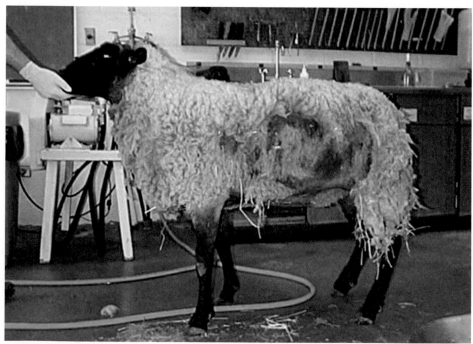

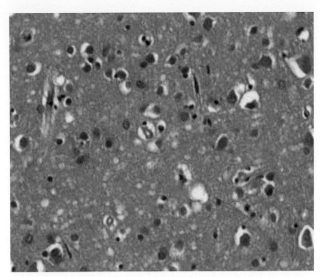

**Figure 17.11** Light micrograph of a stained thin section of brain from a cow with BSE. The cavities (center) contain plaque composed of the prion protein, the presumed cause of the disease. (Courtesy of R. Williams. http://www.ncbi.nlm.nih.gov/pmc/articles/PMC3238497/. https://creativecommons.org/licenses/by/4.0/legalcode.) doi:10.1128/9781555818517.ch17.f17.11

irradiated without their infectivity being destroyed suggested that they probably contain no nucleic acid; in other words, they aren't viral in nature.

In 1982, Stanley Prusiner suggested that TSEs were caused by infectious particles made entirely of protein (no nucleic acid) and dubbed these particles "prions." This idea met with heavy resistance, at least some of which was dogmatic. But very quickly it was discovered that plaque was composed of a protein (protease-resistant protein [PrP]) normally found in the central nervous system, but in a different conformation (Fig. 17.12). The normal conformation of this protein (PrP$^c$) is soluble and primarily alpha-helical, and is a fragment of a larger preprotein released by a specific proteolysis. The pathological conformation of this protein (PrP$^{Sc}$) is insoluble, mostly beta-sheet, and protease resistant. Amazingly, the introduction of small amounts of PrP$^{Sc}$ into a solution of PrP$^c$ in vitro causes the PrP$^c$ to change into PrP$^{Sc}$. In other words, PrP$^{Sc}$ catalyzes the transformation of PrP$^c$ into PrP$^{Sc}$. It looked like the "prion" had been found!

If the idea of a protein with "normal" and "abnormal" conformations, in which the abnormal conformation has the ability to convert normal protein to its abnormal form, seems odd, then think of it this way. If you have crystals of salt and a container of supersaturated salt solution, these represent two conformations of the same chemical, sodium chloride. If you add a single salt grain to the supersaturated solution, much of the dissolved salt is converted to crystals. One of these new crystals could then be transferred to another container of saturated brine, causing the dissolved salt in this new container to crystalize in turn. Thought of this way, TSEs are not so much an infection as a transmissible metabolic chain reaction.

In the years since the prion theory was proposed, the study of TSEs was focused on PrP and prions—but there were lingering doubts. The argument about

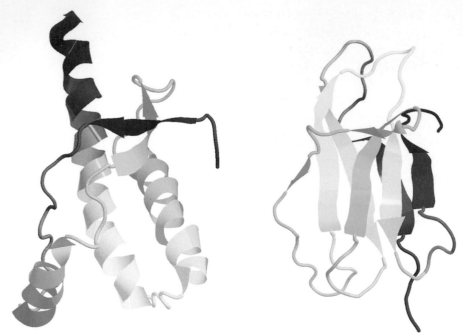

**Figure 17.12** Ribbon diagrams of PrP$^c$ (normal, left) and PrP$^{Sc}$ (diseased, right). doi:10.1128/9781555818517.ch17.f17.12

the cause of TSEs was loud and heated over the years, for the simple reason that nobody had proven definitively what the infection was. It was clear that PrP$^{Sc}$ was associated with TSEs, but were they the cause or just a symptom? In microbiology, the gold standard of proof in identifying the cause of a disease is Koch's postulates:

1. The microorganism must be isolated from a diseased organism and grown in pure culture.
2. The cultured microorganism should cause disease when introduced into a healthy organism.
3. The microorganism must be reisolated from the inoculated, diseased experimental host and identified as being identical to the original specific causative agent.

But diseases like TSEs aren't easy to prove by using Koch's postulates, and the inability to do so resulted in lingering doubt about prions.

Recently, these doubts have largely been laid to rest. Over the past few years, the generation of TSE in previously healthy, wild-type mice by exposure to PrP$^{Sc}$ generated in vitro from recombinant protein (made in *E. coli*) has been demonstrated.

Just as important has been the demonstration that it is possible to generate various "forms" of PrP$^{Sc}$ that generate different kinds ("strains") of TSE. This is important because the major weakness in the prion hypothesis was the fact that TSEs come in different "strains" that are faithfully transmitted. There are over two dozen strains of scrapie, for example, and they are quite easily distinguished—there are itchy strains, walking in circles strains, dementia strains, downer strains, etc. This can be explained in the prion hypothesis only if the PrP protein can exist as multiple disease structures, each of which induces (templates?) the same structure on PrP$^{Sc}$ molecules it encounters. This has now been demonstrated, and all of this is reviewed in C. Soto, The prion hypothesis: the end of the controversy? *Trends in Biochemical Sciences* **36**:151–158, 2010.

## Questions for thought

**1.** Are viruses alive? (There is a great *Scientific American* article with this title in the December 2004 issue.) How would you argue your point to a critical audience?

**2.** How would you go about attempting to figure out the origin of a particular virus (choose one)?

**3.** What do you think the limit might be for how small and simple a virus could be? Why not smaller?

**4.** What do you think the limit might be for how big and complex a virus might be? Why not bigger?

**5.** Do you find the near fulfillment of Koch's postulates in the case of TSEs convincing? Why or why not?

A thermal pool in Yellowstone National Park.

# SECTION III   *Microbial Populations*

Most of what we know about the microbial world was (and is being) learned by examination of cultivated organisms. But most microbial species are not readily cultivated; the often-quoted number is "less than 1%," and the real number is probably many, many orders of magnitude less than this. This means inevitably that we end up examining unrepresentative organisms if we rely on cultivation. Inferences from these species are unlikely to be realistic; imagine how poorly we would understand plant diversity if we could study only a handful of weeds.

Molecular phylogenetic analysis can solve this problem, because cultivation is not a required part of the process. Molecular phylogenetic analysis can be performed on isolated (cultivated or not) material or even on microbial populations taken directly from the environment.

Molecular phylogenetic analysis of uncultivated organisms is similar to analysis of cultivated species but starts with DNA extracted directly from a purified sample or from an environmental sample, rather than from a pure culture (facing illustration). In these cases, the small-subunit rRNA polymerase chain reaction product is likely a mixture of sequences, and so it is generally necessary to separate the products before sequencing (often by cloning) so that each sequencing reaction mixture contains a single sequence; otherwise no clean sequence data are generated. It is then common to design specific fluorescently labeled oligonucleotide probes (phylogenetic probes) based on the resulting sequences and use fluorescent in situ hybridization (FISH) to identify the organisms that each sequence represents in the original population. This FISH analysis can be used to confirm the identity of any specific organism in the population, to determine the morphology of a specific organism, or as a handle on the organism for attempts to grow the organism in culture.

In this section, this process is described in order of increasing scale: first the identification of uncultivable but more or less physically isolatable organisms (chapter 18), sequence-based surveys or censuses of complex populations (chapter 19), FISH analysis of complex populations (chapter 20), and methods to assess complex populations that do not rely on obtaining sequence data: molecular "fingerprinting" (chapter 21). Finally, the issue of what to do with a microbial census is addressed (chapter 22): how we determine which component in our census is responsible for what process we know is going on in that environment. This linking of phylotype and phenotype is usually the ultimate goal of any microbial survey.

Flow diagram of the molecular phylogenetic analysis of microbial populations. doi:10.1128/9781555818517.sIII.f1

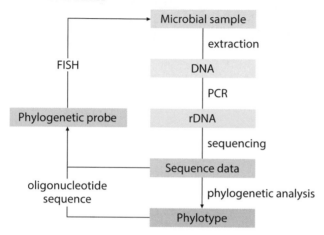

# 18 Identification of Uncultivated Organisms

The most common use of molecular microbial phylogenetics is in the identification of a cultivated organism in-hand. However, it is very common to be able to perform much the same kind of analysis on organisms that cannot be cultivated but can be isolated more or less cleanly from other organisms. Here are some examples of uncultivated species characterized by molecular phylogenetic analysis.

- The pink filamentous organism that grows in the outflows of neutral-pH hot springs in Yellowstone National Park. Initial attempts to grow this organism in the 1960s resulted instead in the cultivation of an entirely different pink bacterium: *Thermus aquaticus*. This is the example reviewed in detail in this chapter.

- The sulfur-oxidizing endosymbionts of the giant tubeworm, the vent clam, and the scaly snail. These organisms are found only at deep-sea hydrothermal vents and have no functional digestive tracts; they are fed entirely by their sulfur-oxidizing proteobacterial endosymbionts.

- Magnetotactic bacteria, which can be isolated by taking advantage of their magnetotaxis but cannot generally be grown in culture.

- The causative agent of bacillary angiomatosis, a close relative of the causative agent of cat scratch disease, *Bartonella quintana*. This knowledge led directly to the cultivation of the organism, now known as *Bartonella bacilliformis*, using culture methods used for *B. quintana*.

- *Cenarchaeum symbiosum*, a psychrophilic (it grows optimally at 10°C) crenarchaeal symbiont of a marine sponge of the genus *Axinella*. The symbionts are related to abundant crenarchaeal species found in marine and soil environments, although none have been cultivated.

- *Epulopiscium fishelsoni*, a gut symbiont of some marine herbivorous fish (surgeonfish, doctorfish, and tangs). For a long time the largest known bacterium (a firmicute, as it turns out), it was originally thought to be a protist, perhaps a ciliate, based on size (big enough to see clearly with the naked eye), but did not seem to have any organelles.
- Endosymbiotic bacteria in sap-sucking and wood-eating insects. These insects contain an unusual organ, the bacteriome, consisting of cells called bacteriocytes. These cells contain endosymbiotic bacteria, which provide the insect with the vitamins and essential amino aids that are absent in their nutritionally poor diets. This is an obligate relationship; the insect cannot survive without the bacteria (antibiotic treatment is lethal), and the bacteria have never been cultivated outside of the insect. However, they have been characterized using small-subunit ribosomal RNA (SSU rRNA) analysis; most are γ-proteobacteria, but some are β-proteobacteria and some are bacteroids. Very often, the insect contains a mixture of endosymbionts. One such insect endosymbiont, *Buchnera aphidicola*, has had its entire genome sequenced even though it cannot be cultivated except in aphids.

And many, many, more . . .

## ■ STUDY AND ANALYSIS
### The Pink Filaments of Yellowstone

Described in the 1960s by Thomas Brock, there is an abundance of pink filamentous growth in many neutral-pH Yellowstone hot springs, including Octopus Spring (Fig. 18.1). Early attempts to label these filaments by feeding them radioactive organic compounds failed, as did early attempts to extract nucleic acids, causing some to suggest that they were not alive but were just the dead remains of mesophilic groundwater organisms that had been regurgitated by the hot spring. *Thermus aquaticus* (cultures of which are often also pink) was isolated as a by-product of attempts to cultivate these filaments, but it turns out that *T. aquaticus* is a minor constituent of this microbial community.

The following two papers describe first the phylogenetic identification of the pink filamentous organism and then the cultivation and characterization of this organism based on this information.

**Reysenbach A-L, Wickham GS, Pace NR.** 1994. Phylogenetic analysis of the hyperthermophilic pink filament community in Octopus Spring, Yellowstone National Park. *Appl Environ Microbiol* **60:**2113–2119.
PubMed ID: 7518219    PubMed Central ID: PMC201609

*You should have a copy of this paper at hand before proceeding.*

*Question being asked:* What is the phylogenetic identity (phylotype) of the pink filamentous organism?

**Figure 18.1** Octopus Spring, Yellowstone National Park.
doi:10.1128/9781555818517.ch18.f18.1

In this 1994 paper by Anna-Louise Reysenbach and Gene Wickham from Norman Pace's laboratory, the pink filaments were identified without cultivation by molecular phylogenetic analysis.

Pink-filament biomass (Fig. 18.2) was collected following growth for several weeks on square, flat cotton filter pads placed in situ, to make sure the organisms actually grew in that environment rather than being groundwater organisms brought, dead, to the surface in the hot spring. These "furnace filters" quickly became colonized by the pink filamentous growth (Fig. 1 of the paper).

DNA was isolated from a washed pink-filament sample, and SSU rDNA was amplified by polymerase chain reaction (PCR). Interestingly, only universal or bacterium-specific primers yielded PCR products; archaeon-specific primers did not amplify anything. This probably is because the archaeon-specific primers were not particularly good (being based, at the time, on only a few archaeal sequences) and because bacteria rather than archaea really do predominate in this environment. The PCR products were cloned into a plasmid vector before analysis; cloning was used to separate this population into individual sequences. A total of 35 clones were obtained.

The clones were sorted in groups of close relatives by using "T-tracts," and a single representative clone from each was sequenced; there is little reason to sequence a slew of identical or nearly identical clones. This was necessary because, at the time of these experiments, sequencing took a lot more work (by orders of magnitude) than it does today. (T-tracts are no longer used; in today's

**Figure 18.2** Pink filaments in the outflow of Octopus Spring.
doi:10.1128/9781555818517.ch18.f18.2

automated sequencing systems, all four reactions are run in the same tube and run on a single lane of a gel.) "T-tracts" are standard DNA-sequencing reactions, but instead of performing the reactions to identify all four bases (A, G, C, and T) in the sequence, only the "T" reactions are performed. These are run side by side on a gel, and the patterns of T's in the sequences (a sort of fingerprint) are sorted to identify sets of identical or nearly identical sequences. A representative clone from each unique type (there were only three types) was fully sequenced.

Most of the sequences (26 of the 35 clones obtained) consisted of a novel sequence designated EM17, related to what was at the time a newly isolated organism, *Aquifex pyrophilus*, a hyperthermophilic hydrogen oxidizer isolated from a submarine hydrothermal vent north of Iceland (Fig. 2 of the paper). Two other sequences showed up less often among the clones: EM19 (2 of the 35 clones) is a less close relative of *Aquifex*, and EM3 (7 of the 35 clones) is a relative of *Thermotoga*.

Consistent with EM17 being related to *Aquifex* are some details of the secondary structure inferred from the SSU rRNA sequence (Fig. 4 of the paper). The authors show this in a hypervariable region of the SSU rRNA. In our previous discussion about molecular phylogenetic clocks, this region might represent a "second hand." Not only does the sequence of the RNA change very quickly over evolutionary history, but so does its secondary structure. It is clear that the structure of this region is much like that of *Aquifex pyrophilus*, a 6-bp proximal helix and a 5-bp stem-loop flanking an asymmetric internal loop with conserved sequence ("CUC" and "A"). Notice how the sequences in the helices are quite different, but all changes are consistent with the secondary structure. The only difference in the secondary structures is the size of the terminal loops. In contrast, the same region in *Thermotoga* or *Escherichia* is very different in both sequence and structure.

The question then was: Which of these three sequences, if any, are from the pink filamentous organism? To determine this, a fluorescently labeled oligonucleotide probe complementary to a species-specific (highly variable) region of the EM17 SSU rRNA sequence was used to probe environmental samples for the EM17 organism; this is called fluorescent in situ hybridization (FISH) (Fig. 3 of the paper). The EM17 probe fluorescently labeled the pink filaments, but not other inhabitants of the sample, showing that the EM17 SSU rRNA sequence is that of the pink-filament microbe. The EM19- and EM3-specific probes failed to hybridize to anything observable in the samples.

So, in conclusion, the authors have determined the phylotype of the pink filaments, which turns out to be a relative of *Aquifex*.

## A note about informal clone names

By the way, the "EM" in the sequence names stands for "Electric Monk," the haywire labor-saving device described in *Dirk Gently's Holistic Detective Agency* by Douglas Adams. When the Electric Monk is introduced in the story, he is stuck because he believed (believing is, after all, the function of an Electric Monk) that everything around him is a uniform shade of pink, making it impossible to distinguish any one thing from any other thing. The only connections between the Electric Monk and these clones are (i) the color pink and (ii) that the authors of the paper had a copy of this book with them during the long drive to Yellowstone for their fieldwork. Informal strain designations can come from anything—such as the student's initials, their cat's name, their love interest, or a sports team.

Huber R, Eder W, Heldwein S, Wanner G, Huber H, Rachel R, Stetter KO. 1998. *Thermocrinis ruber* gen. nov., sp. nov., a pink-filament-forming hyperthermophilic bacterium isolated from Yellowstone National Park. *Appl Environ Microbiol* **64:**3576–3583.

PubMed ID: 9758770    PubMed Central ID: PMC106467

*You should have a copy of this paper at hand before proceeding.*

*The task:* Cultivate the YNP pink filamentous organism (EM17) on the basis of its phylotype, and isolate it as a pure culture from a single cell.

In 1994, the phylotype of the pink filamentous organism common in Yellowstone neutral-pH hot springs had been determined; it was closely related to *Aquifex*. In the 4 years since, additional genera of organisms in this group had been identified: *Hydrogenobacter* and *Calderobacterium*. All of the cultivated members of this phylogenetic group are physiologically similar, being thermophilic hydrogen oxidizers. So, in an attempt to cultivate the pink filamentous organism EM17, a series of enrichment cultures based on the known conditions in Octopus Spring and the optimal growth conditions for growth of *Aquifex* were made. The chemical composition of Octopus Spring was mimicked in detail and was provided with both hydrogen and oxygen in low concentrations of 3% each. These enrichments were inoculated with washed pink filaments from Octopus Spring.

The progress of enrichment cultures was assessed by performing microscopic examination (to monitor the growth of the pink filaments) and by testing samples by FISH for hybridization with the EM17 probe. Although none of the enrichment cultures resulted in pink filamentous growth, some enrichments yielded good growth of a rod-shaped pink bacterium that hybridized strongly to the EM17-specific probe.

In order to isolate this organism in pure culture, a 1,064-nm (about 1-μm)-wavelength focused infrared laser was used as "optical tweezers" in a one-of-a-kind microscope to capture single cells that looked like the ones that hybridize to the EM17-specific probe (Fig. 18.3). Note that FISH analysis kills cells (they are fixed in glutaraldehyde), so you can only test cultures after the fact, you cannot test the individual cells before separating them out and growing them.

**Figure 18.3** Isolation of a single cell by using optical tweezers.
doi:10.1128/9781555818517.ch18.f18.3

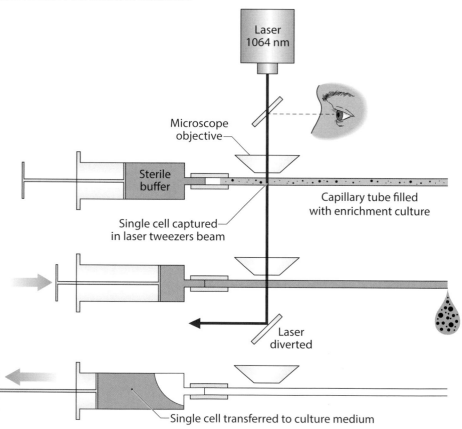

These individual cells were injected into fresh culture medium for growth. Of course, many cells were collected and cultured in this way, and the resulting growth, if any, was tested using the EM17 probe. Some of these cultures did grow, and when tested, one (at least) strongly hybridized to the EM17-specific probe.

With a pure culture in hand, the SSU rRNA sequence of this organism was determined and found to be nearly identical to the EM17 sequence (Fig. 7 of the paper). But it was not exactly identical. This is typical; most microbial populations are collections of closely related organisms, specialized for different microenvironments. Enrichment cultures typically capture only the fastest-growing, robust types (i.e., the weeds), whether they are the predominant strain or not (usually not). The organism was named *Thermocrinis ruber* ("hot red hair").

*Thermocrinis ruber* grows as long rods in suspension cultures (Fig. 2 and 4 of the paper), but when cultivated in an artificial "creek" culture vessel, with a turbulent overflow, it grows as nice pink filaments that are just like those seen in the wild (Fig. 1 and 3 of the paper).

This paper is a monograph: the formal description and naming of a newly isolated organism. At the end of the paper is a list of a series of more or less standard tests used to describe the general properties of the organism.

## Questions for thought

1. More and more diseases that were previously thought to be noninfectious are being discovered to be caused by bacterial or viral infection. What kind of "noninfectious" diseases can you think of that might eventually be found to have an infectious cause?

2. Why do you suppose *T. ruber* grows as rods when cultivated in standard culture conditions, but as filaments in the artificial (or natural) spring?

3. The products of the PCR from DNA isolated from the pink filaments were cloned before analysis. What would have happened if the PCR product DNA had been sequenced directly? In other words, what would the sequence data have looked like?

4. Why do you suppose microbiologists had such a hard time showing (or believing) that the *T. ruber* mats were actually living?

5. The authors of the first paper seem to have been surprised by their lack of detection of archaea in the pink-filament sample. Why do you suppose this is? What could they have done to investigate this further?

# 19

# Sequence-Based Microbial Surveys

A typical molecular phylogenetic analysis starts with genomic DNA isolated from a pure culture of an organism, or an enriched sample, as in the case of *Thermocrinis ruber* in chapter 18. The small-subunit ribosomal RNA (SSU rRNA) sequence obtained by polymerase chain reaction (PCR) amplification is used to determine the place of that organism in the "big tree," i.e., its phylotype.

It is also possible, however, to start a molecular phylogenetic analysis with DNA extracted directly from an environmental sample instead of a pure or even enriched culture. The PCR amplification products in this case are a collection of SSU rRNA sequences representing (perhaps) the population of SSU rRNA genes in the DNA, in turn representing the population of organisms in the original sample. The SSU rRNA sequences are separated by cloning, and then SSU rRNA sequences from each clone are determined to survey the microbial inhabitants of an environment. This can be a relatively simple survey, as in the case of the *T. ruber* pink filaments, or more thorough surveys, as in the example of the hydrothermal vent scaly snail microbial symbionts described below.

More modern methods push this approach over the limit, obtaining not dozens, not hundreds, not thousands, not even tens of thousands, but hundreds of thousands of sequences from environmental samples. This is possible because sequencing technology is improving very, very quickly. This chapter includes a great example of this kind of analysis of the human microbiome.

Whether using the old-fashioned clone-and-sequence approach or the more modern high-throughput sequencing approach, molecular phylogenetic

doi:10.1128/9781555818517.ch19

surveys are far superior to the older cultivation-based approaches and are used a lot.

However, it is important to keep in mind that several aspects of PCR/cloning/sequencing of microbial populations limit its quantitative interpretation. In fact, bias exists in every step of the process, and so estimation of the relative abundance of organisms from molecular phylogenetic surveys is not quantitative. Some cells cannot be "opened" using standard methods (or are not opened as efficiently), some DNA can be selectively lost to some extent in purification (perhaps tightly bound by protein), some sequences are easier than others from which to amplify rRNA genes (rDNAs) (depending on primer target sequences, PCR conditions, etc.), different organisms have different rDNA copy numbers, some clone more readily than others, and so forth. Nevertheless, even a qualitative molecular phylogenetic survey of the microbial population is very useful and is far more informative than examination of whatever happens to grow on plates.

---

■ **STUDY AND ANALYSIS**
## The Scaly Snail and Human Microbiomes

**Goffredi SK, Warén A, Orphan VJ, Van Dover CL, Vrijenhoek RC.** 2004. Novel forms of structural integration between microbes and a hydrothermal vent gastropod from the Indian Ocean. *Appl Environ Microbiol* **70**:3082–3090.
PubMed ID: 1518570   PubMed Central ID: PMC404406

*You should have a copy of this paper at hand before proceeding.*

*Questions being asked:* Is this snail sustained by autotrophic microbes, and if so, what are they? Is the composition of the body scale biofilms unusual and likely to be responsible for the metal-sulfide surface coating?

Deep-sea hydrothermal vents are hot springs found on the ocean floor at the boundaries between tectonic plates. These environments are rich in microbial and animal life, subsisting on the oxidation of sulfide and hydrogen from the hydrothermal fluid. A wide range of animals have adapted to these unique environments. Many of these feed on the sulfide-oxidizing bacteria or each other, but many of them have formed symbioses with the sulfide-oxidizing bacteria. Perhaps the best known of these is the giant vent tube worm *Riftia* (Fig. 19.1), whose gills absorb environmental oxygen, carbon dioxide, and sulfide, which are carried to a specialized organ full of endosymbiotic sulfide-oxidizing autotrophic bacteria where they are digested by the bacteria, providing the worm with the nutrients it needs. *Riftia*, like other vent animals with similar endosymbionts, lacks a digestive tract.

In this paper, the authors describe a new deep-sea hydrothermal vent animal, a snail about 5 cm in diameter. This animal was found in the Indian Ocean in a hydrothermal vent field at a depth of about 2,400 m (Fig. 19.2).

This snail lacks an operculum (the plate the snail uses to cover the opening of its shell when retracted) but instead has a body covered in small scales

**Figure 19.1** A clump of *Riftia* near a deep-sea hydrothermal vent. (From Dworkin M, Falkow S, Rosenberg E, Schleifer K-H, Stackebrandt E [ed]. 2006. *The Prokaryotes*, 3rd ed., with kind permission from Springer Science and Business Media. Courtesy of Irene Lucile Garcia Newton.) doi:10.1128/9781555818517.ch19.f19.1

**Figure 19.2** Google Earth view of the Indian Ocean. The site of the hydrothermal vent field described in this paper is circled, at the intersection of tectonic plates. doi:10.1128/9781555818517.ch19.f19.2

**Figure 19.3** *Crysomallon squamiferum*, the deep-sea hydrothermal vent scaly snail. See also Figure 10.24. (Courtesy of Cindy Van Dover.) doi:10.1128/9781555818517.ch19.f19.3

hardened by a plating of iron sulfides (Fig. 19.3 and Fig. 1 of the paper). This animal does not retract into its shell, but when it contracts its foot, it is protected by an arsenal of hardened scales.

In addition to its unusual body covering, this snail has only the remnants of a digestive tract, so vestigial that it seems unlikely to be functional. The digestive tract never contains anything like food, only scraps of iron sulfides; the radula, a sort of a tongue that snails use to scrape up the food they eat, is almost nonexistent. On the other hand, the snail has very large esophageal glands; these are the glands that usually secrete lubricating mucus into the esophagus. The esophageal gland contains what look like bacteriocytes (host cells with vacuoles full of bacteria) (Fig. 2 of the paper). The esophageal gland and foot are very heavily vascularized, and presumably sulfide and oxygen are absorbed from seawater by the foot and transported directly to the esophageal gland. (The authors of the paper also examined the gills, because other vent mollusks that harbor chemo-autotrophic symbionts do so in the gills, but apparently not so in this case.)

The first step, then, was to identify the esophageal gland symbiont. This is not really a microbial survey but, rather, the identification of a specific organism, a lot like the Yellowstone pink-filament analysis in chapter 18. Glands were dissected from six frozen animals, DNA was extracted, and SSU rRNA sequences were amplified by PCR using bacterium-specific primers. These were cloned, and 29 clones were tested by restriction fragment length polymorphism, but it turned out that all of the clones were alike; in other words, as far as they could tell, these glands contained only a single bacterial type. This was sequenced and analyzed phylogenetically, and it was found to be a γ-proteobacterium related to other sulfide-oxidizing endosymbionts and related free-living vent species (Fig. 4 of the paper—find the black arrow).

The next question, then, is whether these endosymbionts really are autotrophic feeders of the animal. This was addressed by examining the isotope fractionation of the animals. Primary production by sulfide oxidation results in a large isotope effect: a lower fraction of heavy carbon and nitrogen isotopes in tissues. Metabolic processes eventually reduce this isotope effect, so that if the snail gets its carbon and nitrogen directly from sulfide-oxidizing bacteria, it should be enriched for light isotopes, whereas if it grazes on these and digests them, or if it is a predator, this isotope effect would be reduced. As Table 1 of the paper shows, the scaly snail has a greater isotope effect than either grazers (*Lepetodrilus*) or predators (*Phymorhynchus*) of the same environment. So it really does look as though this is a chemoautotrophic snail.

Next, the authors wanted to look at the snail's scales (Fig. 3 of the paper), and in particular their unusual iron sulfide plating. These scales look like a highly modified operculum and contain the expected concholin tissue that would normally coat the interior of a snail's shell or operculum. The scales are covered and permeated with a plating of iron sulfides, and this is in turn coated with a microbial biofilm.

DNA was extracted from the scale-associated biofilm, and SSU rDNA was amplified and cloned, yielding 155 sequences originating from two snails (Fig. 4 and 5 of the paper). Most of these (67%) were ε-proteobacteria; this group is predominant in most vent environments. There were also a number (15%) of sequences related to sulfate-reducing δ-proteobacteria and to γ-proteobacteria related to known sulfur-oxidizing organisms, many of which are symbionts of vent animals. They also got a few *Chloroflexi* (green nonsulfur [GNS]), *Spirochaete*, and *Bacteroid* (*Cytophaga-Flavobacterium-Bacteroides* [CFB]) sequences.

The surface of the shell is only lightly covered in bacteria, looking like the stuff covering a wide range of other vent animals. The diversity of bacteria on the shell (Fig. 4 of the paper) is much lower than that on the scales, and consists of a haphazard collection of sulfur-oxidizing or -reducing ε- or γ-proteobacteria. These are probably free-living organisms rather than symbionts, just using the shell as a surface to grow on. Keep in mind that although the shell is a similar color to the scales, it is made of regular calcium carbonate (aragonite) and is not coated in iron sulfides.

The external scale symbionts (the authors use the term "epibionts") are unique to the snail's scales; they are not common in this environment. Given their relationship to known sulfur oxidizers and reducers, it seems likely that this biofilm, or some organisms in this biofilm, is responsible for this iron sulfide mineralization. Since these bacteria could not be found either on the foot or on the shell of the same animals, their growth on the scales must be facilitated somehow by the animal. On the other hand, there is some iron sulfide *within* the tissue of the scales and the foot, suggesting that maybe the animal itself is capable of this mineralization. This seems unlikely, but at this point it is not certain which organism(s) are responsible for this mineralization.

So, in conclusion, this scaly snail is an autotrophic animal, absorbing oxygen and sulfide from its vent environment and feeding them to γ-proteobacterial endosymbionts in an enlarged, specialized esophageal gland. This snail lacks a typical operculum and is instead covered in protective scales plated with iron sulfides that are probably mineralized by a complex bacterial biofilm.

### A note about chimeras

Although the authors do not mention it, any time you do this sort of population analysis, you need to screen your sequences to eliminate any chimeras, i.e., sequences from two (or more) organisms joined together (Fig. 19.4). These types of sequences show up in most SSU rRNA PCR experiments from natural mixed populations. They arise when a molecule of DNA polymerase stalls partway through a PCR cycle (Fig. 19.5). If the resulting truncated DNA strand ends in a conserved part of the rDNA, it can anneal to another DNA molecule, whether or not it is from the same organism, and serve in the next PCR cycle as a primer for the synthesis of a DNA molecule that comes from one organism at one end and another organism at the other end.

Chimeras can be identified using three standard methods:

- Making phylogenetic trees based independently on the 5′ and 3′ ends of each sequence. If the two trees disagree significantly, it is probably a chimeric sequence.

- Drawing the secondary structure of the RNA. If the base pairs involving the 5′ and 3′ sequences do not work, it is probably a chimera.

- Using the CHECK_CHIMERA function and the Ribosomal Database Project, or related computational tools. CHECK_CHIMERA compares the similarity of a sequence along its length to other sequences in the database; a "break" in this similarity, where the sequence begins to look less and less like one sequence and more like another, indicates that the sequence is probably a chimera.

Any sequence that seems to be a chimera is, of course, discarded from the analysis.

**Costello EK, Lauber CL, Hamady M, Fierer N, Gordon JI, Knight R.** 2009. Bacterial community variation in human body habitats across space and time. *Science* **326:**1694–1697. (Also see the supplement to this paper.)
PubMed ID: 19892944

*You should have a copy of this paper and the supplementary data for the paper at hand before proceeding.*

*Purpose:* To begin to determine how human-associated microbial communities vary from place to place on the body, from person to person, and over time.

Although microbial ecologists usually focus on other environments, one microbial environment of special importance to us is the human body. The collective microbial community of the human body is sometimes referred to as the human "microbiome." Microbiologists know a lot more about human microbiology

**Figure 19.4** The chimera decorating an Apulia dish at the Louvre (Source: Wikimedia Commons.)
doi:10.1128/9781555818517.ch19.f19.4

**Figure 19.5** The generation of chimeras by the priming of PCR products from abortive reaction products. The red and blue sequences represent different organisms.
doi:10.1128/9781555818517.ch19.f19.5

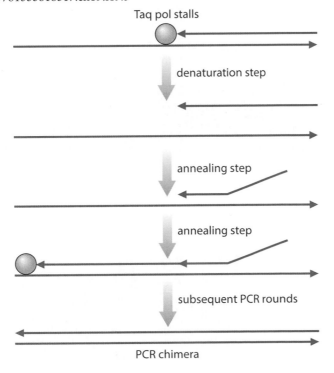

Taq pol stalls

denaturation step

annealing step

annealing step

subsequent PCR rounds

PCR chimera

than any other system, and particularly about organisms that cause serious acute disease. But this is the tip of the iceberg; there are more than 10 times as many bacterial cells in and on the human body as there are human cells! Given that all the human cells are of the same species while the microbial populations are of countless species, the genetic information content of a human is dominated by microbial genes. Very basic questions remain, such as the following:

- How much microbial diversity is there in the human body?
- How much do the microbial communities differ in different parts of the body?
- How much variation is there in "normal" microbial communities?
- How much do our microbial communities change over time?
- How do these communities contribute to our overall health?

Think of it this way. Picture an old-growth tree (Fig. 19.6). It would be futile to try to understand the health and function of this tree by studying only its leaves,

**Figure 19.6** An old-growth tree community. (Courtesy of Piotr Skubisz/Fotolia.) doi:10.1128/9781555818517.ch19.f19.6

limbs, trunk, roots. and wood, or even its genome; you would get some information, but to really understand this tree, you have to think of it as an ecosystem, and include the animals, plants, fungi, and microbes that cover its surface (both above and below ground) and fill its nooks and crannies. If the tree were unhealthy, a good place to start might be to examine all of these symbionts. But in the absence of an obvious cause of disease, such as kudzu or tuberculosis, what do you look for? What is "normal"? How can you distinguish normal from problematic? In general, in human microbiology, we don't have a clue. *This* is the purpose of this paper.

In this series of experiments, the authors took samples from 27 locations on nine healthy humans (three female and six male) on four separate days over 3 months (Fig. S1 [supplement 1] of the paper).

DNA was isolated from each sample, and a region of bacterial SSU rRNA was amplified by PCR (Fig. 19.7). The PCR primer sets for each reaction had a unique "barcode" sequence on the forward primer; these barcodes allow the exact source of each sequence to be identified directly from the sequence. This allows the authors to mix as many samples as they wish into a single 454 sequencing run (see below). They ended up with 815 different samples and an average of 1,315 classifiable sequences per sample, totaling just over a million total usable sequences.

As you might predict, nearly all (92.3%) of the sequences were from four bacterial phyla: the *Actinobacteria*, *Firmicutes*, *Proteobacteria*, and *Bacteroidetes* (Fig. S2 of the paper). Most of the rest are cyanobacteria (from pollen chloroplasts) and some fusobacteria (primarily in the mouth).

The data are sorted by body site in Figures S12 and S13 of the paper.

To compare populations against each other, the authors use a metric called Unifrac (which stands for "unique fraction"). It starts with a reference phylogenetic tree and then, for each population, determines which branches in this tree lead to sequences in that sample. Two populations are compared by determining the fraction of branch length that is shared between the populations and the fraction of branch length that is unique (i.e., the unifrac) to each population (Fig. 19.8).

Population similarities can be visualized graphically by calculating all pairwise unifrac distances and then generating a tree either from these distances or by principal-component analysis (PCoA or PCA) (Fig. 19.9). Note that in the case of trees, each leaf of the tree does not represent a specific sequence but instead represents the entire population of sequences, and distance in the tree represents how similar or distant these populations are.

Their metric for "diversity" is much simpler; for any one community, they just add up all of the branch lengths leading to sequences that appear in that sample and divide by the number of sequences obtained.

Note that in both the unifrac and the diversity metrics, sequences greater than or equal to 97% identical are considered to be the same thing; this is sometimes called a "molecular species," or operational taxonomic unit.

Thus, they sorted the fractions of each kind of sequence in each body habitat, per individual, across the time points. These data are summarized graphically in Fig. S4 of the paper. In the graphs at the top, each bar represents the

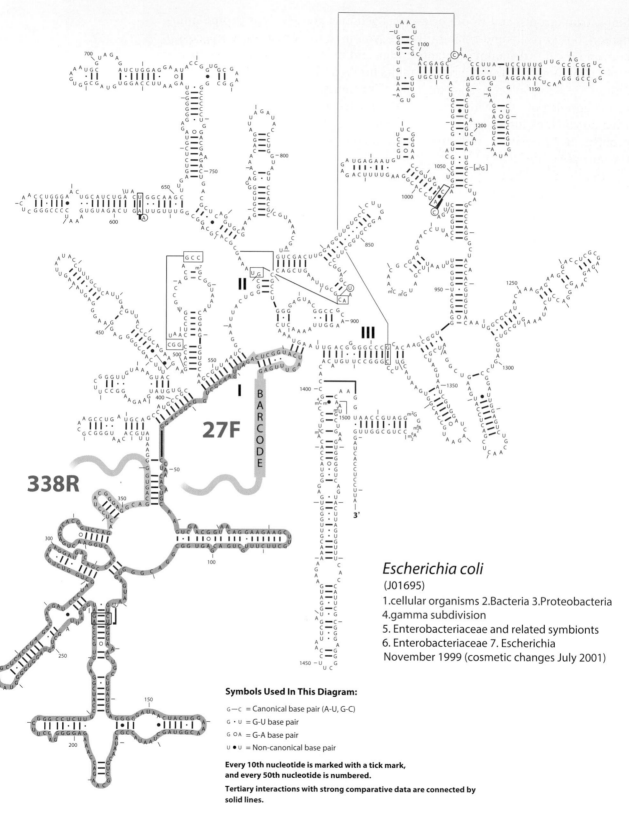

*Escherichia coli*

(J01695)

1.cellular organisms 2.Bacteria 3.Proteobacteria
4.gamma subdivision
5. Enterobacteriaceae and related symbionts
6. Enterobacteriaceae 7. Escherichia
November 1999 (cosmetic changes July 2001)

**Symbols Used In This Diagram:**

G—C  = Canonical base pair (A-U, G-C)

G • U  = G-U base pair

G ОА  = G-A base pair

U • U  = Non-canonical base pair

Every 10th nucleotide is marked with a tick mark,
and every 50th nucleotide is numbered.

Tertiary interactions with strong comparative data are connected by
solid lines.

Citation and related information available at http://www.rna.icmb.utexas.edu

**Figure 19.7** Locations of the primers and amplified regions of the SSU rRNA. The primer and target sequences are shown in blue, and the amplified sequence obtained is highlighted in green. Primer 27F is the forward primer, ending (the 3′ end of the primer) at position 27 of the *E. coli* SSU rRNA. Primer 338R is the reverse primer, ending at position 338. Secondary structure: SSU rRNA. (*E. coli* SSU rRNA secondary structure courtesy of Robin Gutell.) doi:10.1128/9781555818517.ch19.f19.7

**Figure 19.8** Graphical description of Unifrac. (Reprinted from Lozupone C, Knight R, *Appl Environ Microbiol* **71**:8228–8235, 2005, with permission.) doi:10.1128/9781555818517.ch19.f19.8

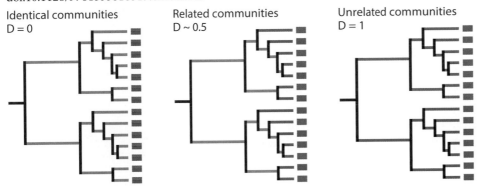

Identical communities
D = 0

Related communities
D ~ 0.5

Unrelated communities
D = 1

**Figure 19.9** Graphical descriptions of how principal-component analysis (PCoA) and clustering (very basic trees) are used to compare populations of sequences from a microbiome. (Reprinted from Lozupone C, Knight R, *Appl Environ Microbiol* **71**:8228–8235, 2005, with permission.) doi:10.1128/9781555818517.ch19.f19.9

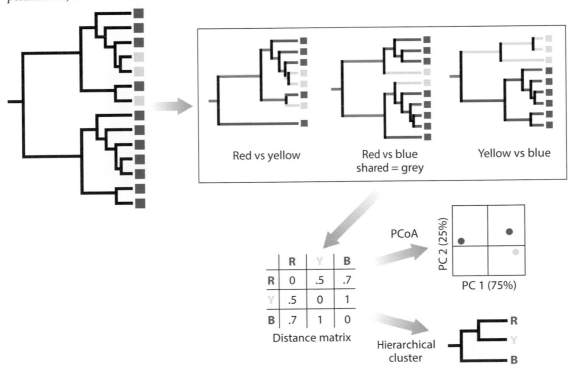

Red vs yellow

Red vs blue
shared = grey

Yellow vs blue

PCoA

PC 2 (25%)

PC 1 (75%)

|   | R | Y | B |
|---|---|---|---|
| R | 0 | .5 | .7 |
| Y | .5 | 0 | 1 |
| B | .7 | 1 | 0 |

Distance matrix

Hierarchical cluster

R
Y
B

fraction of sequences from any particular phylogenetic group from a specific person. The lower graphs are by date. Look, for example, at the oral cavity and the gut portions of the diagram. Notice that these two body sites are very different in composition: a mixture of *Streptococcus, Veillonella, Prevotella, Pasteurellaceaea*, and *Neisseria* (and so on) in the mouth, but predominantly *Bacteroides* in the gut. Notice also the big differences in the fraction of organisms in different people (upper graphs); e.g., male #4 has a mouth full of *Veillonella* and *Prevotella* whereas male #3 has a mouth full of *Neisseria* (*Neisseria* species are common oral symbionts, not just agents of sexually transmitted diseases; I suspect this subject was relieved to be reminded of this). In contrast, the populations are much less variable over time (lower graphs).

Therefore, human microbial populations are defined primarily by body site. This can also be visualized in the PCA plots (Fig. 1 of the paper). As you can see (Fig. 1B to D), sorting by person, time, or even sex (although there is some discrimination here) is not helpful, but sorting by body location shows that the mouth, gut, and external auditory canal (EAC) form nice, distinct clusters. Figure 1E and F show that there is much more variation between body sites than between people or over time. Figure 1G shows this in a tree of microbial populations by location, and the supplementary data include a number of graphs, trees, and plots going through this in detail.

The authors also spend some time showing that some sites are more diverse than others (Fig. 2 of the paper). Figure 2A shows that some parts of the skin (generally the arms, hands, and legs) contained the most diverse of populations. The gut and mouth were also very diverse. In contrast, other parts of the skin were not very diverse, nor were the EAC or labia minora. (A more detailed version of this graph can be found as Fig. S14 to S16 of the paper.) Figure 2B shows that the tongue was very consistent whereas the forehead was more variable and the forearm was highly variable. The changes in diversity over time are shown in Fig. 2C.

This information is graphed as *rarefaction curves*, a statistical representation in which you plot from your data the average amount of diversity (remember: sum tree branch length) with increasing number of sequences sampled. For example, in the first graph, for "F1 palm," with 500 randomly chosen sequences from the data, the average sum length of the tree of these sequences is about 17. If you double the number of sequences used to 1,000, the total tree length increases to 26. With large enough numbers of sequences, these should flatten out to horizontal; this would mean that you have collected enough sequences to get everything. Notice, however, that the curves do not approach an asymptote; they do flatten out but not toward a slope of zero. This means that you should expect to continue to get new sequences no matter how much you increase your data sample size. In other words, the microbial diversity of these samples is a bottomless pit!

One interesting finding was that gut communities varied a lot between individuals but were relatively constant over time. This is a huge community; there are certainly more microbes in the gut than everywhere else on the body put together. The authors believe that these stable differences in microbial gut communities might have a big impact on health, including nutrition, immunity, and especially obesity.

On the other hand, the mouth generally harbored a very diverse microbial population but was much more consistent from person to person or from time to time than were other sites of the body. This means that disease related to problematic oral microbiology might be easier to identify than at other sites of the body; "normal" is more readily defined.

Skin populations also vary with site and could be clustered into three types: the head (dominated by propionibacteria), the arms (also lots of propionibacteria, but less predominantly so), and the trunk and legs (staphylococci and corynebacteria).

There is a second interesting implication of the observation that there is so much variation in human microbial populations and that this variation is stable: your microbiome is unique to you, and you might leave traces of this wherever you go (on doorknobs, cell phones, keyboards, and so on). In a more a recent paper, this same group showed that they could match computer keyboards and mice to their users just from the microbial populations left behind, and they could do so weeks after their last use.

In the supplementary material, they demonstrate that they could readily distinguish individuals by their composite microbiomes (Fig. S8 of the paper). Note that in all cases, days 1 and 2 are related, as are days 3 and 4; remember that 2 is one day after 1, and 4 is one day after 3, but these pairs are 3 months apart.

The authors then go on to ask whether these differences in skin populations are the result of differences in environmental conditions or just historical contingency (who got there first). Much of the rest of the paper, which is not reviewed here, deals with "transplants," in which areas of skin are disinfected and populated artificially with populations from other sites or other people.

---

## High-throughput sequencing technology

The authors of this paper were able to collect so many sequence data because they used a modern high-throughput sequencing method: 454 sequencing. 454 sequencing is a relatively new technique that allows the researcher to get hundreds of thousands of sequences in a single experiment. The main drawbacks of this method are that (i) the sequences obtained are, at least for now, relatively short (less than 500 bp) and (ii) it is expensive (per run, not per sequence) and requires one-of-a-kind technology. Another popular high-throughput sequencing technology is Illumina sequencing. Illumina sequencing yields even more sequence data than 454 sequencing, but in shorter bits of about 100 nucleotides (as I write this).

454 sequencing starts with a standard PCR amplification, using primers that generate a relatively short (<500-bp) product. Each of the primers has extra sequences on the end, one for annealing to beads and the other serving as a primer-binding site for the sequencing reactions.

The PCR product DNA is denatured and mixed with an excess of small beads coated with oligonucleotides that bind to one end of the PCR product. Because there is an excess of beads, most of the beads get only one (or no) DNA molecule annealed to them. The slurry of beads in PCR mixture is then emulsified with oil. Each bead ends up in a droplet of aqueous PCR solution in this oil emulsion; each of these aqueous droplets acts as a microscopic PCR tube.

The emulsion is run through a set of PCR cycles as usual. The PCR product generated in each drop anneals to the oligonucleotide tags on the bead, so each bead ends up covered in DNA with a single sequence. The emulsion can then be broken (separating the oil and aqueous phases) and the DNA-covered beads separated out.

The beads are then washed over a silicon wafer honeycomb picotiter plate. Each cell of the honeycomb is big enough for only one bead, so the honeycomb ends up filled with one bead per well. Each of these serves as a sequencing reactor. The wells are filled with very small beads coated in two enzymes: sulfurylase and luciferase.

The honeycombs are then percolated with reaction mixture of the sequencing primer, DNA polymerase, APS (adenosine 5′-phosphorothioate), and luciferin, followed by sequential cycles of dGTP, dATP, dCTP, and dTTP, over and over again. When the next nucleotide in a sequence matches the deoxynucleoside triphosphate (dNTP) added, the base is added to the growing DNA chain, releasing pyrophosphate. Sulfurylase replaces the thio group on APS with the pyrophosphate, creating ATP, which is used by luciferase to react with luciferin to generate a pulse of light.

In other words, the well "blinks" a flash of light to signal that the nucleotide in the sequence matches the current dNTP. If there is a stretch of two or more of the same base, proportionally more pyrophosphate is generated and the flash of light is proportionally stronger. So, as the dNTPs flow past in repeated cycles, each well signals the sequence of the DNA on that bead in a kind of Morse code. A magnifying charge-coupled device (CCD) video camera watches the honeycomb, collecting the data from all of the wells over time, and a computer then separates the data from each spot to read the sequences.

The name "454" is said to come from the address of the startup company that developed this method, although others have suggested that it comes from the notion that this is the temperature at which money burns. Refinement of the technology is increasing the length of the sequence reads, and there is no conceptual reason why they could not ultimately be as long on average as those in traditional sequencing methods.

## Questions for thought

1. Sulfide is very toxic to animals (largely because is displaces oxygen in the blood and in the mitochondria), and yet these vent animals swim in it. How do you think they can be so resistant?

2. Can you think of any way to determine whether it is the snail or the biofilm, or some combination of both, that creates the iron sulfide plating on the scales?

**3.** The scaly snail endosymbionts are in the esophageal gland. Given that these are egg-laying animals, how do you think the snail's offspring get their endosymbionts? How would you test this?

**4.** Likewise, how do you think young snails get their (presumably) protective biofilm of armor-plating organisms?

**5.** The authors of the scaly snail paper go to great lengths to remind the reader that the relative number of sequences they obtained is not a good measure of the relative abundance of the different kinds of organisms. How, then, could you use the sequence data to either count the organisms in or on the snail, or identify them specifically among the complex biofilm community?

**6.** Is there some confirmation experiment the authors have failed to perform to show that the single bacterial SSU rRNA sequence they obtained really is from the organism they can see in the bacteriocytes of the esophageal gland?

**7.** Given a 200- to 270-bp sequence and 1 g (ca. $10^{12}$ cells) of bacterial biomass, which would do you think would run out first on the rarefaction curves above, the number of actual SSU rRNA molecules in the sample or the number of sequence possibilities?

**8.** What do you think the data would look like if the authors chose to compare people from different parts of the world? Of different races? Or other animals? Old versus young? Lean versus obese?

**9.** What systematic biases do you think this approach might have?

**10.** The data in this paper are overwhelming and are rather hard to interpret. Can you think of a better way to present the data more clearly?

**11.** How useful do you think this sort of a census is? What kind of problems can you think of that you could solve with this approach?

# 20 Fluorescent In Situ Hybridization Surveys

## Fluorescent in situ hybridization

We have already talked a little about fluorescent in situ hybridization (FISH), but it perhaps bears a general review. In this method, cells in an environmental sample are fixed with glutaraldehyde, then treated to make them permeable (e.g., with toluene) and mixed with an oligonucleotide probe that contains a fluorescent tag such as Texas Red or acridine orange. The probe finds matches in the DNA and/or RNA of the permeabilized cells and sticks to it. Unannealed probe is washed out, and the sample is examined by fluorescence microscopy. If enough probe accumulates in a cell, i.e., if it contains the target for which the probe is designed, it should be fluorescent.

The most commonly used probes for FISH in the microbial world target the small-subunit ribosomal RNA (SSU rRNA). FISH probes targeting SSU rRNA sequences are sometimes called phylogenetic probes because they target specific phylogenetic groups (Fig. 20.1). There are two reasons why SSU rRNA is a good target for FISH. First, it allows you to search using phylogenetically relevant sequences; often the SSU rRNA sequence is all you know about an organism. This also allows you to tune the range of organisms to be labeled; the more conserved the target region of the SSU rRNA, the wider the phylogenetic range of cells that can be labeled. Second, because there are thousands of ribosomes in each cell, a lot of probe can bind to each cell, yielding a strong fluorescent signal. In fact, generally only metabolically active cells contain enough ribosomes to be labeled with an SSU rRNA probe.

doi:10.1128/9781555818517.ch20

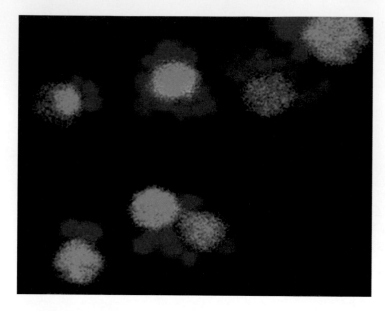

**Figure 20.1** *Nanoarchaeum equitans* stained with a specific Texas Red-labeled FISH probe (red), and its host *Igniococcus* stained with an acridine orange (green)-labeled specific FISH probe. (Courtesy of Reinhard Rachel, Harald Huber, Reinhard Wirth, and Michael Thomm, University of Regensburg.) doi:10.1128/9781555818517.ch20.f20.1

## Confocal laser scanning microscopy

Confocal laser scanning microscopy (CLSM) is a method that allows the collection of three-dimensional (3-D) data from a fluorescence microscope sample. The laser light is focused on a single point in the field of view (about 1 μm in diameter), and the resulting fluorescence from that point is visualized and quantitated by the detector. The laser focal point is scanned back and forth across the field of view to collect a single-plane image about 1 μm thick: a bitmap, just like a digital image. The focal plane is then moved slightly up or down in the sample and another plane image is collected . . . and again, over and over. These focal-plane images are stacked one on top of the other, and the result is a 3-D digital image of the fluorescence in the sample. An easy way to view these 3-D images is to compress them vertically, to create a 2-D view, as if the fluorescence microscope had a depth of field so deep that the entire sample was in focus in the same view.

■ **STUDY AND ANALYSIS**
**FISH survey of *Chloroflexi* in wastewater sludge**

Björnsson L, Hugenholtz P, Tyson GW, Blackall LL. 2002 Filamentous *Chloroflexi* (green non-sulfur bacteria) are abundant in wastewater treatment processes with biological nutrient removal. *Microbiology* **148**:2309–2318.
PubMed ID: 12177325

*You should have a copy of this paper at hand before proceeding.*

*Purpose:* To design phylum- and subdivision-specific oligonucleotide probes for the *Chloroflexi* and evaluate them on sludge samples by using FISH, in order to determine the abundance, morphology, and spatial distribution of *Chloroflexi* in activated sludges.

Wastewater is commonly treated to remove a majority of the organic and nitrogenous (and sometimes phosphate) material in the form of sludge; the remaining water is disinfected (chlorine, chloramine, ozone, and ultraviolet [UV] light are the most common disinfection agents in the United States) and released into some waterway (Fig. 20.2). First the raw wastewater is passed through a rough grating to remove large foreign objects (e.g., condoms, guns, dead bodies, and alligators). It then undergoes an aerated digestion; in this step, dissolved organics are taken up by microbes and converted into biomass or converted to $CO_2$ by aerobic respiration. Then the wastewater is allowed to settle; the bacteria collect in flocs and settle into a relatively thin layer at the bottom of the pool. The clarified water is pumped out from the top of the settled pool, disinfected, and released into the environment. The settled sludge is pumped into a cesspool (called a "lagoon" in one of the most egregious of Orwellian misuses of the English language) for anaerobic digestion. During anaerobic digestion, a large part of the biomass is converted by microbial activity to $CO_2$ and $CH_4$, and the residue is applied to the soil as fertilizer.

A problem that can occur in the settling stage is the overgrowth of filamentous bacteria that trap air and prevent floc settling; the result is bulking (if settling results in a thick, water-laden layer) or foaming (if flocs float to the surface).

It is commonly thought that mesophilic members of the *Chloroflexi*, relatives of *Herpetosiphon*, are major contributors to wastewater bulking. However,

**Figure 20.2** Typical wastewater treatment process. doi:10.1128/9781555818517.ch20.f20.2

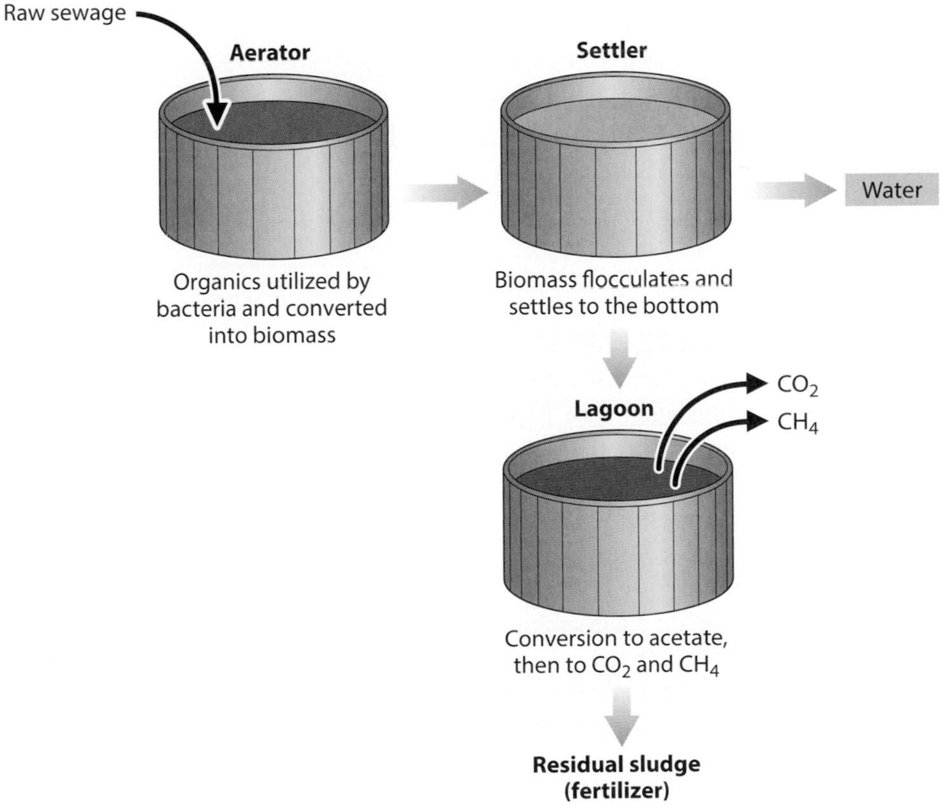

Raw sewage

**Aerator**
Organics utilized by bacteria and converted into biomass

**Settler**
Biomass flocculates and settles to the bottom

Water

**Lagoon**
Conversion to acetate, then to $CO_2$ and $CH_4$

$CO_2$
$CH_4$

**Residual sludge (fertilizer)**

the vast majority of filamentous bacteria in wastewater sludges have not been cultivated or identified; they are usually categorized only by microscopic morphology or staining.

The samples used in the survey described in this paper come from activated sludge from 10 wastewater treatment plants in Queensland and New South Wales, Australia, and two laboratory-scale reactors (Table 1 of the paper). In preliminary experiments, the authors had previously cloned and sequenced SSU rRNAs from six members of the *Chloroflexi* from their laboratory-scale reactors; these are the "sludge clone SBR" sequences. (Notice the discussion in the first paragraph of the Results that one of their sequences was a chimera.)

The authors designed a series of three phylogenetic probes to examine *Chloroflexi* in normal wastewater after aerobic digestion (Table 2 of the paper). These probes were designed from an alignment of all of the *Choroflexi* SSU rRNA sequences available at the time, including their SBR sequences. This alignment was used to generate a phylogenetic tree (Fig. 1 of the paper), and the sequences that the probes ought to "hit" are shown on the right side of the figure. The authors also include a previously described *Chloroflexi* PCR primer for use as a probe (GNSB-941) and also a probe mixture (EUBMIX) that is designed to hit as many members of the *Bacteria*, but not *Archaea* or *Eukarya*, as possible. All of the *Chloroflexi* probes were tested against organisms that were perfect matches (positive controls) and single mismatches (negative controls) to demonstrate their specificity. For use in FISH, these probes were labeled with fluorescein isothiocyanate (FITC) (blue), Cy3 (green), or Cy5 (red). The abundance of *Chloroflexi* was assessed subjectively on a scale of 0 (none) to 6 (excessive).

Notice that phylogenetic probes are *not* nice and neat; none of the probes hit all members of the targeted group and nothing else. Probes GNSB-941 and CFX1223 are the broadest, hitting most of the *Chloroflexi*, and mixed together give very good coverage except for a small group of sequences that, by hard luck, includes one of their SBR clones (SBR2022). Probe CFX784 hits most of subdivisions 1a and 1b and not much else, and probe CFX109 hits most of subdivision 3 and nothing else.

The first result the authors describe is just a phylogenetic analysis of the *Chloroflexi* sequences (Fig. 1 of the paper). In this phylogenetic analysis, the authors agree with the previous subdivision of the *Chloroflexi* into four subdivisions (1, 2, 3, and 4), plus the *Thermomicrobia*. The cultivated species (*Chloroflexus, Roseiflexus, Oscillochloris,* and *Herpetosiphon*) are predominantly in subdivision 3, but most of the uncultivated "environmental" sequences are in subdivision 1.

The authors then used these probes with samples of wastewater sludge (Table 1 of the paper). The columns in the table under "Abundance" are from the FISH data: "General" are the data for filaments that hybridize with the EUBMIX probe, "Chloroflexi" are filaments that hybridize with the GNSB-941 and CFX1223 probe mix, "Subdivision 1a,b" are those that hybridize with probe CFX784, and "Subdivision 3" are those that hybridize with probe CFX109. The authors point out that nearly everything that hybridized with any of the *Chloroflexi* probes was filamentous, consistent with the phenotype of cultivated species.

The majority of filaments seen in the activated sludge were members of the *Chloroflexi*, and the differentially labeled probes were able to sort many of these into subcategories on the basis of signal color (Fig. 2 of the paper). Most of the filaments they saw seem to be from subdivision 3, even though most environmental clones are from subdivision 1. The use of confocal microscopy allowed the authors to examine these flocs in three dimensions.

The authors concluded that their probes were a lot better than traditional staining and microscopy for seeing and identifying *Chloroflexi* in sludge flocs. This is probably true, but it should be noted that they did not actually do any head-to-head comparisons to validate this.

Interestingly, although they had a lot of *Chloroflexi* in these sludges, the filaments were inside the flocs, not bridging them (which is what seems to lead to bulking), and none of these sludges were suffering from bulking. The authors went on to suggest that perhaps *Chloroflexi* are actually important for good floc formation and macromolecule degradation; in other words, they might be good for the wastewater treatment process rather than problematic.

This paper is a great example of the right way to use SSU rRNA-based FISH to study a microbial community, but it also points out the uncertainties and the work involved; this represents a huge investment of time and effort for the analysis of a few samples. Ecologists need to study samples taken from many locations in an environment and at various times. It is really not yet reasonable in most circumstances to do this using the FISH approach.

## Questions for thought

1. Can you think of any target other than SSU rRNA that might make a good target for FISH?

2. Given that FISH combined with CLSM allows 3-D phylogenetic analysis of an environment, can you imagine the 4-D (three in space and one in time) analysis of a sample?

3. What are the differences in the data obtained from a molecular phylogenetic analysis of a microbial population using FISH versus the clone-and-sequence approach? Under what circumstances would you choose one of these approaches over the other?

4. The authors gloss over the fact that phototrophic organisms, decaying plant material, and a wide range of other stuff in wastewater are inherently fluorescent. How do you think this would affect FISH analysis? How do you think you might get around this, or accommodate it?

# 21

# Molecular Fingerprinting of Microbial Populations

Molecular phylogenetic surveys give a much larger picture of the diversity of an environment than the older cultivation-dependent methods, but a real drawback is the time, expense, and energy required. This might not be the case much longer, as sequence-based methods continue to evolve. Nevertheless, at least for now these molecular surveys are almost always performed one at a time, a single snapshot of a single time and place in the environment. However, microbial ecologists know that a lot of the important microbiology is in the variation in population from one place to the next (measured at scales ranging from micrometers to thousands of kilometers) or from one time to another (measured in scales of minutes to thousands of years). How, for example, do you survey the dynamics of the microbial population in each layer of a Yellowstone microbial mat, from its source pool to its drainage into a cold stream, during the diurnal and seasonal cycles? How do you monitor the changes in microbial population in the subsurface as a gasoline plume forms from a leaky storage tank?

The usual answer is to use "fingerprinting" methods that provide patterns that can be used to distinguish populations without necessarily sequencing anything (although it is useful to be able to get sequences, or identifications of organisms, from the bands that make up the patterns in some cases). A relatively old-fashioned approach is denaturing gradient gel electrophoresis (DGGE). The emerging alternative is terminal restriction fragment length polymorphism (t-RFLP). This chapter discusses these technologies and examples of how they can be used.

We also address real-time polymerase chain reaction (PCR), not because it is a fingerprinting method, but because it is often used in conjunction with these methods to look at the numbers of specific kinds of organisms in a population.

doi:10.1128/9781555818517.ch21

# Denaturing gradient gel electrophoresis

DGGE starts out like most molecular phylogenetic analyses do, by the isolation of DNA from environmental samples, followed by PCR of small-subunit ribosomal RNA (SSU rRNA) genes (SSU rDNA). Rather than cloning and sequencing from this pool of DNA molecules, however, this method separates them into a pattern of unique bands based on their denaturation properties.

DGGE is carried out with polyacrylamide gels in which the concentration of urea and formamide increases from the top to the bottom in the gel; i.e., the gel contains a gradient of denaturants (Fig. 21.1). (Remember that denaturation of DNA means separation of the two strands.) The PCR-amplified SSU rDNA is loaded into wells at the top of the gel, where the concentration of urea and formamide is too low to denature it. As the SSU rDNA migrates down the gel during electrophoresis, the concentration of urea and formamide to which the DNA is exposed increases until, at some point, it is high enough to denature the DNA. At this point, the SSU rDNA band essentially stops moving (it slows drastically). Because every unique SSU rDNA sequence has a slightly different denaturation point, the different DNAs denature at different levels of the gel and separate into distinct bands despite the fact that the DNAs in all of the resulting bands are essentially the same size.

**Figure 21.1** Diagrammatic representation of the DGGE process. Note that this diagram shows how electrophoresis progresses over time; this process occurs in each well of a gel. doi:10.1128/9781555818517.ch21.f21.1

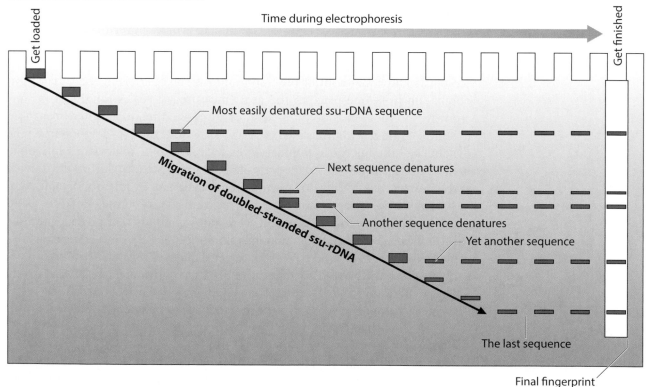

Time during electrophoresis

Get loaded

Get finished

Most easily denatured ssu-rDNA sequence

Migration of doubled-stranded ssu-rDNA

Next sequence denatures

Another sequence denatures

Yet another sequence

The last sequence

Final fingerprint

A technical improvement that has become standard in this method is the incorporation of a long tail of GC base pairs to the end of one of the PCR primers. This "GC clamp" keeps the denatured strands of the SSU rDNA from becoming completely separated, effectively doubling the length of the denatured single-stranded DNA (slowing it to a near stop), and so that the two strands do not begin to separate into two distinct bands in the gel, confusing the issue.

Another version of DGGE is temperature gradient gel electrophoresis (TGGE), in which the denaturant is temperature instead of (or in addition to) urea and formamide. The electrophoresis unit is designed so that the gel is heated in a controllable fashion, usually to a higher temperature at the bottom than at the top.

DGGE or TGGE gels are stained and visualized as usual. Each visible band represents an abundant organism in that environment. The pattern of bands is a "fingerprint" of the microbial population of the environment. The intensity of each band represents the abundance of the organism, at least to some extent, and can be monitored from place to place or time to time. To identify the organism represented by each band, you can excise the band from a gel, reamplify it by PCR, and sequence it.

### ■ STUDY AND ANALYSIS
### Response of Community Structure in a Geothermal Spring to Thermal Stress

**Yim LC, Hongmei J, Aitchison JC, Pointing SB.** 2006. Highly diverse community structure in a remote central Tibetan geothermal spring does not display monotonic variation to thermal stress. *FEMS Microbiology Ecology* **57**:80–91.

PubMed ID: 13523374    PubMed Central ID: PMC2537641

*You should have a copy of this paper at hand before proceeding.*

*Question being asked:* How do the bacterial communities in various hot springs and temperature zones differ?

The authors attempted in this study to examine two aspects of diversity in this Tibetan hot spring: how the diversity varies with temperature, and how the organisms differ from those of similar hot springs elsewhere in the world. For the first, they hypothesized that the populations do not vary directly (monotonically) with regard to thermal stress; in other words, the populations do not get incrementally simpler with each increase in temperature. For the second, they compared the sequences they got with those available in GenBank, including sequences from hot springs all over the world, to see if they could detect any distinctness of their sequences; in other words, were they seeing "phylogeographic" groups?

The issue of geographic groups of microbes, i.e., the notion that certain bacterial species or strains might have a geographic range, in the same way that macroscopic plants and animals do, is an important one that is usually ignored in microbiology. It is usually assumed that the local environment controls the microbial species present and that there is enough exposure of every environment to every kind of organism that all places where the same local environment exists will have the same kinds of microbes. In the introduction of the paper, the authors described several cases where this was and was not true for hot spring organisms.

If this Tibetan site looks familiar, it should. It is a lot like the typical neutral-pH low-sulfide hot springs of Yellowstone National Park (Fig. 21.2).

So, in this study, the authors took samples in transects across thermal gradients. They used PCR primers specific for *Bacteria*, *Archaea*, *Cyanobacteria*, and *Chloroflexi* to amplify SSU rRNA genes and then separated them by DGGE (Fig. 1 of the paper). Each distinct band from each sample was excised from the gel, reamplified, and sequenced so that the phylotype of the organism it represents could be determined. They also got a rough idea of what the organisms were by identifying their closest relatives in GenBank using BLAST. A table of these results is given in the online supplement for the paper. Most of the remainder of the paper is a series of phylogenetic trees based on these sequences (Fig. 2 through 5 of the paper).

**Figure 21.2** Geyser at the Daggyai Tso geothermal region of Tibet. (Courtesy of Michael Searle.) doi:10.1128/9781555818517.ch21.f21.2

The authors made a point of saying that 19% of their sequences were archaeal, but keep in mind that they used distinct archaeal and bacterial primers for their PCRs. In fact, they used the percentage of their total numbers of sequences all the time as an implicit measure of abundance in the environment. However, it is the relative intensities of the bands, not the number of times they obtained a particular sequence, that is relevant. Even so, it is unclear whether their PCRs are designed to be quantitative; real quantitation would require either fluorescent in situ hybridization (FISH) or real-time PCR, which they have not done.

The last figure of the paper (Fig. 6) is an assessment of diversity along the temperature gradient. They used several measures of diversity and attempted to show that the most diverse samples were from the 63 to 70°C temperature samples, not the lowest temperatures; in other words, they were trying to show that diversity does *not* decrease evenly with increased thermal stress. This seems to me like a straw man argument, an intentionally weak hypothesis created specifically to be refuted. I do not think any ecologist would argue that diversity should decrease quantitatively with environmental stresses, only that diversity tends to be lower in harsher environments. This decrease in diversity is expected to be most obvious near the limits (well above 70°C, in the case of high temperature), and there is no reason to expect it to be a direct relationship (monotonic). You can see this in Yellowstone hot springs, for example: the color changes take place stepwise, as bands (not smoothly), as the temperature changes.

Unfortunately, I think their final point is also weak: the notion that the sequences they see are distinct from anything in GenBank and therefore represent novel species found only in this area. Where is the control experiment for this? This conclusion would require obtaining similar samples from other similar hot springs worldwide, some close to one another and some remote, and *then* seeing if different sequences are specific to specific geographic ranges.

## Terminal restriction fragment length polymorphism

t-RFLP is similar to DGGE in that it generates fingerprints of a population, but, unlike DGGE, the bands can be assigned to specific organisms directly, without the need for sequencing.

Imagine the simplest case of a pure culture of an unknown organism. You amplify the SSU rRNA gene with some set of primers, e.g., 515F and 1492R, and one of the primers (515F in this example) is fluorescently labeled. You digest this SSU rDNA with several different restriction enzymes and separate the products by size on a sequencing gel (Fig. 21.3). The exact sizes of the labeled fragments could be compared to a database of fragments of a wide range of SSU rRNA sequences that would be generated from PCR products from those primers digested with those enzymes. If the restriction enzymes were carefully

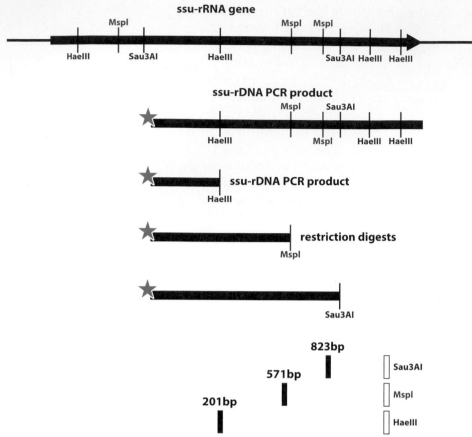

**Figure 21.3** How t-RFLP works. The restriction map of an example SSU rRNA gene is shown at the top. PCR amplification using a fluorescently labeled (star) primer (in this case at about position 500) and an unlabeled primer (in this case near the 3′ end of the SSU rDNA) yields a specifically labeled PCR product. Digestion of this PCR product, in separate reactions, with various restriction enzymes produces a pattern of bands that can be used to identify the origin of the sequence. doi:10.1128/9781555818517.ch21.f21.3

chosen, a computer program should be able to sift through the database and identify your organism based on the observed (from the gel) sizes of the labeled fragments. For example, there might be 100 organisms whose SSU rDNA, if amplified with labeled 515F and unlabeled 1492R and digested with HaeIII, should give a 201-bp fragment. There might also be another 100 organisms that would have a 571-bp MspI fragment. But there is only one name on both lists; that's your organism. This identification might be verified by the presence of a predictable 823-bp Sau3AI fragment.

This should be pretty easy, but now imagine doing the same thing with a population of organisms from a natural environment. Now you have several abundant organisms, some more common than others, creating a pattern of bands in each digest. However, the computer can, if the experiment is properly set up, sift through the peaks and determine what mixture of organisms would create that pattern of bands (Fig. 21.4).

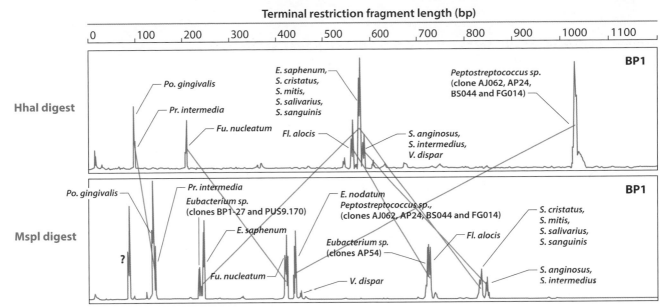

**Figure 21.4** Deconvolution of sequence identities from multiple restriction digests in a t-RFLP experiment. (Based on Sakamoto M, Huang Y, Ohnishi M, Umeda M, Ishikawa I, Benno Y, *J Med Microbiol* **53**:563–571, 2004.) doi:10.1128/9781555818517.ch21.f21.4

The ability to sift through the microbial population by using t-RFLP is limited by the choice of primers (what kind of organisms they amplify SSU rDNA from), the ability to choose the optimal restriction enzymes, and the database of SSU rRNA sequences. t-RFLP is an emerging technology, so there is plenty of room for improvement in all of these aspects, but this approach is very useful and incredibly promising. Probably the ultimate limitation will be PCR primers; this is a limitation shared by all of the molecular phylogenetic approaches discussed.

## Real-time PCR

Although not a fingerprinting technique, real-time PCR is often used in conjunction with fingerprinting methods.

Real-time PCR (sometimes abbreviated rt-PCR, which is unfortunately readily confused with reverse transcriptase PCR, also rt-PCR) is a method used to quantitate the number of target genes in a DNA sample. In real-time PCR, the concentration of DNA product is continuously measured during PCR amplification, usually by the addition of SYBR green, a fluorescent dye that binds double-stranded DNA and is fluorescent when bound to DNA (Fig. 21.5). This fluorescence is measured during amplification with a fluorometer built into the PCR machine. In a real-time PCR experiment, the fluorescence increases more or less exponentially at first, then plateaus as the reagents are exhausted. The point at which each PCR crosses some threshold (best set relatively low) can be plotted against a standard curve to determine the concentration of PCR target in the original sample.

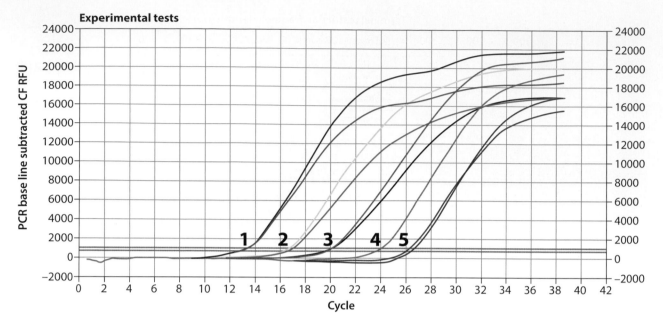

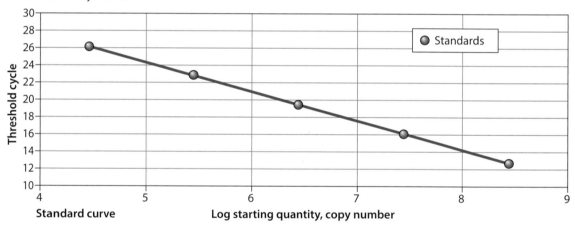

Correlation coefficient: 0.999   Slope: −3.353   Intercept: 41.054   Y = −3.353 X + 41.084
PCR efficiency: 98.7%

**Figure 21.5**  Real-time PCR. The top panel shows the result of a real-time PCR experiment; each line represents a single PCR and its accumulation of DNA. The bottom panel shows a standard curve based on gene copy number and how long the real-time PCR took to cross a threshold level of DNA product; this can be used to measure gene copies in an unknown sample. (Reprinted from Yang J-L, Cheng A-C, Wang M-S, Pan K-C, Li M, Guo Y-F, Li C-F, Zhu D-K, Chen X-Y, *Virol J* **6:**142, 2009. http://creativecommons.org/licenses/by/4.0/) doi:10.1128/9781555818517.ch21.f21.5

## Changes in Oral Microbial Profiles after Periodontal Treatment

**Sakamoto M, Huang Y, Ohnishi M, Umeda M, Ishikawa I, Benno Y.** 2004. Changes in oral microbial profiles after periodontal treatment as determined by molecular analysis of 16S rRNA genes. *Journal of Medical Microbiology* **53:** 563–571.

PubMed ID: 15150339

*You should have a copy of this paper at hand before proceeding.*

*Question being asked:* How does the periodontal microflora change after treatment?

In this study, the authors used t-RFLP to study subgingival plaque from three periodontal disease patients before and after treatment (lessons in oral hygiene and very complete teeth cleaning involving scaling and planing both above and below the gums). Samples were taken from the subgingiva of three or four teeth before and 3 months after treatment.

Although the authors used real-time PCR and SSU rDNA clone libraries as well as t-RFLP, let's review the t-RFLP data first.

t-RFLP began with PCR amplification from the samples by using 6-carboxyfluorescein (6-FAM)-labeled 27F (bacterium specific) and 1492R (universal). Samples of the PCR products were digested with HhaI (GCG^C) and MspI (C^CGG) and separated by size on a DNA sequencing machine.

Figure 1 of the paper shows an example of their data from the HhaI digests. At this point, these are viewed as fingerprints; the identities of the organisms represented by the peaks are not important. The figure shows that there are differences before and after treatment. These are pretty subtle for patient A but clear for patients B and C.

Figure 3 of the paper gives a better example of how the data are examined. In the top panel (a) are the HhaI digests from one site of one patient, before (top) and after (bottom) treatment. The bottom panel (b) shows the same data from the MspI digests. Notice that after treatment (this is patient B), the *Peptostreptococcus*, *Porphyromonas gingivalis*, and *Prevotella intermedia* (known problem organisms) all disappear or diminish. The other organisms, which the authors showed (using the same methods) to be common in healthy individuals, remain abundant.

Given that t-RFLP is a new technology, the authors used real-time PCR and more traditional SSU rDNA clone libraries to verify the results of their t-RFLP. They discuss these at some length in the paper. First they used real-time PCR with species-specific primers to determine the numbers of cells of several specific species in each sample (Table 4 of the paper).

Notice that they were looking specifically for some common oral spirochetes. Notice also that although patient A did not have much change in the abundance of these problem organisms, patients B and C showed dramatic decreases in them all. However, notice that these particular organisms (the spirochetes) did not show up in the t-RFLP experiments; this is probably why the authors also did these real-time PCRs.

They also cloned sequences from their PCRs and counted species found before and after treatment (Table 5 of the paper). This is the traditional SSU rRNA microbial survey method. They looked at a relatively small number of clones (90 before and 88 after treatment in a single tooth of a single patient), but even here you can see pathogens decreasing or disappearing and being replaced by commensals. However, given the small numbers (rarely more than a handful of any one organism), this table really should have included some statistics to show whether these changes are significant.

The take-home message is that these experiments confirm the observations of the t-RFLP: known harmful organisms are reduced and commensal ones remain constant or increase long after periodontal treatment.

## Questions for thought

**1.** How well do you think the computer would be able to identify members of a population from t-RFLP patterns if SSU rRNA sequences from some of the organisms are not in the database? Would this just result in unidentified peaks, or could it foil the ability of this method to identify organisms it *does* have in the database?

**2.** What might be the advantages or disadvantages of labeling one primer versus the other in a t-RFLP experiment? What about labeling both primers? How would you do this experiment?

**3.** What might you do if the restriction enzymes you use do not identify the organisms definitively? In other words, what if a set of bands could be one or more of several organisms?

**4.** How might you deal with the fact that many of the identifications in a t-RFLP experiment are likely to be from uncultivated organisms?

**5.** Why do you think the real-time PCR experiment was able to measure changes in the spirochete populations but these populations were not detected by t-RFLP?

**6.** Can you think of an environment that might be interesting to examine across space and/or time by DGGE technology?

**7.** What limitation(s) of molecular phylogenetic analysis does DGGE or t-RFLP not improve upon?

# 22 Linking Phenotype and Phylotype

Microbial ecologists have the tools to quantitatively decipher the processes taking place in microbial communities. They can monitor nitrogen, sulfur, carbon, phosphorus, reducing equivalents, energy, and so forth as they are transformed from one form to another. For example, it is reasonable to measure the steady-state levels of all of the forms of nitrogen in a nitrogen-cycling ecological community, including the inputs, outputs, and flux rates among all of the nitrogen compounds.

The limiting factor in microbial ecology these days is that essentially all we know about the organisms that make up these communities comes from studying the less than 1% of microbes that are readily cultivated. It should be clear at this point that taking even a very basic quantitative census of a microbial population is not straightforward, but it is possible, with much work, to get a semiquantitative assessment of the makeup of the population, especially the organisms that are abundant in the population. The next task is to link these bits of information together so as to identify the organism(s) responsible for specific steps of an ecological transformation.

For example, suppose you have an environment in which a certain process is taking place and a phylogenetic census of the organisms in a sample from this environment, how would you determine which organisms in your census are the ones that carry out the process you are interested in? Or, coming from the other direction, suppose you know an organism is abundant in an environment, how do you determine its ecological niche? The trick is to find or make a connection between the metabolic process and phylogenetic information. In this chapter we discuss two approaches to this problem: genomics and stable-isotope probing.

doi:10.1128/9781555818517.ch22

# The genomic or metagenomic approach

One way of linking microbial processes to specific organisms uses the genomic or metagenomic approach. In the case of "genomics," if you have an organism in culture and want to know what it's capable of, you can sequence its genome and find out. More commonly, you are not working with cultivated organisms and so you might use metagenomics: fishing large fragments of DNA out of environmental samples that contain both phylogenetic information (a copy of the small-subunit ribosomal RNA [SSU rRNA] gene) and phenotypic information (genes for phototrophy, carbon fixation, etc.). If both types of genes come from a single fragment of DNA, they must be from the same organism, therefore linking the phylotype (from the SSU rRNA sequence) and phenotype (from the metabolic gene) of that organism.

---

■ **STUDY AND ANALYSIS**
## Bacterial Rhodopsin: a New Type of Marine Phototrophy

Béjà O, Aravind L, Koonin EV, Suzuki MT, Hadd A, Nguyen LP, Jovanovich SB, Gates CM, Feldman RA, Spudich JL, Spudich EN, DeLong EF. 2000. Bacterial rhodopsin: evidence for a new type of phototrophy in the sea. *Science* **289**: 1902–1906.

PubMed ID: 10988064

---

*You should have a copy of this paper at hand before proceeding.*

**Question being asked:** What is the ecological role of the SAR86 rDNA sequence group?

In this study, the authors already knew from molecular phylogenetic analysis of ocean water that members of the SAR86 group of γ-proteobacteria are abundant worldwide, but they had no idea about the role of these microbes in the ecology of the ocean. Thus, again, the question is linking a particular phylogenetic group (the known in this case) to a particular ecological process (the unknown in this case).

In attempt to find out what SAR86 is doing in the ocean, they made a cosmid bank of DNA isolated from seawater. These cosmid clones contain DNA fragments more than 100 kbp in length. They screened these cosmids by hybridization to identify those that contained a gene (SSU rDNA) for SSU rRNA, so they could identify the organism the DNA fragment came from. One such cosmid clone (EBAC31A08) proved to be a member of the SAR86 group, based on phylogenetic analysis of this SSU rRNA sequence, and the authors therefore sequenced the entire 130-kbp DNA fragment, hoping to find genes that would provide clues about the metabolism of the organism (Fig. 22.1). This is not as hopeless as you might think at first; most γ-proteobacteria have a genome size of 2 to 4 Mbp and have about four to seven rRNA operons, so on average about 20% of the genome is within 130 kbp of an rRNA gene. Another way to look at it is that the average size of a protein-encoding gene is about 1 kbp, so they expected to get about 130 genes; the chances are good that 130 genes chosen randomly from a bacterial genome would give clues about its phenotype.

# Uncultured marine gamma proteobacterium EBAC31A08 BAC sequence

GenBank: AF279106.2

GenBank  FASTA

Link To This Page | Feedback

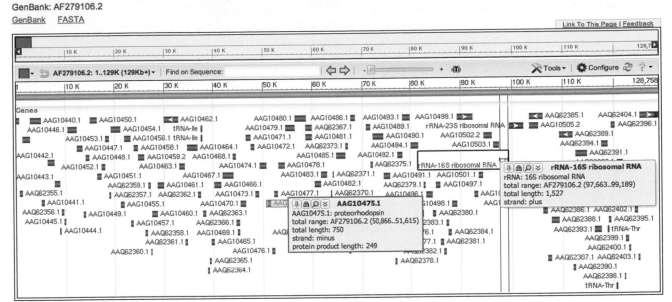

**Figure 22.1** NCBI Genome Browser view of the SAR86 cosmid EBAC31A08 sequence. Each red rectangle is a putative protein-encoding gene sequence; each blue rectangle is a putative RNA-encoding sequence. The SSU rRNA, from which the phylotype of the sequence can be determined, is just before the 100-kbp mark. The proteorhodopsin gene (AAG10475) at just after 50 kbp is highlighted. doi:10.1128/9781555818517.ch22.f22.1

All the same, they were successful beyond any reasonable expectation. One of the genes they found in their cosmid sequence seemed to be a gene encoding a rhodopsin, presumably originating by horizontal transfer from a halophilic archaeon. Because this rhodopsin was found in a proteobacterium, they refer to it as proteorhodopsin.

Halophilic members of the *Archaea* have three distinct rhodopsins. The main one is bacteriorhodopsin, the light-driven proton pump used to capture energy (via the proton motive force) from light. Another is halorhodopsin, a light-driven chloride pump used to bail Cl⁻ out of the cell (cells predominantly use organic anions such as glutamate rather than chloride). The third is a sensory rhodopsin, used along with a signal transduction protein to signal the presence of sufficient light to justify turning on the genes for the other rhodopsins and retinal (the cofactor for all of these rhodopsins) biosynthesis. Rhodopsin genes have also been transferred to eukaryotes at least once; the mold *Neurospora* has a rhodopsin it uses for light sensing to control its diurnal cycle. It may also be that the opsins used in vision in animals originated from one of the archaeal genes. Could this member of the SAR86 family use this rhodopsin to grow phototrophically? Is light harvesting in the ocean based on two kinds of photosynthesis instead of just one?

In Fig. 1 of the paper, the SSU rRNA tree is shown to demonstrate that the genome fragment comes from an organism of the SAR86 group. The rhodopsin tree is an attempt to determine the likely function of the proteorhodopsin; is its sequence related to proton pumping, chloride pumping, or sensory rhodopsins? The tree suggests that it might be in the sensory rhodopsin group, but this is not a strong association. (In the end [see below], this turns out to be misleading; it is not a sensory opsin.) Also note that the sequence is not related to the *Neurospora* rhodopsin gene NOP1.

In Fig. 2 of the paper, they show that the predicted secondary structure of the proteorhodopsin is consistent with that of a bona fide opsin and contains the conserved amino acids needed to bind its cofactor, retinal. In Fig. 3 of the paper, they show that when they expressed this protein in *Escherichia coli* and added retinal (*E. coli* does not make retinal), the cells quickly turned a hue of red (absorbance maximum of 520 nm) consistent with a functional rhodopsin. In other words, the protein as expressed by *E. coli* was correctly folded and inserted into the membrane in a form that correctly bound retinal.

But does it pump protons? Figure 4 of the paper shows that *E. coli* cells expressing rhodopsin and with retinal added pumped protons from inside to outside (as measured by the change in pH of the solution) when and only when provided with light. They used tetraphenylphosphonium uptake by rhodopsin/retinal-containing vesicles to measure the electrical potential generated: $-90$ mV, which is consistent with a strong proton pump.

As shown in Fig. 5 of the paper, they dissected the reaction cycle photometrically to show that it looks like a proton or chloride pump rather than a sensory rhodopsin. Sensory opsins have a slow reaction cycle, $>300$ milliseconds. The longer the recycling time, the stronger is the signal sent to its associated transducer/regulator protein. This is why the rod cells in your eyes allow you to see better in the dark; their opsins have a longer reaction cycle than those of the cone cells. Proton or chloride pump rhodopsins, however, have very fast reaction cycles, $<20$ milliseconds, so they are ready to absorb another photon and pump another ion as quickly as possible.

Panel A shows absorbance changes in rhodopsin-containing *E. coli* over time after being pulsed with a 532-nm laser. Absorption increased at 400 nm for a very short time ($<5$ milliseconds); this is the activated retinal (the "M" intermediate). The increase in absorbance at 590 nm, which increased over the same time scale as the M-intermediate decayed and in turn decayed with a half-life of 15 milliseconds, is another intermediate in the light cycle of retinal called the "O" intermediate. The decrease in absorbance at 520 nm returned to normal in the 15-millisecond timescale as well, so the O-to-ground-state transition seemed to be the rate-limiting step of the photocycle, as it is in proton-pumping bacteriorhodopsins. The timescale of the cycle was clearly that of a pump rather than sensory opsin.

Therefore, proteorhodopsin seems to be a functional, light-driven proton-pumping rhodopsin in a group of uncultivated γ-proteobacteria that make up as much as 10% of the biomass of the ocean surface water. In a subsequent paper, Ed Delong's laboratory showed that they could readily detect this rhodopsin (spectroscopically) in bacteria from ocean water and demonstrated that it actually works in cells isolated directly from the ocean, that the rhodopsin is present in large enough amounts per cell to provide for its energy needs, and that deep-water and surface water organisms produce rhodopsins that are tuned to absorb at different wavelengths of light based on what is available to them at that depth. Furthermore, they showed by cloning and sequencing genome chunks from other SAR86 organisms that different SAR86 types have different proteorhodopsin genes, presumably tuned by wavelength.

It therefore appears that the SAR86 organisms are phototrophic using rhodopsin! As abundant as they are in the ocean, this represents a huge ecological impact. But are they also photo*synthetic*? That is, can they fix carbon? Are they primary producers? So far, the answer seems to be "no." No cultivated SAR86 organism since isolated (there are only a few, and they are not easy to work with) can fix carbon. So they presumably use organic carbon for growth, but use light and proteorhodopsin for energy. This means, of course, that they are photoheterotrophs, like most purple nonsulfur bacteria.

But this is not the end of the story. It turns out that a lot of marine bacteria from many phylogenetic groups have proteorhodopsin genes, apparently acquired by horizontal transfer. This gene seems to move around a lot by horizontal transfer. Phototrophy via rhodopsins may turn out to be an important part of energy collection. Given that many of these other organisms are probably capable of carbon fixation, this may also represent an important, and previously unsuspected, component of primary production in the ocean.

## The stable-isotope probing approach

If isotopically enriched substrates are fed to an organism, the biomass of that organism becomes enriched in those same isotopes. Some of the labeled molecules are phylogenetically informative, e.g., the rRNAs or DNA, and so by separating and sequencing these molecules it should be possible to determine what, in a mixed culture, is eating the labeled substrate. Because this process usually uses stable rather than radioactive isotopes, it is called stable-isotope probing (SIP) (Fig. 22.2).

For example, if [$^{13}$C]phenol is fed to an environmental sample in which phenol is being degraded (for example, in a wastewater aerobic digester), the organisms that take up and use this phenol will end up with $^{13}$C-labeled SSU rRNAs. These can then be separated from the unlabeled ($^{12}$C) SSU rRNAs in cesium

**Figure 22.2** Overview of the SIP process. (Second panel reprinted from Gallagher EM, et al, *Appl Environ Microbiol* **76**:1695–1698, 2010, with permission; fourth panel reprinted from Moreno AM, et al, *Appl Environ Microbiol* **76**:2203–2211, 2010, with permission.) doi:10.1128/9781555818517.ch22.f22.2

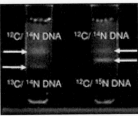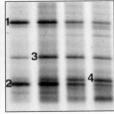

density gradients because of their higher density, amplified by polymerase chain reaction (PCR), cloned or separated by denaturing gradient gel electrophoresis (DGGE), and sequenced. This process has been used a great deal to identify (i) organisms responsible for degrading pollutants (using the $^{13}C$-labeled pollutant compound), (ii) primary producers (using $^{13}CO_2$), (iii) organisms in carbon cycle bottlenecks (e.g., propionate turnover in rice paddy sediment), and (iv) organisms in key steps of the nitrogen cycle (using $^{15}N$-labeled nitrate, nitrite, or ammonium).

## How cesium tetrafluoroacetate density gradients work

The key to this process is to get the heavy isotope label into the organisms and then separate labeled from unlabeled molecules: molecules with phylogenetic information. How is this separation by density performed? Usually by using cesium density gradient ultracentrifugation (Fig. 22.3).

Solutions of cesium salts form stable density gradients when subjected to very high gravitational fields in a centrifuge for many hours; these gradients are more concentrated (and therefore more dense) at the bottom and less concentrated (less dense) at the top. In other words, although still in solution, cesium ions "settle" somewhat at very high $g$ forces. If the solution also contains macromolecules, these float up or sink down to form bands where the density of the macromolecule matches the density of the surrounding cesium solution. RNA binds lots of cesium and is therefore fairly dense in cesium solutions, about 1.7 g/cm³. DNA binds less cesium and so has a lower density

**Figure 22.3** A Beckman TLA 100.2 rotor, capable of carrying as many as 10 4-ml samples at 435,000 × $g$ at 100,000 rpm. doi:10.1128/9781555818517.ch22.f22.3

of about 1.5 g/cm$^3$, whereas protein does not bind much cesium at all and so has a density of 1.1 to 1.2 g/cm$^3$. Within these regions, specific molecules separate slightly based on small differences in their density, because of differences in their base composition or isotopic composition. In other words, molecules (in this case RNAs) with a higher fraction of heavy-isotope atoms settle into bands that are just a little bit lower in the gradient than those with lighter isotopes.

The tetrafluoroacetate salt is used rather than chloride or sulfate (often used for other kinds of cesium density gradients) because it is more soluble. To band RNAs (with a density of about 1.7 g/cm$^3$) in approximately the middle of a centrifuge tube, the density of the solution at the bottom might be about 1.8 g/cm$^3$, well above the solubility of CsCl or even (depending on the details) that of Cs$_2$SO$_4$. If the solubility is exceeded, some of the salt precipitates at the bottom of the tube, initiating a chain reaction of precipitation that results in solid salt forming a large, heavy knob at the bottom of the tube, unbalancing the rotor. These rotors typically spin at 50 to 100 krpm, in a vacuum because the outer parts of the rotor are traveling supersonically. An off-balance rotor will rip the centrifuge apart, disintegrate, come off the motor spindle, or fail in any of several equally catastrophic ways. Ultracentrifuges have armor plates on all sides for this very reason, but a rotor failure is nevertheless dangerous, expensive, and messy.

■ STUDY AND ANALYSIS
## Identification of Ciliate Grazers of Autotrophs in Activated Sludge by Stable Isotope Probing

Moreno AM, Matz C, Kjelleberg S, Manefield M. 2010. Identification of ciliate grazers of autotrophic bacteria in ammonia-oxidizing activated sludge by RNA stable isotope probing. *Applied and Environmental Microbiology* **76**:2203–2211.
PubMed ID: 20139314    PubMed Central ID: PMC2849243

*You should have a copy of this paper at hand before proceeding.*

*Purpose:* To identify the ciliate grazers that control populations of primary producers in ammonia-oxidizing wastewater sludge.

Although SIP experiments usually target bacteria or archaea that grow directly on specific organic or nitrogenous compounds, in this case the authors were looking one step up the food chain, to the unicellular eukaryotes that eat these bacteria, and specifically those that eat the primary producers in this environment, i.e., those that incorporate carbon from $CO_2$. The particular environment in which they were interested is an ammonia-oxidizing wastewater activated sludge. In this environment, carbon is the primary limiting nutrient rather than usable nitrogen ($NH_4^+$, $NO_3^-$, $NO_2^-$) or phosphorus, and so carbon fixation predominates. The resulting microbial growth incorporates the ammonia, either directly or after being oxidized to nitrite or nitrate. Unicellular eukaryotes (protists), mostly ciliates, are present in large numbers in activated (aerobic) sludge, where they graze on the resulting lush bacterial growth. The authors describe conflicting data about whether protist grazing has positive or negative effects on microbial growth or nitrogen cycling. Of particular interest to the authors, however, is whether protists in this environment have feeding preferences or whether they feed indiscriminately.

The first step in this process was to take a census of the protists in this sludge. The authors used the same approach we talked about before; they isolated DNA from this environment, amplified SSU rRNA, in this case using eukaryote-specific primers. Individual sequences were separated by cloning and sequenced (Table 1 of the paper). This process works just as well with microbial eukaryotes as it does with bacteria or archaea, although it has been used much less frequently.

They sequenced only 29 clones but estimated that they covered almost two-thirds of the diversity of the sample because the sequences were dominated by a single type, the amoeba *Arcella*. They obtained several other amoebas (*Chaos*, *Euglypha*, *Balamuthia*, and *Stenamoeba*), some ciliates (*Zoothamnium*, *Epistylis*, and *Acineta*), and a small number of cercozoans (*Cercomonadida*) and apicomplexians (*Eimeriidae*) (Fig. 22.4). Only one sequence was unidentifiable, i.e., less than 90% similar to any characterized type.

The authors then expended some effort to prove they could label and recover protist SSU rRNA from this predominantly bacterial ecosystem. They started by seeding their sludge with $^{13}$C-labeled *Tetrahymena* (an easy-to-grow ciliate that they fed with $^{13}$C-labeled bacteria) at $10^4$/ml (Fig. 1 of the paper) and showed they could get the corresponding SSU rRNA band in the heavy-isotope fractions of cesium trifluoroacetate (TFA) gradients. For this analysis, they separated total RNAs (which were mostly rRNA) by density in cesium TFA gradients, fractionated these rRNAs (the numbers are fraction numbers), performed reverse transcriptase PCR with protist-specific primers, and separated the products by DGGE.

They went on to show that they could perform heavy-isotope labeling of grazing-protist SSU rRNA from labeled bacteria by seeding the sludge with $^{13}$C-labeled *E. coli* at $10^8$ ml and allowing grazing for 16 hours (see Fig. 2 of the paper). On the DGGE gel, all of the protists appeared to be labeled to some extent, but the most heavily labeled (and so the predominant grazers) were *Epicarchesium*, *Spumella*, and *Zoothamnium*. One type, the common amoeba *Hartmannella*, seemed specifically *not* to be labeled.

They showed (see Fig. 3 through 5 of the paper) that carbon (carbonate) fixation and acetate utilization in this sludge were limited by ammonium concentration, and that carbon fixation could be increased by adding ammonium sulfate. This might be expected in this environment, which is conditioned on the feeding of nitrogen-rich "input." The implication is that ammonia-oxidizing bacteria are growing and fixing carbon in the process.

The experiment they were building up to involved testing the protists that were labeled when they fed $^{13}CO_2$ (actually bicarbonate) to the sludge (Fig. 6

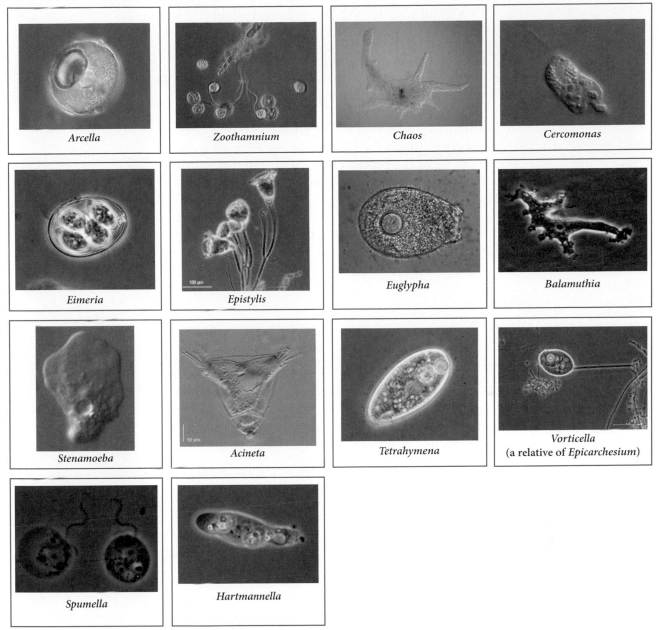

**Figure 22.4** Images of the unicellular eukaryotes identified in this paper. [Sources: *Arcella*, http://commons.wikimedia.org/wiki/File:Arcella_sp.jpg, http://creativecommons.org/licenses/by-sa/2.5/legalcode; *Zoothamnium*, Frank Fox, http://de.wikipedia.org/wiki/Zoothamnium, http://creativecommons.org/licenses/by-sa/3.0/de/legalcode; *Chaos*, Tsukii Yuuji, http://en.wikipedia.org/wiki/Chaos_(genus), http://creativecommons.org/licenses/by-sa/2.5/legalcode; *Cercomonas*, http://en.wikipedia.org/wiki/Cercomonadida, http://creativecommons.org/licenses/by-sa/2.5/legalcode; *Eimeria*, Agricultural Research Service, USDA; *Epistylis*, *Acineta*, *Tetrahymena*, *Spumella*, and *Hartmannella*, D.J. Patterson, courtesy of http://microscope.mbl.edu; *Euglypha*, Eugen Lehle, http://commons.wikimedia.org/wiki/File:Euglypha.jpg, http://creativecommons.org/licenses/by-sa/3.0/legalcode; *Balamuthia*, DPDx, Centers for Disease Control and Prevention; *Stenamoeba*, Geisen S, et al., *Eur J Protistol* **50**(2):153–165, 2014, with permission from Elsevier; *Vorticella*, Donata Dubber and Nick Gray.] doi:10.1128/9781555818517.ch22.f22.4

of the paper) along with some ammonia. The only lanes shown are those of the heavy-isotope-labeled RNA after 0, 6, and 10 hours of labeling. $^{13}CO_2$ was being fixed by (presumably) autotrophic ammonia oxidizers, which were being eaten by the protists. The only bands that were darker in the $^{13}$C-labeled lanes than in the light ($^{12}$C) lanes were those corresponding to *Epistylis*. The implication is that this ciliate predominates in the grazing of ammonia-oxidizing autotrophs.

In contrast, as a control they fed the sludge with [$^{13}$C]acetate (rather than $^{14}CO_2$) under nitrate-reducing (denitrifying) conditions (lots of nitrate, no oxygen), which presumably would serve as a general carbon source for heterotrophic bacteria (see Fig. 7 of the paper). They found no specific labeling of protists (the $^{12}$C and $^{13}$C lanes were the same).

The implication is either that there are no protists that specifically feed on denitrifiers or that protist feeding is no longer prevalent under anaerobic conditions (this would not be surprising).

## Questions for thought

**1.** Proteorhodopsin seems to be used in a major form of phototrophy in the ocean. How could this have been missed all these years?

**2.** Do you think *E. coli* carrying the proteorhodopsin gene could grow photochemotrophically if given retinal in the growth medium? How would you know it is using the organics only for fixed carbon rather than energy? If you did this, would you have created a new species of *Escherichia*?

**3.** Do you know of any other examples where the whole ecological niche of an organism is defined by one or a few genes acquired horizontally?

**4.** How would you go about trying to cultivate one of the members of the SAR86 group?

**5.** Given that you cannot grow SAR86, how might you go about trying to determine whether it is autotrophic (i.e., can fix carbon) and therefore photosynthetic?

**6.** Although rRNA is more readily labeled than DNA with $^{13}$C in SIP experiments, can you think of any reasons why it might be more useful to be able to isolate the DNA (rather than the rRNA) of organisms that can eat the labeled substrate?

**7.** Can you think of systems in which it might be a mistake to assume that the rRNAs that are labeled after addition of a labeled growth substrate represent the organisms that use the substrate directly?

**8.** Other than carbon or nitrogen, what other elements do you think you could use in SIP experiments?

**9.** Why do you think the authors got such poor labeling of protists with [$^{13}$C] acetate? What do you think they would find if they performed PCR with bacterium-specific primers to look for [$^{13}$C]acetate-labeled bacteria instead of protists?

**10.** What difference do you think might result from a protist feeding on planktonic (free-floating) versus attached bacteria? What kind of protists do you think would feed predominantly on planktonic bacteria? On sessile bacteria?

*Nothing in biology makes sense except in the light of evolution.*
THEODOSIUS DOBZHANSKY

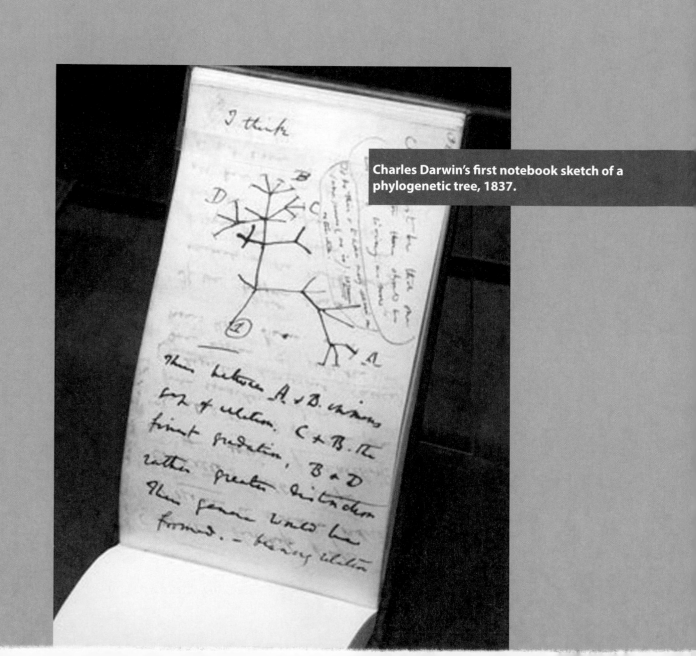

Charles Darwin's first notebook sketch of a phylogenetic tree, 1837.

# SECTION IV *Conclusion: The Phylogenetic Perspective*

Any discussion of a topic as broad and deep as microbial diversity must be organized on the basis of some underlying perspective. In this book, the vantage point from which all of microbial diversity is viewed is the phylogenetic perspective. Other perspectives are possible and are very useful. Medical microbiology views the microbial world from the perspective of its influence on microbe-human interactions and human health. Environmental microbiology views the microbial world from the perspective of biogeochemical processes and ecosystems. Industrial microbiology and biotechnology views the microbial world from the perspective of the utility of microbes as biological tools. Microbial physiology views the microbial world from the perspective of biochemical and regulatory networks. Microbial genomics views the microbial world from the perspective of information systems, processing, and expression.

However, the organizing principle of biology is evolutionary theory. The phylogenetic perspective is the view of biological diversity as the outcome of evolutionary history. This perspective is not exclusive of any other perspective on microbiology, but instead enriches these other perspectives.

We all implicitly understand the importance of this perspective in our personal lives. What is the best way to *really* get to know someone? You can spend time with them, talk and have experiences together. But real understanding comes when you meet their family. The various facets of a person's personality become clear when viewed in various forms in their relatives. Although we all seek the approval of our relatives when we introduce them to a new friend or love interest, the real anxiety of introducing them to the family comes mostly from the fact that it lays us bare to this person.

So it is also in biology. Upon encountering an unfamiliar organism, in the laboratory or in nature, your first question should always be "What *kind* of organism is this?" In other words, what is this organism's address in the "tree of life"? This will not be your last question, but it should be your first question. And when working with organisms, familiar or not, always keep in mind what *kind* of organism it is.

# 23 Genomics, Comparative Genomics, and Metagenomics

The availability of the complete genome sequences of microbial species has completely transformed microbiology over the past 20 years. It is difficult to imagine today, but not so long ago a major part of the work in most microbiology research laboratories involved cloning and sequencing experiments. Having complete genomes has, to a large extent, freed up this time and energy for more interesting questions. It has also started the process of changing our perspective of genomes from a collection of genes to that of an organized whole. The complete genome sequence of an organism provides a plethora of information about it; in many cases, it is now easier to figure out how an organism makes a living by sequencing its genome than doing the array of biochemical and microbiological tests required to get the same information.

## Genomics

### How to sequence a genome

The traditional way to sequence a microbial genome starts with DNA isolated from a pure culture of the organism. This DNA is sheared by passing a solution of the DNA through a small hole; this tears the DNA into random fragments. The DNA is fractionated by size, and fragments of 3 kbp are collected and cloned into a plasmid vector. Plasmid DNA is isolated or amplified, and each end of the DNA fragment is sequenced. A typical sequencing reaction yields 500 to 1,000 bp of sequences, resulting in two lengths of sequence data separated by 1 or 2 kbp. Typically, enough clones are sequenced to give about 10 times as many sequence data as the genome is long, e.g., 40 Mbp of data for a 4-Mbp genome.

doi:10.1128/9781555818517.ch23

454 or Illumina sequencing works in essentially the same way, except that the sequence reads are shorter (ca. 300 and 100 nucleotides, respectively, as I write this) but there are a lot more of them.

The next step is to assemble the raw sequence reads into large chunks. This is a matter of finding sequences that overlap, putting them in order, and then finding new sequences that overlap this, and repeating the process over and over again, walking down the sequence.

Imagine separating 10 copies of a book into random ~30-letter pieces. You could take any string of characters, find all copies of it in your collection, and then order them by aligning the common part. Then you could take each end of this new, longer phrase and search for all of the matches to these in other chunks of the text, merging them. For example:

```
Listen. Billy Pilgrim has co.....................................................................................
.....n. Billy Pilgrim has come unst.............................................................................
...........y Pilgrim has come unstuck in t.......................................................................
.....................as come unstuck in time. Bi.................................................................
..................... unstuck in time. Billy has gon.............................................................
..................... in time. Billy has gone to sleep a ........................................................
.....................y has gone to sleep a senile widower a......................................................
..................... to sleep a senile widower and awaken.......................................................
.....................e widower and awakened on his we...........
.....................................................................kened on his wedding day.
```

. . . can be assembled into:

```
Listen. Billy Pilgrim has come unstuck in time. Billy has gone to sleep a senile widower and awakened on his wedding day.
```

As long as you have lots of sequence, you can assemble most of the genome (or your book) this way into very large nonoverlapping chunks, called "contigs" (contiguous sequences).

These contigs, however, cannot usually be assembled into a single complete sequence because of repeated sequences. If a repeated sequence is longer than the individual sequence reads, the assembly process results in several contigs leading into and coming out of these repeated sequences, and it is not clear which "lead ins" go with which "coming outs." In our book analogy, if our average sequence read is 30 letters, what if our book contains more than one copy of the phrase "My name is Yon Yonson, I work in Wisconsin"? We end up with several large chunks of text beginning or ending within "My name is Yon Yonson, I work in Wisconsin", but do not know how these large chunks fit together:

```
... the song that goes My name is ...............................................................................
........... that goes My name is Yon Yo..........................................................................
.....................y name is Yon Yonson, I work in ............................................................
.........................................nson, I work in Wisconsin, I wor........................................
..................................... Wisconsin, I work in a lumberm.............................................
```

but then we get

```
.......................................................................ork in a lumbermill there. Some...................
```

and

```
.......................................................................work in a lumbermill there. The ...................
```

This means that this is a repeated sequence, but we do not know which of these next fragments goes with the one we were working with. Both of them can be continued; this is a fork in the road. Presumably a different sequence that ends in "I work in a lumbermill there" will be found later; in this book, there are actually three of them. Coming in from "upstream" (the forward direction) we have

```
...and his Pall Malls. My name is Yon Yonson, I work in Wisconsin...
... the song that goes My name is Yon Yonson, I work in Wisconsin...
...r name? And I say, 'My name is Yon Yonson, I work in Wisconsin...
```

and coming out from "downstream" (the reverse direction) we have

```
... My name is Yon Yonson, I work in Wisconsin, I work in a lumbermill there. Sometimes I try to...
... My name is Yon Yonson, I work in Wisconsin, I work in a lumbermill there. The people I meet ...
... My name is Yon Yonson, I work in Wisconsin... And so on to infinity. Over the years, people ...
```

There are three repeats in our "genome" (from *Slaughterhouse-Five*, by Kurt Vonnegut), and so the question is, in what order are the three fragments of the book assembled? Which upstream goes with which downstream? Each fragment (if the book were circular like most microbial genomes) would begin and end in the repeated phrase. (In this example, two of the repeats are longer than the third; this happens a lot in biology as well.)

You can see that as long as the text chunks (sequence reads) are at least a bit longer than the repeat, it is possible to distinguish each repeat by its flanking text:

```
...and his Pall Malls. My name is Yon Yonson, I work in Wisconsin, I work in a lumbermill there. Sometimes I try to...
... the song that goes My name is Yon Yonson, I work in Wisconsin, I work in a lumbermill there. The people I meet ...
...r name? And I say, 'My name is Yon Yonson, I work in Wisconsin... And so on to infinity. Over the years, people ...
```

This is one reason for sequencing the ends of 3-kbp fragments in traditional genome sequencing: it enables repeats up to about 3 kbp to be spanned; a fragment may end in a repeat, but there is also another associated chunk of sequence about 3 kb away, and these can be used to match up the repeats as long as the repeated sequence is less than about 3 kb.

The 3-kbp span covers most of the common repeats in microbial genomes, but not all. Large transposons can be longer than this, as well as the rare case of multiple copies of some gene or gene cluster, but the common cases are the ribosomal RNA (rRNA) operons. Each of these is about 5 kbp and occurs up to a dozen times in a bacterial genome (averaging about five copies). So the final task in assembling a genome is to bridge any remaining gaps in the sequence and join the contigs across these rRNA operons or any other large repeats.

For example, in Fig. 23.1, the sequence assembly ends up with three contigs, each ending in an rRNA operon. Joining contigs is usually done by making polymerase chain reaction (PCR) primers pointing toward each end of the contigs and running all possible combinations of PCRs. Because the operons have a specific orientation, there are not as many possibilities as you might think. Which PCR amplifications give you products will tell you how they need to be assembled, and this can be confirmed by sequencing each of these PCR products. With this done, the genome sequence is "closed." Notice that in our

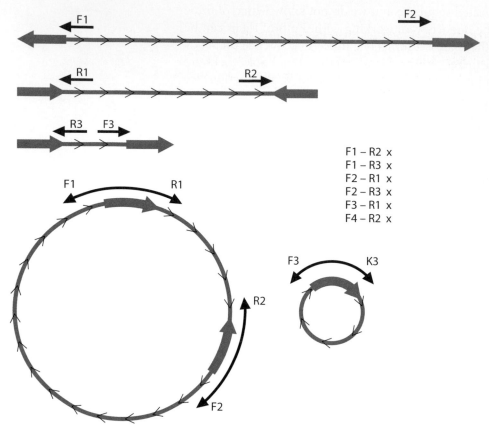

F1 – R2 x
F1 – R3 x
F2 – R1 x
F2 – R3 x
F3 – R1 x
F4 – R2 x

**Figure 23.1** Assembling contigs into complete genomes. Primers (black) directed outward from each end of all contigs (red) are used in PCR amplifications; products are generated only across the junctions joining contigs in the genome. These PCR products would be sequenced for confirmation of the overlap. doi:10.1128/9781555818517.ch23.f23.1

example, the result is not a single circular chromosome like you might have been expecting, but *two* circular chromosomes!

The next step, just as important but far more time-consuming than the first, is to annotate the genome sequence; in other words, identify the genes and what they encode, as far as possible. In the case of protein-encoding genes, this starts by identifying all of the strings of codons that follow the general rules of protein-encoding sequences. These are called open reading frames (ORFs). Each ORF is then translated, and the encoded amino acid sequences are compared to the database of identified sequences to see if it looks like a sequence of known function. Sometimes these can be assigned definitively, sometimes tentatively. Sometimes the identification is very specific (e.g., "fetal alpha-2 hemoglobin"), and sometimes it is only general (e.g., "dehydrogenase of unknown function"). Very often a gene can only be said to fall into a class of conserved genes but the function of this gene may not be known. RNA genes (e.g., transfer RNA [tRNA], rRNA, signal recognition particle [SRP] RNA, ribonuclease P [RNase P] RNA, and so on) are usually identified using BLAST or specific RNA sequence-structure model search programs, e.g., tRNAscan.

■ STUDY AND ANALYSIS

## Evidence for Lateral Gene Transfer between Archaea and Bacteria

Nelson KE, Clayton RA, Gill SR, Gwinn ML, Dodson RJ, Haft DH, Hickey EK, Peterson JD, Nelson WC, Ketchum KA, McDonald L, Utterback TR, Malek JA, Linher KD, Garrett MM, Stewart AM, Cotton MD, Pratt MS, Phillips CA, Richardson D, Heidelberg J, Sutton GG, Fleischmann RD, Eisen JA, White O, Salzberg SL, Smith HO, Venter JC, Fraser CM. 1999. Evidence for lateral gene transfer between Archaea and Bacteria from the genome sequence of *Thermotoga maritima*. *Nature* **399**:323–329.

PubMed ID: 10360571

*You should have a copy of this paper at hand before proceeding.*

*The task:* To report the complete genome sequence of *Thermotoga maritima*. The reason for focusing on this organism is to try to understand bacterial ancestry and early evolution.

In this study, the authors sequenced the entire genome of *Thermotoga maritima*. This was an early genome sequence, reported only 4 years after the very first complete genome sequence was published (that of *Haemophilus influenzae* strain Rd in 1995), so a lot of the paper is devoted to basic information about the genome: size, number of genes, and so forth, compiled in Table 1 and Fig. 1 of the paper. The genome is a single circle of 1.86 Mbp, about average for a bacterial genome, with 1,877 identified ORFs, a full complement of tRNAs and rRNAs (a single rRNA operon).

Figure 2 of the paper is the huge diagram of the entire genome, with each gene shown. Table 2 of the paper is a list of all of these genes, categorized by function where they can be identified. Take a moment to survey these in the original paper and stand in awe; that is largely the purpose of these diagrams.

About half of the ORFs are similar enough to characterized genes from other organisms to be identifiable with some level of confidence. The authors use these (along with the known properties of *T. maritima*) to infer the metabolic pathways it uses (Fig. 3 of the paper). A large fraction of the genome is taken up by genes encoding the uptake and metabolism of sugars and oligopeptides as part of the organism's heterotrophic lifestyle. Most of the uptake pumps are adenosine triphosphate (ATP) dependent, and mostly ATP-binding cassette (ABC) transporters. It can do a lot of sugar metabolism and can synthesize at least nine of the amino acids it needs, along with many of the usual cofactors and vitamins. Glycogen is probably the primary energy storage compound.

Notice that there is no electron transport chain and the "ATP synthase" is shown hydrolyzing ATP to pump protons out rather than allowing protons in for the synthesis of ATP. This is how many organisms that do not have the electron transport chain make a proton gradient to drive their active-transport systems, which *Thermotoga* obviously has a lot of. There is also a pyrophosphatase-driven proton pump. There is some electron transport, however, in which $Fe^{3+}$ or sulfur, rather than protons, serves as the terminal electron acceptor; this would prevent hydrogen accumulation which inhibits growth profoundly for hydrogen-generating organisms. This electron transport is presumably carried out using flavoproteins and iron-sulfur proteins; no cytochromes are encoded in the genome.

Global gene expression is probably regulated by four alternative sigma factors.

Two aminoacyl-tRNA synthetases are absent, those for charging tRNAs with glutamine and asparagine. These tRNAs are charged instead by using the glutamate and aspartate synthetases, and then amidases (identifiable in the genome) convert tRNA-glutamate to tRNA-glutamine, and tRNA-aspartate to tRNA-asparagine. These can either be used directly in translation or released from the tRNA and used in central metabolism; the tRNAs and charging enzymes have become part of the biosynthetic pathways for these amino acids.

## *Horizontal transfer of genes*

This might be a good time to review our previous discussion on horizontal gene transfer; see chapter 7.

The most interesting thing about the *T. maritima* genome sequence was a big surprise; there is good evidence for significant numbers of "foreign" genes from archaea in the *Thermotoga* genome. The authors argue that up to about a fourth of the genome was acquired from other sources recently enough that its foreignness can still be detected; others would reduce this to about 5%. Either way, it is clear that a significant fraction of the DNA in this organism does not share the vertical ancestry with the small-subunit rRNA (SSU rRNA); the organism is, at least to some extent, an evolutionary mosaic. The implication is that maybe *all* bacteria are as well.

How do the authors deduce that certain regions or genes originated somewhere else?

1. They argue that 24% of the genes in *Thermotoga* are more similar in sequence to their homologs in archaea rather than in other bacteria. About one-fifth of these genes are clustered in 15 distinct "archaea-like" regions associated with repeated elements (repeats are a sign of potential mobile elements).

2. There are about 50 regions in the genome (many of these overlap with the ones above) in which the base composition (the ratios of G, A, T, and C) is significantly different from that in the rest of the genome, and these include another 3-bp repeated sequence.

3. Although the authors do not mention it, another form of evidence of lateral transfer is the clustering of genes that use unusual codon bias. In other words, although each organism generally prefers certain codons over others for the same amino acid, regions that came recently from another organism follow the codon bias of that source.

The major source of disagreement about their evidence for horizontal transfer comes from item 1 in the list. This is summarized in Table 4 of the paper. For example, of the 71 conserved (i.e., present in bacteria, archaea, and eukaryotes) genes in *Thermotoga* for amino acid biosynthesis, 49 are most similar (closest hit in BLAST) to a bacterial gene but 20 are more similar to an archaeal gene and 2 are more similar to a eukaryotic gene. They conclude that these genes were horizontally transferred from archaea. But let's look at the tree (Fig. 23.2). *Thermotoga* is apparently a primitive, early branch in the bacterial domain, and archaea are likewise primitive. It is easy to see that although *Thermotoga* is a bacterium, variation in the length of branches (evolutionary rates) when the trees are generated with different gene sequences could cause the closest match to jump from bacteria to archaea. This is where most of their overestimation of the number of archaeal genes acquired by horizontal transfer comes from.

**Figure 23.2** Placement of *Thermotoga maritima* in rRNA-based trees, relative to what this might look like in the case of protein-coding sequences with different evolutionary rates (branch lengths) in different kinds of organisms. The top tree shows the general relationship between genes in bacteria (B), archaea (A), and eukarya (E), in which the relationship of genes in *Thermotoga* to these organisms is in question. In the middle tree, the protein has evolved relatively slowly in bacteria, and so the closest match to the *Thermotoga* homolog would correctly be identified as bacterial. In the bottom tree, the protein has evolved relatively quickly in other bacteria, and as a result the closest match to the *Thermotoga* homolog is in archaea, even though the protein is more closely related by ancestry to its bacterial homologs. doi:10.1128/9781555818517.ch23.f23.2

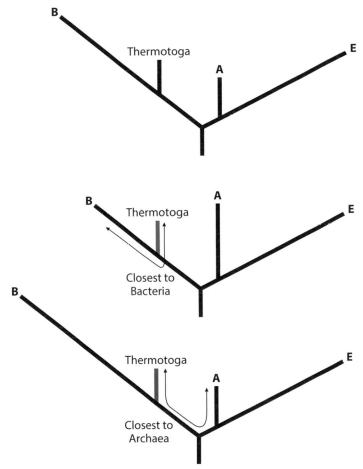

The second most important argument the authors use for horizontal transfer is the regions of different base composition (ratios of G, A, C, and T) in the genome. Suffice it to say that there are many reasons why regions might have a different base composition that have nothing to do with horizontal transfer. They point out that the biggest skew in base composition is in the rRNA operons, which are enriched in G+C, as might be expected of RNA-encoding genes in a thermophile. This is related to the function of the sequence; it is not evidence for horizontal transfer. Their statistical analysis of base composition skew is very simplistic; with the more recent availability of many genomes, these base composition skews largely turn out to be red herrings.

But these are arguments about how much of the genome was acquired by horizontal transfer; even the most skeptical would agree that the evidence for about 5% of the genome coming recently from elsewhere is pretty good. This is a lot of DNA, encoding many important functions that directly affect the phenotype of the organism.

Notice, however, that the authors specifically point out that even with their generous (too generous) 24% of acquired genes, these do not (generally) include genes for central information processing; the "central dogma" genes, including replication, transcription, and translation. Instead, they are predominantly metabolism/phenotype genes. There is pretty good evidence that regardless of horizontal transfer of metabolic/phenotypic genes, the central information-processing genes can be viewed as a core of the genome that is inherited vertically and so represents the main phylogenetic lineage of the organism.

In the end, the authors argue pretty strongly against the use of any single gene to create phylogenetic trees; the obvious target for this comment is the common use of the SSU rRNA gene. They go on to talk about a series of trees generated by using other genes which give trees different from those of SSU rRNA, although they fail to consider whether these other sequences are good phylogenetic clocks. They are clearly talking from the perspective of microbial species as transient amalgams of genes from a common microbial gene pool. The notion is that microbes (or at least bacteria and archaea) do not really have meaningful phylogenies, that each gene has its own history, but that the organism as a whole has no unique evolutionary history. This represents the opposite extreme of the notion that the SSU rRNA phylogenetic tree is the only meaningful representation of an organism's evolutionary history. Both views are, of course, false, and the truth lies somewhere in between. *Where* in between is not yet clear.

## Comparative genomics

The complete genome sequence of an organism gives a great deal of basic information about it. But deeper insight requires a comparative approach; comparing the genomes of related organisms gives information about how, and why, they differ phenotypically. For example, *Escherichia coli* strains vary from completely innocuous to extremely virulent. The difference in their virulence lies in the differences in their genome sequences.

■ STUDY AND ANALYSIS
**Evolution of Chlamydiae**

Horn M, Collingro A, Schmitz-Esser S, Beier CL, Purkhold U, Fartmann B, Brandt P, Nyakatura GJ, Droege M, Frishman D, Rattei T, Mewes HW, Wagner M. 2004. Illuminating the evolutionary history of chlamydiae. *Science* 304:728–730.

PubMed ID: 15073324

*You should have a copy of this paper at hand before proceeding.*

*The task:* To compare the genome sequences of UWE25 and its more specialized animal-pathogenic relatives to learn about how the ancestors of these animal pathogens were preadapted to the pathogenic lifestyle.

This is a genome sequence paper, although you might get all the way through it and still not be sure if you don't go to the online supplementary data, where the actual sequencing and annotation are described. That's the way it is; it used to be that you would triumphantly herald the "Complete Genome Sequence of (fill in name here)!," but now you need to actually have something interesting to focus on other than just the sequence: anything interesting, no matter how desperately you have to look. This should not be so hard—after all, you had some reason (or at least an excuse) in the grant proposal that funded the sequencing—but sometimes you have to wonder. Not so in this case.

This is a great example of comparative genomics, where you learn about an organism or group of organisms (in this case the pathogenic chlamydiae) by comparing its genome with those of related organisms with different properties. In this case, we would like to know how chlamydiae evolved from free-living bacteria to obligately parasitic, almost virus-like, intracellular pathogens. The genome sequences of four human-pathogenic chlamydiae had already been determined when this paper was published, but this was the first available "environmental" *Chlamydia* sequence. The idea was to compare this relatively primitive symbiont of amoebas to the more specialized animal parasites in an attempt to see how the animal parasites have adapted to the lifestyle from their less specialized ancestors and to see how the ancestors of the parasites were "preadapted" by their relationship with unicellular eukaryotes. In other words, the primitive relative is acting as a surrogate for the unavailable primitive ancestor. The authors understand the dangers of this—hence the focus on horizontal transfer and recent changes—but it bears keeping in mind that UWE25 is *not* an ancestor of pathogenic chlamydiae but, rather, a primitive cousin.

The organism UWE25 (since named *Protochlamydia amoebophila*) grows symbiotically in amoebas of the genus *Acanthamoeba*. These amoebas are common environmental organisms, although some are opportunistically pathogenic to humans; *Acanthamoeba castellanii* is commonly found in tears and can cause infections in the eyes, especially those of contact lens wearers. UWE25 cannot be grown in pure culture, but the amoebas (fed with *E. coli*) infected with UWE25 can be grown, and then the UWE25 elementary bodies (EBs) can be physically isolated from the culture medium. In this study, genomic DNA isolated from these EBs was sequenced in the usual way, although a "gap" of about 5 kbp (ca. 0.5% of the genome) remained unsequenced, despite some effort (details are not given). This nearly complete genome was annotated in the usual ways, which, by the way, overlook all of the RNA genes except rRNAs and tRNAs (no effort is made to find the genes for RNase P RNA, transfer-messenger RNA [tmRNA], SRP RNA, 6S RNA, or any of the other small RNA sequences).

The genome of UWE25 is clearly less reduced than those of the pathogenic chlamydiae. It is about 2.4 Mbp long, with over 2,000 protein-encoding genes: twice the size of those of the animal-pathogenic chlamydiae and as large as those of many free-living organisms.

UWE25 relies on its host for many of its amino acids but retains the ability to make some of them (glycine, serine, glutamine, and proline). Interestingly, it lacks the genes for tryptophan synthesis; the ability to synthesize tryptophan is a virulence factor in the pathogens, somehow affecting sensitivity to interferon, and tryptophan is the only "standard" amino acid they can produce. Because it can make these few amino acids, UWE25 retains a complete tricarboxylic acid (TCA) cycle, from which these amino acid synthetic pathways originate.

UWE25 retains much more sugar metabolism than the animal pathogens, which get all of their phospho sugars and pentoses from their host.

UWE25 can apparently also make a proton gradient by using an abbreviated proton-pumping electron transport chain. The authors interpret this to mean that UWE25 can use this gradient to make ATP, but my guess is that it uses this gradient to run its active transport uptake systems, from which it makes its living from the host. These pumps would also be needed by the pathogens, which would presumably energize them with a proton gradient generated by ATPase run in reverse. The pathogens have the genes to make small amounts of ATP from a very short electron transport chain that does not make a proton gradient; it just serves to dump electrons onto the terminal electron acceptor. Both UWE25 and the pathogens have the genes for ATP generation by substrate-level phosphorylation. However, UWE25, like the pathogens, also has the ATP/ADP antiport, so it presumably gets the bulk of its ATP by "energy parasitism."

Interestingly, UWE25, like its pathogenic relatives, has a type III secretion system. The authors use a phylogenetic tree (Fig. 2 of the paper) to argue that these genes in UWE25 have been there for a while (not horizontally acquired), suggesting that this system would also have existed as a preadaption to the pathogenic lifestyle in the common ancestor of the chlamydiae.

These genes are strict virulence factors in pathogens; their use in UWE25 is not known, but presumably it involves secreting proteins into the host for some purpose. The most likely candidate is a chlamydial protease-like activity factor (CPAF) (the gene for which is in UWE25), which in the animal pathogens is involved in degrading transcription factors required for major histocompatibility complex (MHC) ("self") antigen synthesis. Amoebas do not have MHCs, so exactly what CPAF is doing is unclear.

UWE25 also contains another complex, a type IV secretion system, presumably for pumping some other protein into its host. These proteins are not found in the animal pathogens, and the unusual G+C content of these genes suggests that they *might* be recent transplants from some other organism.

## Metagenomics

If a "genome" is the genetic composition of an organism, then a "metagenome" is the genetic composition of a population of organisms. Genomes are sequenced shotgun-style, i.e., by sequencing large random collections of DNA from an organism, and then assembling the snippets of sequences to get the genome sequence. Not so long ago, the task of sequencing an entire microbial genome was considered monumental, taking months or years and millions of dollars. Now genome sequencing is done as a service, taking days or weeks (if you include annotation) and costing thousands of dollars. In this context, it is feasible in the case of simple populations to sequence the metagenome and end up assembling most of the genomes in the population, or at least the most abundant members of the population (Fig. 23.3).

In a way, we have been talking about metagenomics for a while; the process of sequencing SSU rRNAs from microbial populations (molecular surveys) is a gene-specific form of metagenomics.

**Figure 23.3** Genomics versus metagenomics. (DNA panels: Bruce Chassey Laboratory, National Institute of Dental Research; Genome panel: Wegmann U et al., *J Bacteriol* **189**:3256–3270, 2007, with permission; Metagenome panel: Venter JC et al., *Science* **304**:66–74, 2004, reprinted with permission from AAAS.) doi:10.1128/9781555818517.ch23.f23.3

**Genome**: the complete genetic complement of an organism

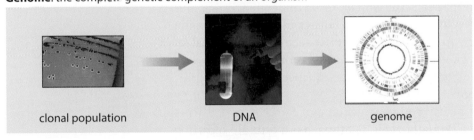

clonal population          DNA          genome

**Metagenome**: the complete genetic complement of an environment

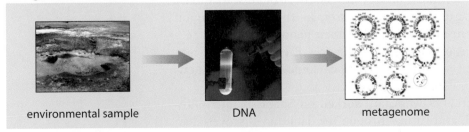

environmental sample          DNA          metagenome

■ STUDY AND ANALYSIS
## Environmental Genome Shotgun Sequencing

Venter JC, Remington K, Heidelberg JF, Halpern AL, Rusch D, Eisen JA, Wu D, Paulsen I, Nelson KE, Nelson W, Fouts DE, Levy S, Knap AH, Lomas MW, Nealson K, White O, Peterson J, Hoffman J, Parsons R, Baden-Tillson H, Pfannkoch C, Rogers YH, Smith HO. 2004. Environmental genome shotgun sequencing of the Sargasso Sea. *Science* **304**:66–74.

PubMed ID: 15001713

*You should have a copy of this paper at hand before proceeding.*

*The task:* To obtain a broad sampling of the genetic potential of a complex microbial environment. Also, to patent all of the sequences and somehow later make money for the investors based on this ownership.

Many of the first bacterial genome sequences were generated by The Institute for Genome Research (TIGR), a company led at the time by the visionary Craig Venter. The human genome sequence is about 1,000 times larger than a typical bacterial genome sequence; again, a monumental task, and again the driving force behind getting it done was Craig Venter. So, what do you do next if you are Craig Venter? Sequence a metagenome! The metagenome chosen was near-surface water collected from the Sargasso Sea, because so much microbial ecology research had already been done on this environment.

The paper (and supplementary data) starts with summary tables, intended to blow your mind away with the sheer scale of these sequences (Tables 2, S4, and S5 of the paper). Then the authors expend some effort with the bits of the sequences they can assemble into coherent parts (a *very* small fraction of the whole). The first is a collection of contigs that are chunks of a population of *Prochlorococcus* (Fig. 2 of the paper). The *Prochlorococcus* sequences are not all the same; there is the expected heterogeneity (alleles of genes) and some rearrangements of gene order. At least five types of sequences are detectable, each presumably representing an individual strain. The authors were also able to pull out essential complete *Shewanella* genome sequences, probably representing two major strains (Fig. S3 of the paper).

Finally, they pulled out a nearly complete genome of *Burkholderia*. This is odd in at least two ways. First, *Burkholderia* is a pathogen or symbiont of mammals. It has never been detected in seawater, much less any of the Sargasso samples that have been studied over many, many years. And yet there seems to be a lot of it in the metagenomic data, enough to assemble the thousands upon thousands of individual sequence reads into a complete genome sequence. Second, unlike the *Prochlorococcus* or *Shewanella* sequences, there is no significant heterogeneity in the *Burkholderia* sequences. They are clonal within experimental error.

And so a group of researchers who had worked on Sargasso microbiology for a while were skeptical, and sifting through the data they discovered an important aspect of the *Burkholderia* sequences that the authors of this paper had overlooked: all of the *Burkholderia* sequences—every single one—came from the DNA extracted from material collected on a single filter. Furthermore, the sequence turns out to be *B. cepacia* group K, a strain associated only with human infections from contaminated medical implants and not able to grow in seawater. Finally, the authors later confirmed that although all of the other filters were new when used to collect the Sargasso samples, the single filter from which all of the *Burkholderia* sequences came had been recycled and was of unknown origin. This indicates that the *Burkholderia* sequences are almost certainly contaminants.

In addition to these nearly complete genome sequences, they got some complete plasmid sequences (Fig. 4 of the paper). Some of these plasmids are large and contain housekeeping genes and so ought probably to be considered small chromosomes.

The researchers were particularly interested in rhodopsins, and they got a lot of rhodopsin sequences (Fig. 7 of the paper, which is legible only when magnified from the electronic file). They analyzed these phylogenetically to get some idea of where they might have come from, and it was clear that they were from more than SAR86-like γ-proteobacteria, where they were first seen in bacteria (chapter 22). How do they know? They determined this by looking at the flanking genes and treeing them out in groups. They found that the flanking genes were from a wide range of organisms, all kinds of proteobacteria, bacteroids, and many other groups. So the proteorhodopsin gene is probably being moved around by lateral gene transfer a lot.

Lastly, they tried to sort out what kinds of organisms might be in the samples based on the rRNA genes in their data, as well as other conserved, phylogenetically informative sequences (Fig. 6 of the paper). Although there are some differences, these are grossly consistent and similar to what others have seen in other surveys of this environment.

---

## Questions for thought

**1.** How do you think genome sequencing would deal with linear chromosomes?

**2.** What would you get in a genome-sequencing project if your original DNA was contaminated by small amounts of DNA from some other organism, for example, *Pseudomonas* from the distilled-water jug, or *Sargassum* from agar?

**3.** Many organisms are not particularly stable genetically in culture. How would changes in genome sequence or structure (arrangement) in your original sample show up in your data? What if the organism you thought was a pure culture was actually a coculture of two unrelated organisms living symbiotically?

**4.** Do you think *Bacteria* and *Archaea* have a meaningful genealogy in the face of lateral gene transfer? What observation would sway your conclusions either way?

**5.** What do you think a bacterial species is? How would you distinguish between the meaning of the terms "species" and "strain" or "genus"? How is this similar or different in plants and animals?

**6.** In what way(s) are lymphocytes and amoebas similar that might contribute to their acting as host to these *Chlamydia* parasites/symbionts?

**7.** The authors of the UWE25 paper talk a lot about genes in chlamydiae that are most similar to those of cyanobacteria, chloroplasts, and plants. What could this mean?

**8.** The chlamydiae we know about infect animals (mammals, birds, reptiles, and insects) and amoebas. Where would you look for whole new kinds of chlamydiae? How would you go about this?

**9.** Can you think of a metagenome you could sequence that might be much simpler in composition, and therefore would require less sequencing to cover? How many data do you think would be required to "finish" the Sargasso metagenome?

**10.** If you could sequence a metagenome, what would it be? Why?

**11.** What kind of useful genes might there be in these metagenomic data? What kind of useful variants of well-known genes might there be in this metagenome? How would you go about identifying these sequences?

**12.** What do you think you would get if you did the same kind of metagenomic analysis with sub-0.1-$\mu$m filtrates? 100- to 10-$\mu$m filtrates?

# 24 Origins and Early Evolution

One of the reasons to study biological diversity is to answer, at least a little bit, the biggest question of them all: *Where did we come from?* In this chapter, we trace backward down the tree of life, starting with the last common ancestor.

Because we have access to modern living things and traces of previous living things (fossils), we can infer something about this last common ancestor by using the phylogenetic approach we have used all along in the previous chapters. Even so, there is already some real uncertainty about the nature of this creature.

Before the last common ancestor, we are working mainly by conjecture. It should be absolutely clear that at this point, we largely leave the realm of real science because the notions covered in this chapter are, for the most part, both unverifiable and unfalsifiable, at least for now. And so, for the record:

> *How did life originate, and what was it like before the last common ancestor? We don't know!*

The rest is hand-waving. But it is interesting hand-waving, and it is the last bit of this book, so let's forge ahead.

## The timescale

We have already dealt at some length about how most of biological diversity today is microbial. It is worse than this: throughout *most* of Earth's history, life has been *entirely* microbial.

doi:10.1128/9781555818517.ch24

Figure 24.1 is a copy of the geological timescale in its usual format. The problem with this view of geological time is that it is a backward logarithmic scale, and so it dramatically emphasizes recent times over ancient times. It is done this way simply because we have a lot more information about recent geological time; the further back we go, the less we know, and so the lower the resolution of our view. In other words, our view of the past is foreshortened.

In an attempt to try to bring these "deep timescales" into a more realistic focus, Table 24.1 shows the timescale of life events (very approximate) compared to the length of a typical ~46-ft-long classroom. Notice that life seems to have originated and evolved into recognizable forms very quickly, within about 200 million years. After this, for the next 3 billion years (80% of Earth's history), life was entirely microbial. During this time the concentration of oxygen in the atmosphere was not high enough to be breathable. Remember this when people dismiss other planets for being "uninhabitable" or say that even if there could be life there, it would "only" be microbial. That was true for *this* planet over most of its history. And, of course, microbes *still* rule the ecosystems; in the words of Stephen J. Gould, famous paleontologist and evolutionary biologist, "*This* is the Age of Bacteria—as it ever has been and always will be."

**Figure 24.1** Geological time scale. (Source: Geological Society of America.)
doi:10.1128/9781555818517.ch24.f24.1

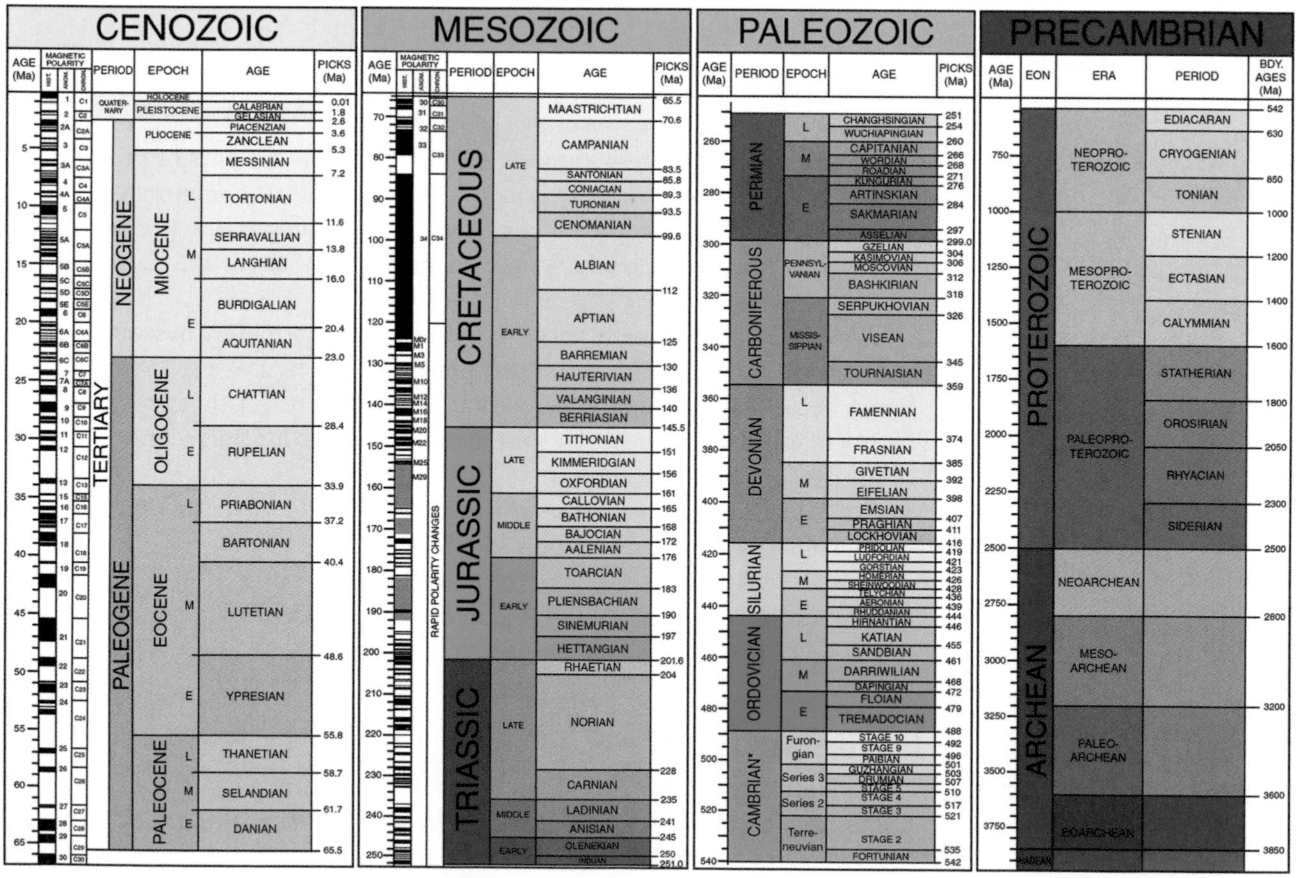

## CENOZOIC

| PERIOD | EPOCH | AGE | PICKS (Ma) |
|---|---|---|---|
| QUATERNARY | HOLOCENE | | 0.01 |
| QUATERNARY | PLEISTOCENE | CALABRIAN / GELASIAN | 1.8 |
| NEOGENE | PLIOCENE | PIACENZIAN | 2.6 |
| NEOGENE | PLIOCENE | ZANCLEAN | 3.6 |
| NEOGENE | MIOCENE (L) | MESSINIAN | 5.3 |
| NEOGENE | MIOCENE (L) | TORTONIAN | 7.2 |
| NEOGENE | MIOCENE (M) | SERRAVALLIAN | 11.6 |
| NEOGENE | MIOCENE (M) | LANGHIAN | 13.8 |
| NEOGENE | MIOCENE (E) | BURDIGALIAN | 16.0 |
| NEOGENE | MIOCENE (E) | AQUITANIAN | 20.4 / 23.0 |
| TERTIARY / PALEOGENE | OLIGOCENE (L) | CHATTIAN | 28.4 |
| TERTIARY / PALEOGENE | OLIGOCENE (E) | RUPELIAN | 33.9 |
| TERTIARY / PALEOGENE | EOCENE (L) | PRIABONIAN | 37.2 |
| TERTIARY / PALEOGENE | EOCENE (L) | BARTONIAN | 40.4 |
| TERTIARY / PALEOGENE | EOCENE (M) | LUTETIAN | 48.6 |
| TERTIARY / PALEOGENE | EOCENE (E) | YPRESIAN | 55.8 |
| TERTIARY / PALEOGENE | PALEOCENE (L) | THANETIAN | 58.7 |
| TERTIARY / PALEOGENE | PALEOCENE (M) | SELANDIAN | 61.7 |
| TERTIARY / PALEOGENE | PALEOCENE (E) | DANIAN | 65.5 |

## MESOZOIC

| PERIOD | EPOCH | AGE | PICKS (Ma) |
|---|---|---|---|
| CRETACEOUS | LATE | MAASTRICHTIAN | 65.5 / 70.6 |
| CRETACEOUS | LATE | CAMPANIAN | |
| CRETACEOUS | LATE | SANTONIAN | 83.5 |
| CRETACEOUS | LATE | CONIACIAN | 85.8 |
| CRETACEOUS | LATE | TURONIAN | 89.3 |
| CRETACEOUS | LATE | CENOMANIAN | 93.5 / 99.6 |
| CRETACEOUS | EARLY | ALBIAN | |
| CRETACEOUS | EARLY | APTIAN | 112 |
| CRETACEOUS | EARLY | BARREMIAN | 125 |
| CRETACEOUS | EARLY | HAUTERIVIAN | 130 |
| CRETACEOUS | EARLY | VALANGINIAN | 136 |
| CRETACEOUS | EARLY | BERRIASIAN | 140 / 145.5 |
| JURASSIC | LATE | TITHONIAN | 151 |
| JURASSIC | LATE | KIMMERIDGIAN | 156 |
| JURASSIC | LATE | OXFORDIAN | 161 |
| JURASSIC | MIDDLE | CALLOVIAN | 165 |
| JURASSIC | MIDDLE | BATHONIAN | 168 |
| JURASSIC | MIDDLE | BAJOCIAN | 172 |
| JURASSIC | MIDDLE | AALENIAN | 176 |
| JURASSIC | EARLY | TOARCIAN | 183 |
| JURASSIC | EARLY | PLIENSBACHIAN | 190 |
| JURASSIC | EARLY | SINEMURIAN | 197 |
| JURASSIC | EARLY | HETTANGIAN | 201.6 / 204 |
| TRIASSIC | LATE | RHAETIAN | |
| TRIASSIC | LATE | NORIAN | 228 |
| TRIASSIC | LATE | CARNIAN | 235 |
| TRIASSIC | MIDDLE | LADINIAN | 241 |
| TRIASSIC | MIDDLE | ANISIAN | 245 |
| TRIASSIC | EARLY | OLENEKIAN | 250 |
| TRIASSIC | EARLY | INDUAN | 251.0 |

(vertical note: RAPID POLARITY CHANGES)

## PALEOZOIC

| PERIOD | EPOCH | AGE | PICKS (Ma) |
|---|---|---|---|
| PERMIAN | L | CHANGHSINGIAN | 251 |
| PERMIAN | L | WUCHIAPINGIAN | 254 / 260 |
| PERMIAN | M | CAPITANIAN | 266 |
| PERMIAN | M | WORDIAN | 268 |
| PERMIAN | M | ROADIAN | 271 / 276 |
| PERMIAN | E | KUNGURIAN | |
| PERMIAN | E | ARTINSKIAN | 284 |
| PERMIAN | E | SAKMARIAN | |
| PERMIAN | E | ASSELIAN | 297 / 299.0 |
| CARBONIFEROUS | PENNSYLVANIAN | GZELIAN | 304 |
| CARBONIFEROUS | PENNSYLVANIAN | KASIMOVIAN | 306 |
| CARBONIFEROUS | PENNSYLVANIAN | MOSCOVIAN | 312 |
| CARBONIFEROUS | PENNSYLVANIAN | BASHKIRIAN | 318 |
| CARBONIFEROUS | MISSISSIPPIAN | SERPUKHOVIAN | 326 |
| CARBONIFEROUS | MISSISSIPPIAN | VISEAN | 345 |
| CARBONIFEROUS | MISSISSIPPIAN | TOURNAISIAN | 359 |
| DEVONIAN | L | FAMENNIAN | 374 |
| DEVONIAN | L | FRASNIAN | 385 |
| DEVONIAN | M | GIVETIAN | 392 |
| DEVONIAN | M | EIFELIAN | 398 |
| DEVONIAN | E | EMSIAN | 407 |
| DEVONIAN | E | PRAGHIAN | 411 |
| DEVONIAN | E | LOCKHOVIAN | 416 |
| SILURIAN | L | PRIDOLIAN | 419 |
| SILURIAN | L | LUDFORDIAN | 421 |
| SILURIAN | L | GORSTIAN | 423 |
| SILURIAN | M | HOMERIAN | 426 |
| SILURIAN | M | SHEINWOODIAN | 428 |
| SILURIAN | E | TELYCHIAN | 436 |
| SILURIAN | E | AERONIAN | 439 |
| SILURIAN | E | RHUDDANIAN | 444 |
| ORDOVICIAN | L | HIRNANTIAN | 446 |
| ORDOVICIAN | L | KATIAN | 455 |
| ORDOVICIAN | L | SANDBIAN | 461 |
| ORDOVICIAN | M | DARRIWILIAN | 468 |
| ORDOVICIAN | M | DAPINGIAN | 472 |
| ORDOVICIAN | E | FLOIAN | 479 |
| ORDOVICIAN | E | TREMADOCIAN | 488 |
| CAMBRIAN* | Furongian | STAGE 10 | 492 |
| CAMBRIAN* | Furongian | STAGE 9 | 496 |
| CAMBRIAN* | Furongian | PAIBIAN | 499 |
| CAMBRIAN* | Series 3 | GUZHANGIAN | 501 |
| CAMBRIAN* | Series 3 | DRUMIAN | 503 |
| CAMBRIAN* | Series 3 | STAGE 5 | 507 |
| CAMBRIAN* | Series 2 | STAGE 4 | 510 |
| CAMBRIAN* | Series 2 | STAGE 3 | 517 / 521 |
| CAMBRIAN* | Terreneuvian | STAGE 2 | |
| CAMBRIAN* | Terreneuvian | FORTUNIAN | 535 / 542 |

## PRECAMBRIAN

| EON | ERA | PERIOD | BDY. AGES (Ma) |
|---|---|---|---|
| PROTEROZOIC | NEOPROTEROZOIC | EDIACARAN | 542 / 630 |
| PROTEROZOIC | NEOPROTEROZOIC | CRYOGENIAN | 850 |
| PROTEROZOIC | NEOPROTEROZOIC | TONIAN | 1000 |
| PROTEROZOIC | MESOPROTEROZOIC | STENIAN | 1200 |
| PROTEROZOIC | MESOPROTEROZOIC | ECTASIAN | 1400 |
| PROTEROZOIC | MESOPROTEROZOIC | CALYMMIAN | 1600 |
| PROTEROZOIC | PALEOPROTEROZOIC | STATHERIAN | 1800 |
| PROTEROZOIC | PALEOPROTEROZOIC | OROSIRIAN | 2050 |
| PROTEROZOIC | PALEOPROTEROZOIC | RHYACIAN | 2300 |
| PROTEROZOIC | PALEOPROTEROZOIC | SIDERIAN | 2500 |
| ARCHEAN | NEOARCHEAN | | 2800 |
| ARCHEAN | MESOARCHEAN | | 3200 |
| ARCHEAN | PALEOARCHEAN | | 3600 |
| ARCHEAN | EOARCHEAN | | 3850 |
| HADEAN | | | |

**Table 24.1** Time scale of the history of life in comparison with a 46-ft-wide classroom

| Time | Scale | Events |
|------|-------|--------|
| 4.6 Bya | 46 ft (14 m) | Earth formed by accretion in the developing solar system |
| 3.8 Bya | 38 ft (~12 m) | Oldest rocks formed; final sterilizing impacts |
| 3.6 Bya | 36 ft (11 m) | First apparent trace fossil bacteria |
| 2.7 Bya | 27 ft (8.2 m) | Trace oxygen appeared in the atmosphere |
| 2 Bya | 20 ft (6.1 m) | The oxygen catastrophe; [O$_2$] jumped to ~0.01 atm (5% of modern level) |
| | | Snowball Earth |
| 1.2 Bya | 12 ft (3.7 m) | First colonial animals appeared |
| 700 Mya | 7 ft (~2 m) | First multicellular (but still microscopic) animals and plants appeared |
| 540 Mya | 5 ft 5 in. (~1.7 m) | Cambrian explosion; nearly all animal phyla appeared at once; macroscopic life appeared |
| | | O$_2$ jumped to essentially modern levels |
| | | Beginning of the Paleozoic |
| 430 Mya | 4 ft 4 in. (~1.3 m) | Plants began to colonize land |
| 360 Mya | 3 ft 7 in. (1.1 m) | First trees appeared; these were the ones we burn in our cars today |
| | | First insects appeared |
| | | Amphibians invaded land |
| 245 Mya | 2 ft 5 in. (74 cm) | Permian extinction; ~95% of species extinguished |
| | | Beginning of the Mesozoic, the Age of Reptiles |
| 65 Mya | 6 in. (15 cm) | Mass extinction of dinosaurs |
| | | Start of the Age of Mammals |
| 3 Mya | 1/3 in. (0.85 cm) | First hominids (*Australopithecus*) |
| 200,000 ya | 0.024 in. (0.6 mm) | *Homo sapiens* appeared |
| 5,000 ya | 10 μm | Agriculture and first civilizations appeared |
| 235 ya | 500 nm | United States declared independence from Great Britain |
| 20 ya | 500 Å | Today's college students were born |
| | | Cell phones became common |

It is clear that life on Earth has existed for at least 3.5 billion years, and probably more, although not before about 3.8 Bya (billion years ago) because the large impacts that took place before then would have vaporized the oceans and, in some cases, a fair amount of the crust of the planet. Before 3.8 Bya, the planet may have gone through alternating cycles of sterilizing heat from impacts and crushing cold with completely frozen oceans.

In the last 3.8 billion years, life has survived many trials in the history of the planet, some of its own making. Large impacts are now rare, and not of sterilizing energy, but they do occur. The most famous is the K-T impact that probably played a role in the extinction of dinosaurs about 65 Mya, but previous impacts were much larger, such as the Permian impact 240 Mya that extinguished 95% of all macroscopic species. Think for a moment about the fraction of *individuals* that would have to be killed in order to extinguish 95% of *species*. Macroscopic life on Earth came within a hair's breadth of ending on the day the Permian impact occurred.

The temperature of the planet has also changed, and seems to be less stable than previously thought. It seems likely that Earth has looked a lot like Jupiter's moon Europa, with completely frozen-over oceans, perhaps several times over the last 4 billion years, the last time about 0.6 Bya, just before the Cambrian explosion. The planet is in a delicate balance between the energy provided by the Sun and how much of this is retained via greenhouse gases. Today, the major greenhouse gas is carbon dioxide, produced by photosynthesis, without which the Earth would freeze. Prior to about 2 Bya, however, methane was probably the major greenhouse gas, and this methane was apparently also the product of life (methanogenesis).

Another issue is atmospheric oxygen. Although we think of oxygen as an essential for life, high concentrations of oxygen are a recent feature of Earth's atmosphere. Oxygen concentrations before about 2.8 Bya are immeasurably low, less than 0.01% of the current level. Oxygen produced by photosynthesis was absorbed by the planet before 2 Bya, when this buffering capacity was exhausted and the oxygen concentration in the air increased dramatically to as much as 2 to 10% of the current level. This has been described as the greatest climatic change ever to have occurred on Earth. Although animals need oxygen, it is not an innocuous compound; imagine if the Earth's atmosphere suddenly rose to 2% of an equally reactive compound, such as ammonia! Over the last 1 billion years, the oxygen concentration has risen about 10-fold to its current level, most of this change occurring at about the time of the Cambrian explosion.

## Ancient microbial fossils

Fossils are the traces of previous life. Fossils can range from the actual remains of an organism (e.g., an insect preserved in amber) to scant traces of unusual isotope ratios (e.g., in the Greenland banded-iron formations). The further back in time you wish to examine, the harder it gets to obtain and interpret these fossils. Samples are harder to get because of plate tectonics—really old rocks are actually quite rare, since most have been heavily or completely transformed by passage into the mantle. Before the Cambrian explosion around 540 Mya, when all or nearly all of the animal phyla appeared, only microbial (or nearly microbial) fossils exist. In later formations these are common, but *very* old fossils are few and far between. Most of these have been found in Australia and Greenland, where the oldest untransformed deposits are accessible. Here we consider the evidence for microbial life early in Earth's history.

■ **STUDY AND ANALYSIS**
**Microfossils in a 3.2-Billion-Year-Old Sulfide Deposit**

Rasmussen B. 2000. Filamentous microfossils in a 3,235-million-year-old volcanogenic massive sulphide deposit. *Nature* **405**:676–679.
PubMed ID: 10864322

*You should have a copy of this paper at hand before proceeding.*

*The task:* To describe evidence that crystalline structures seen in rocks in a fossilized deep-sea hydrothermal vent represent the fossil remains of organisms from 3.2 Bya.

This paper describes fossil filamentous microbes from the geological remains of a hydrothermal vent that existed over 3.2 Bya and about 1 km below the surface of the Archaean ocean. These are not the oldest microbial fossils known; there are a few from between 3.5 and 3.6 Bya. However, these others are of more typical rods and cocci from shallow sediment deposits, and they are much less visually striking and have a less clear environmental context. There is also reason to be skeptical about the origin of many of these earlier fossils.

The paper contains a great deal of geological and geophysical description that is difficult for a typical biologist (or, for that matter, any biologist) to decipher. Suffice it to say that this site has been described in great detail by previous geologists and there is a clear picture of the environment in which the organisms that became these fossils lived.

This was a deep-sea hydrothermal vent system, much like those that exist today, with hot (ca. 300°C) mineral-laden hydrothermal fluid rising through the fractured rock to the floor of the ocean. As this water neared the top of the sediment, it mixed with cold ocean water, and the minerals (mostly silicates and sulfides) precipitated to create the "massive sulfide deposit." Parts of this deposit probably reached the surface of the ocean floor, to emerge as chimneys and black smokers of the sort we know today. Conversely, it is also true that today, underneath hydrothermal fields there lie massive sulfide deposits of the sort whose remains, in fossil form, are described in this paper.

Of course there were no vent worms, clams, or scaly snails as there are in modern deep-sea vents, since animals did not exist until more than 2.5 billion years in the future. This vent, like many modern vents, contains bitumen deposits, tar, and oil. Although the author does not point this out, these are also signs of life, since they are produced from the transformation of biological organic remains.

The fossils themselves are found on the surface of what appear to have been cracks, fissures, and open spaces in the vent rock (Fig. 3 of the paper). Probably the water flowing through these open spaces was hydrothermal fluid mixed with the surrounding ocean water at less than 100°C, providing a rich supply of geothermal energy and even organic material. These vents, then and now, are prime opportunities for organisms. The hydrothermal fluid might be in chemical equilibrium deep in the crust, but when cooled and mixed with cold ocean water (which is presumably at its own, very different equilibrium), it is no longer in chemical equilibrium, and organisms can step in, facilitate favorable reactions, and capture the energy released.

The filaments are clearly the remains of microbial life. They are nearly, but not quite, uniform in diameter and length and are oriented preferentially along the water channel rather than randomly. The filaments cross crystal boundaries in the rock, and, perhaps most convincingly, they are not branched. The organisms seem to have first nucleated the precipitation of silica around themselves, becoming "petrified," and then later the silica was replaced with pyrite. All of this is standard-issue fossilization. The fossils are threads of pyrite in the original shape of the organisms. The morphology and habitat of the organisms are strikingly similar to those of the filamentous sulfur-oxidizing bacteria that currently inhabit hydrothermal vents, although because there was no oxygen at the time they would most likely have had some other metabolism.

# The last common ancestor

Woese CR. 2000. Interpreting the universal phylogenetic tree. *Proceedings of the National Academy of Sciences of the United States of America* **97**:8392–8396.

If we consider the three-domain phylogenetic tree we have looked at throughout this book, what can we infer about the last common ancestor? It is probably fair to suggest that any features common to all three domains were inherited from the last common ancestor, including:

- DNA, the universal code, and most genes
- Transcription and RNA polymerase
- RNAs of all types (ribosomal RNA [rRNA], transfer RNA [tRNA], etc.)
- Translation and the translational apparatus
- Most proteins and the metabolic pathways they generate
- Membrane and cellular structure

In other words, this covers pretty much everything about the most basic processes of life. Most of biochemical evolution must therefore have predated this last common ancestor.

The phenotype of the last common ancestor was probably thermophilic, because all deep/primitive branches of the tree appear to have been thermophilic. The last common ancestor, if it existed as a single organism, was probably something a little bit like *Thermococcus celer* is today.

However, the last common ancestor may not have been a single organism but, rather, a population of organisms (like "mitochondrial Eve"), or even a "communal" organism. There is theoretical reason to believe that lateral gene transfer may have been pervasive prior to the last common ancestor, and the three-way split between *Bacteria*, *Archaea*, and *Eukarya* may have been life's first real experiment in genealogical diversification.

## *Before the last common ancestor?*

So, if most of evolution (at least biochemical evolution) predates the last common ancestor, let's have an expanded look at the tree (Fig. 24.2). The "progenotes" in the figure refer to organisms in which genotype and phenotype had yet to be tightly linked, i.e., in which translation was inaccurate or not well developed. The last common ancestor may have been a progenote, or alternatively progenotes may have predated the last common ancestor, as shown.

However, the belief that most of evolution (at least biochemical evolution) occurred before the last common ancestor creates a time problem. The Earth's formative meteor/comet bombardment ended ca. 3.8 Bya. Microbial fossils from 3.6 Bya and cyanobacterial fossils (and oxygen isotope signals) from 3.2 Bya have been found, so life apparently originated and much of basic biochemical evolution occurred in a fraction

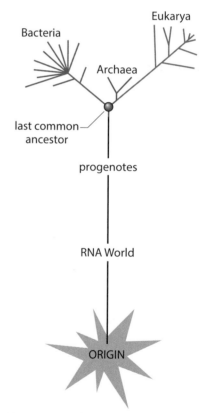

**Figure 24.2** The three-domain tree, the last common ancestor, and how these might be related to progenotes, the RNA world hypothesis, and the origin of life.
doi:10.1128/9781555818517.ch24.f24.2

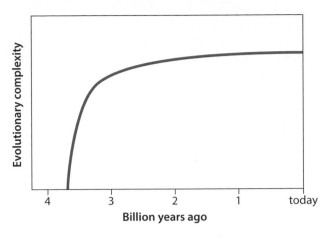

**Figure 24.3** Biochemical evolution may have been very rapid early in evolutionary history and slowed dramatically as life became more complex.
doi:10.1128/9781555818517.ch24.f24.3

of the Earth's history, less than 200 million years. Some people interpret this as evidence that life originated elsewhere in the solar system (or beyond) and came to Earth on meteorites. You can judge the idea of "panspermia" for yourself. However, there is probably no need for life to have originated elsewhere. One can imagine a quick, simple origin and rapid evolution and diversification (Fig. 24.3). As organisms subsequently became increasingly complex, the rate of evolution would be expected to decrease; complex systems evolve more slowly than simple systems. We see this all the time; for example, animal evolution was rapid at its emergence at the Cambrian explosion, and has not changed so much since. Or think about the automobile, or computers, or software. Any new "technology" evolves very quickly when it emerges and explores the available landscape, then slows down dramatically when it gets complex, and especially when the interactions it's involved with get intricate.

## The RNA world hypothesis

Life is based primarily on three major macromolecules: DNA, RNA, and protein. It seems unlikely that these three emerged simultaneously, and so the question is, which was first? What might life have been like at such an early stage?

There are several reasons to think that RNA was most likely the first of these three macromolecules:

1. DNA functions only through RNA intermediates (via transcription)
2. Protein synthesis machinery is very complex and fundamentally RNA based (messenger [mRNA], rRNA, tRNA)
3. RNA can serve as genomes like DNA (information storage, e.g., RNA viruses)
4. RNA can catalyze chemical reactions like proteins (e.g., catalytic RNAs)

Therefore, RNA probably predates both DNA and protein, and therefore there probably was some form of an "RNA world" before the evolutionary invention of either DNA or protein. The alternative is that the three macromolecules, or perhaps RNA and protein, coevolved somehow.

The least obvious of the reasons cited above is the ability of RNA to catalyze chemical reactions. It turns out that not all enzymes are made of protein; some are made of RNA. The three known naturally occurring catalytic RNAs are the large-subunit rRNA (LSU rRNA) (which catalyzes the peptidyl transferase activity of translation), ribonuclease P (RNase P) (which cleaves the 5′ leader from pre-tRNAs), and nuclear spliceosomal RNAs. There are also a number of naturally occurring chemically active RNAs that are almost but not quite catalytic (and therefore not really enzymatic) because they are self-reactive: self-splicing RNAs (group I and II intron RNAs) and self-cleaving RNAs (delta virus RNAs, hammerhead and hairpin RNAs, Varkud mitochondrial RNA), and some others. Most of these self-reactive RNAs can be engineered into true catalysts by separating the catalytic and substrate regions of the RNAs in synthetic forms. In addition, a wide range of reactive and even catalytic RNAs have been generated by in vitro selection.

RNAs can also perform other roles usually played by proteins. They can serve mechanical functions, e.g., the rRNAs that are the works of the translational apparatus; the ribosome is really a nanomachine. RNAs can also perform structural roles and make biochemical decisions and transduce these decisions to other molecules, e.g., riboswitches, attenuators, terminators and antiterminators, antisense RNAs, and replication origin RNAs.

## What could the RNA world have been like?

Two extreme scenarios for an RNA world (if it existed) are the "simple RNA world" and the "complex RNA world" scenarios. Reality (if there was an RNA world at all) might lie anywhere between these two extremes.

### A simple RNA world scenario

In the simplest "one-molecule" RNA world scenario, a single self-replicating RNA might have contained both genetic information (genome) and the ability to replicate itself (function). Perhaps more likely, based on what we see today in RNA viruses, viroids, and virusoids, it might have been a "two-RNA" world in which an RNA genome encoded an RNA replicase. Remember that most RNA viruses replicate through alternating (+) and (−) strand intermediates:

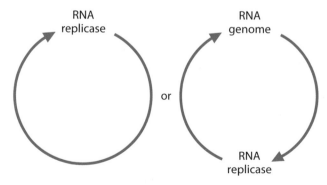

In this scenario, an RNA-based protein synthesis machinery might have emerged right away, presumably to make oligo- or polypeptides that would be better than RNA at catalyzing other chemical reactions. A protein catalyst could then have emerged to replace the simple RNA replicase (we would call this an RNA polymerase). In this scenario, only the genome, replicase, and translational apparatus were ever RNA:

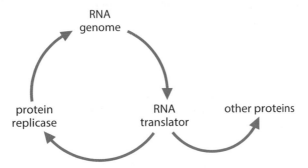

The replacement of the RNA genome with more chemically stable DNA yields the modern system, the DNA/RNA/protein world we live in:

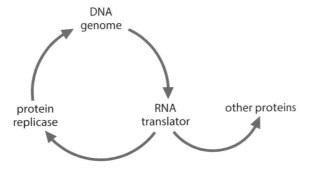

### A complex RNA world scenario

In the complex RNA world scenario, an originally simple RNA world developed into an entire RNA-based metabolic system, with most of the current metabolic pathways and functions catalyzed by RNA enzymes rather than protein enzymes. These RNAs were subsequently replaced one at a time by more efficient protein enzymes, once RNAs acquired the ability to encode and create proteins.

Proponents of this view consider nucleotide cofactors (e.g., adenosine triphosphate [ATP], guanosine triphosphate [GTP], nicotinamide adenine dinucleotide [NAD], flavin mononucleotide [FMN]) to be the vestiges of the original RNA enzymes. In this view, the RNA organisms would be cellular organisms, unlike the simple RNA world scenario in which this is not specified.

### Problems with the RNA world hypothesis: a reality check

Although the RNA world hypothesis is almost universally accepted (even if by default), it has some pretty significant shortcomings that warrant explanation, such as:

- In a simple RNA world, or early in a complex RNA world, where do the activated nucleotides (NTPs) come from for RNA polymerization? The generation of activated monomers is at least as difficult a task as their directed polymerization.

- RNA is far too complex isomerically. There are hundreds of pentose isomers; how do you specifically make D-ribose? The problem is that each nonfunctional isomer is a potential inhibitor of any process in which the functional isomer is likely to engage. The same issue also applies to the bases and how they are assembled into isomerically specific polymers. The number of possible chemical isomers of even a short RNA chain is astronomical.
- RNA is very unstable, and is especially sensitive to metal ions, high temperature, and either high or low pH. How would RNA have been protected from damage, or even plain hydrolysis?

To avoid these problems, some have suggested that early predecessors of RNA may have contained only purines (no pyrimidines; purines are easier to make abiotically) and/or a nonribose backbone (e.g., glycerol or peptide).

But where is the evidence? All of these ideas are pretty nebulous and, at least for now, untestable. So the RNA worldview remains vague and disputed, but it is for now perhaps the best working hypothesis we have.


# The emergence of life

The RNA world hypothesis deals only with the "information" of life: the software. Metabolism, the hardware of life, is a separate issue. In fact, there is reason to believe that the metabolism and information-processing components of life may have originated independently and merged to generate what we think of as life.


## *The Oparin ocean hypothesis: primordial soup*

Perhaps the first realistic suggestion about how metabolism could have originated was the Oparin ocean hypothesis. The idea was that chemical reactions and meteoric influx concentrated organics in the ocean, creating an ocean (or at least tidal pools) of nutrient-rich organic "soup." Spontaneous and random polymerizations created self-replicating molecules that eventually evolved into metabolism and life.


### Sources of organic material

There were two potential sources of organic material in the early Earth: in situ synthesis and meteoric influx.


#### In situ synthesis

Irradiation or electrical sparking of mixtures of methane-ammonia-water-hydrogen cyanide and other possible atmospheric mixes yields lots of organic compounds, including adenine and amino acids. This is the classic Miller-Urey experiment (Fig. 24.4).

The implication of this is that that lightning and cosmic irradiation of an early Earth atmosphere could have generated significant amounts of complex organic compounds from which life may have emerged.

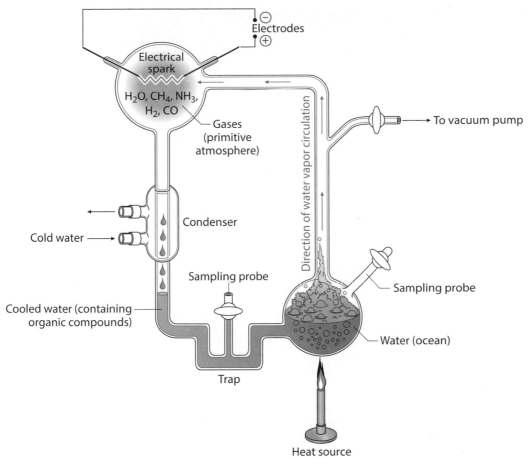

**Figure 24.4** The Miller-Urey apparatus. Mixtures of potential primordial Earth atmospheric gases are circulated in steam and subjected to spark discharges, mimicking lightning. Water-soluble organics generated are trapped and concentrated for analysis. (Source: Wikimedia Commons. http://creativecommons.org/licenses/by-sa/3.0/legalcode) doi:10.1128/9781555818517.ch24.f24.4

### Meteoric influx

Comets and asteroids are rich in carbon and organics. In fact, comets are similar in composition to living things (Table 24.2).

### The origin of metabolic pathways

Metabolism might (according to the Oparin ocean view) be the result of sequential exhaustion of required substrates, followed by the utilization of handy precursors to make what was needed. For example, when compound X was exhausted from the primordial soup, compound Y was used by converting it to compound X:

$$Y \rightarrow X$$

Then when compound Y was exhausted, a related compound, Z, was used by converting it first to compound Y, then to compound X:

$$Z \rightarrow Y \rightarrow X$$

**Table 24.2** Comparison of the compositions of a comet and a human being

| Element or compound | Halley's Comet | Edmond Halley |
|---|---|---|
| H | 55.2% | 63% |
| C | 11% | 9.5% |
| N | 2.4% | 1.4% |
| O | 27.6% | 25.5% |
| $H_2O$ | 50% | 75% |

This process, repeated over and over, would have generated the metabolic pathways we know today.

The Oparin ocean hypothesis was a start, but like any notions about the origin of life it is pretty vague and, at least for now, untestable. And it must be said that there are some significant weaknesses of the hypothesis that need explaining, for example:

- An entire ocean of primordial soup? Really? (Could it have been a tide pool or evaporation pond?)
- Activated precursors are unstable, and so would not likely have accumulated to any significant concentration
- Biochemistry is too complicated stereochemically; there are too many isomers of everything; the useful compounds end up at trivial concentrations, overwhelmed by other isomers that would compete for reactions
- Biological polymerizations are generally dehydration reactions and so are strongly disfavored in aqueous solution (55 M $H_2O$); accumulation of polymers is not going to happen in solution

So the Oparin ocean hypothesis is not a great theory, but it was the best available until . . .

## Wächtershäuser's hypothesis of surface metabolism: primordial sandwich

The basis behind the hypothesis of surface metabolism, developed by Günter Wächtershäuser, is that life originated in an organic scum adhering to mineral surfaces, rather than in solution. The hypothesis comes with a very elaborate set of chemical features that solve many of the problems with the Oparin ocean hypothesis:

- Surfaces can concentrate very low concentrations of solution organics to very high local concentrations by adsorption
- Polymerization is favored on surfaces (low effective water concentration)
- Surface chemistry is stereospecific
- Activated precursors are not required to drive reactions

Pyrite and similar minerals bind anions, especially organophosphates, but let them diffuse in two dimensions on their surfaces. This is consistent with the observation that the most fundamental metabolic pathways use phospho-intermediates. The peripheral pathways are not composed of phospho-intermediates and were presumably invented later.

These surface metabolic reactions may have been catalyzed only by cofactors before the invention of enzymes (either protein or RNA). In this view, metabolism arose before (or at least independently of) biological information processing (RNA or DNA) and even before biological catalysis.

Cellularization, in this scenario, is thought to have occurred when lipid by-products of metabolism later enclosed the surface reaction centers, creating

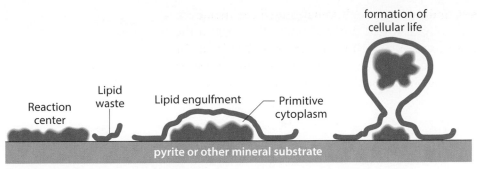

formation of
cellular life

Reaction
center

Lipid
waste

Lipid engulfment

Primitive
cytoplasm

pyrite or other mineral substrate

**Figure 24.5** The primordial sandwich hypothesis, progressing from left to right.
doi:10.1128/9781555818517.ch24.f24.5

a kind of simple cytoplasm enclosed on one side by a lipid membrane and on the other by the mineral surface in a "primordial sandwich." As metabolism became increasingly complex and independent of the mineral surface, these surface organisms could have escaped from their mineral substrate (Fig. 24.5).

The RNA world (if it existed) may have originated within this protobiotic metabolism, or it may have originated independently, and then the two may have merged to create some of the earliest forms that we might recognize as living.

## Questions for thought

**1.** What do you think of the RNA world hypothesis? What are the alternatives? How could you go about testing this?

**2.** Presuming the RNA world hypothesis to be correct, what do you think about the simple versus complex RNA world scenarios? How would you go about testing your hypothesis?

**3.** Make a list of all of the RNA-encoding genes you know about in bacteria (say, in *Escherichia coli*). Of course you would start with rRNA and tRNA, but what other non-protein-encoding RNA genes are there? For any of these, how would you determine whether it is an ancient RNA and so perhaps a relic of the RNA world, or a recently invented RNA?

**4.** The last sterilizing impacts on Earth occurred about 3.8 Bya, and there is good evidence for the presence of microbial life (fossils) at about 3.6 Bya. Do you see this as evidence that life may have originated elsewhere? Why or why not?

**5.** Imagine space aliens visiting Earth in search of intelligent life at some arbitrary point in its history. What are the chances of their finding it? What are the chances of their finding macroscopic life? What are the chances of their finding a "breathable" atmosphere?

**6.** As a microbiologist, you are willing to accept the geologists' description of the history of a rock sample from the isotope data. However, microbial fossils are usually identified on the basis of morphology (what they look like), even though morphology can be severely distorted during fossilization and in the geological transformations of the rock. What, then, would you look for in a potential microbial fossil to convince yourself that it really represents the remains of living things? What kinds of things would convince you that it might not be a real fossil? What kind of rocks would you examine in your search for microbial fossils?

# *Index*

Page numbers followed by f and t indicate figures and tables, respectively.

Archaeal primers, 333
*Archaeoglobus fulgidus*, 76f, 255
*Arthrobacter globiformis*, 179
Asexual organisms, 234, 236
Atmospheric oxygen, 370
Autotrophs, 145, 151, 247, 345–348
*Azotobacter vinelandii*, 152

## B

*Bacillus anthracis*, 169, 170, 171
*Bacillus cereus*, 170–171
Bacteria
    acid-fast, 181
    archaea and eukarya and, 258–259
    branches of, 95–96
    development of, 183–184
    eukaryotes and, 23
    filamentous, 147, 325, 326
    horizontal transfer and, 92, 357–360
    microbial diversity, 6–9
    motility, 197–201
    multicellular behavior in, 184
    phylogenetic tree, 222f
    RNase P RNA sequences and, 76
    rRNA spacer sequence analysis and, 77
    with sheathed filaments, 147
    thermophilic ancestry of, 105
    Tree of life and, 86
Bacterial genomes, 90, 287
Bacterial photosynthesis, 128–131
Bacterial phyla
    with few cultivated species, 225–232
    with no cultivated species, 232–235
    SSU rRNA sequences and, 221, 223
    uncultivated sequences and, 234
Bacterial primers, 310, 332
Bacterial species, concept of, 234, 236
Bacterial vaginitis, 174
Bacteriocytes, 151, 300, 310
Bacteriophages
    bacterial genomes and, 287
    lambda, 286f
    M13, 287
    Mu, 287, 288f
*Bacteroides thetaiotaomicron*, 195
*Bacteroidetes*
    example species, 195–197
    features of, 194–195
    introduction to, 187
    phylogenetic tree, 187f, 194f
    taxonomy, 193
*Bacteroidia*, 193, 195
Banding patterns, 81, 82
Base composition, 79, 345, 358, 360
Bayesian inference, 69

*Bdellovibrio bacteriovorans*, 157, 158f
*Beggiatoa alba*, 151
β-proteobacteria
    chemolithoautotrophs and, 145–146
    features of, 143–145
    heterotrophs and pathogens and, 145
    phylogenetic tree, 144f
    sheathed filamentous, 147
    taxonomy, 143
"Big Tree of Life," 85–91
Biochemical evolution, 372, 373f
Biological diversity, 86–87, 351, 367
*Blastopirellula marina*, 215–216
BLAST searches, 30, 332, 356
Blindness, 212
Blue-green algae, 120–128
Bootstrap analysis, 69–71
*Borrelia recurrentis*, 191–192
Botulism, 171, 172
Bovine spongiform encephalopathy, 292
Branches
    description of, 53
    determining length of, 46–50
*Brocadia anammoxidans*, 216–217
Brock, Thomas, 300
Brown algae, 270, 271
*Buchnera aphidicola*, 151, 300
*Burkholderia* sequences, 364, 365
Buruli ulcer, 181–182

## C

$C_1$-$H_4$MPT reduction, 246
*Caloranaerobacter azorensis*, 61
Calvin cycle, 131–132, 146
Cambrian explosion, 370, 373
Carbon fixation
    *C. aurantiacus*, 115
    *Chlorobi*, 130
    *Chloroflexi*, 129
    *Cyanobacteria*, 130
    green phototrophic bacteria and,
        131–134
    obtaining reducing power for,
        129–130
    SAR86 group and, 343
    SIP process and, 346
Catenated alignments, 78
*Caulobacter crescentus*, 140–141
Cell envelope
    *Archaea*, 238
    gram-negative, 87, 165
    structure of, 166
Cells
    Central Dogma, 11, 14, 91
    DNA and, 11

    isolation of, 304f
    protein and, 11
    RNA and, 11
Cell-to-cell communication, 184
*Cenarchaeum symbiosum*, 240, 299
Cercozoa, 263f, 274, 278, 346
Cesium tetrafluoroacetate density
    gradients, 344–345
CFX1223 probe, 326
Chain of being, 15, 16f, 258
Chain termination sequencing, 31f
Charge-coupled device (CCD), 320
CHECK_CHIMERA function, 312
Chemoautotrophs, 7, 136, 145
Chemoheterotrophs, 7
Chemolithoautotrophs, 145–146
Chimeras, 312, 313f
*Chlamydiae*
    evolution of, 361–363
    example species, 212–213
    features of, 210–212
    phylogenetic tree, 203f, 210f
    taxonomy, 209
    viruses and, 290
Chlamydial protease-like activity factor
    (CPAF), 362, 363
*Chlamydia trachomatis*, 212
*Chlorobi*
    carbon fixation and, 130
    example species, 119–120
    features of, 117–119
    taxonomy, 117
*Chlorobium* bacteria, 8, 117
*Chlorobium limicola*, 119
*Chloroflexi*
    carbon fixation and, 129
    example species, 114–117
    features of, 105, 113–114
    FISH analysis of, 323–327
    phylogenetic tree, 111f, 113f
    taxonomy, 112
*Chloroflexus aurantiacus*, 114–115
Chlorophytes, 263, 267
Chloroplastids, 267, 269
Chloroplasts, 21, 88, 126, 127, 262, 280
Choanoflagellates, 264, 266
*Chondrus crispus*, 269
Chromalveolates
    example species, 271–274
    groups related to, 263f, 270–271
    taxonomy, 270
*Chromatium vinosum*, 57, 153
*Chroococcales*, 120, 121, 122, 126
*Chrysophytes*, 270, 271
Ciliate grazers, 345–348

lipid profiling, 80–81
phenotypic markers, 83–84
RFLP methods, 81–82
scaly snail analysis and, 308–312
serology, 79–80
Sequence data, obtaining, 29–32
Sequence information, availability of, 27
Sequences
    long branches and, 67
    partial, 65
    phylogenetic analysis and, 26–29
Serological method, 79–80
Sexually transmitted diseases, 212, 318
Sheathed filaments, bacteria with, 147
*Shewanella* sequences, 364
Siderophores, 152, 184
Similarity matrix
    bootstrap analysis and, 69, 70, 71
    conversion of, 43–44
    generating, 42, 51
Simple RNA world scenario, 374–375
SIP process, 343–348
Six-parameter model, 64
Skin populations, 319
S-layer, 6, 7f, 205
Small-molecule stability, 107–108
Small-subunit ribosomal RNA (rRNA)
        sequences, *see also* Alternative
        sequences
    bacterial phyla and, 221
    DGGE method and, 330–331
    ES-2 organism and, 57–58
    FISH analysis and, 323–327
    genomic approach and, 340–341
    human microbiome analysis and, 315
    molecular clock, 27–29
    PCR and, 223
    phylogenetic survey, 223, 224t, 307
    phylogenetic tree, 10, 20, 360
    pink-filament biomass analysis and,
        302, 303
    primers and regions of, 317f
    problems related to, 73–74
    scaly snail analysis and, 310–311
    t-RFLP method and, 333–335,
        337–338
    window into alignment of, 32, 33f
*Sphaerotilus natans*, 147
Sphingobacteria, 193, 194, 196
*Sphyraena barracuda*, 266
*Spirilla*, 138, 144, 150, 198
*Spirochaetae*
    example species, 191–193
    features of, 188–189

flagella and, 189–190
introduction to, 187
phylogenetic tree, 187f, 188f
taxonomy, 188
Spirochete motility, 200–201
*Spiroplasma* motility, 201
SR1 phylum, 234, 235f
SSU rDNA, 301, 311, 330, 333, 334, 337
Stable-isotope probing (SIP) approach,
        343–348
Stationary-phase cells, 104, 183
*Stigonematales*, 120, 122, 126
Stramenopiles, 263f, 270–271
*Streblomastix strix*, 283, 284f
Strep throat testing, 79, 80f
*Streptococcus pyogenes*, 79
*Streptomyces antibioticus*, 180, 181f
Streptophytes, 263f, 267
Subgingival plaque, 337
Sudden infant death syndrome, 172
Sulfate reducers, 104, 133, 155
Sulfide deposit, microfossils in, 370–371
*Sulfolobus solfataricus*, 243, 244f
Sulfur-metabolizing thermophiles,
        253–254
Sulfur oxidizers, 145–146, 151, 241, 243
Sulfur-oxidizing bacteria, 308, 311, 371
Sulfur-oxidizing endosymbionts, 299,
        311
Sulfur reduction, 241, 243
Sulfur respiration, 241, 243
Surface metabolism hypothesis, 378–379
Symbionts
    *Enterobacteriaceae*, 28f
    obligately intracellular, 142
    plant, 141
    scaly snail, 311
Symbiosis
    *Chlorobi*, 118, 119f
    *Cyanobacteria*, 122
    scaly snail, 161

## T

Taxonomy
    *Actinobacteria*, 176–177
    *Alphaproteobacteria*, 137
    *Aquificae*, 99
    *Bacteroidetes*, 193
    *Betaproteobacteria*, 143
    *Chlamydiae*, 209
    *Chlorobi*, 117
    *Chloroflexi*, 112
    chromalveolates, 270
    *Crenarchaeota*, 240

*Cyanobacteria*, 120
*Deinococcus-Thermus*, 204
*Deltaproteobacteria*, 154
description of, 21
*Epsilonproteobacteria*, 158
eukaryotes, 262, 263f
*Euryarchaeota*, 245
excavates, 279
*Firmicutes*, 167–168
*Gammaproteobacteria*, 148–149
phenotypic markers for, 83
phylogeny and, 21–22
*Planctomycetes*, 213
plantae, 267
rhizaria, 274
*Spirochaetae*, 188
*Thermotogae*, 102
unikonta, 264
Tenericutes, 172–173
Terminal nodes, 51, 52, 53
Terminal restriction fragment length
        polymorphism (t-RFLP)
    description of, 331
    introduction to, 329
    procedure, 333–335
    study and analysis and, 337–338
Termites, 189, 190f
Texas Red-labeled FISH probe, 323,
        324f
*Thalassia testinum*, 268
*Thermales*
    example species, 208–209
    features of, 206–208
Thermal pool, 296f
Thermal stress, 331–333
*Thermocrinis ruber*, 100–101, 305
*Thermodesulfobacteriales*, 104
*Thermoleophilum album*, 182, 183f
*Thermomicrobium roseum*, 104, 113
Thermophilic bacteria
    *Aquificae*, 98–101
    introduction to, 97–98
    life at high temperatures and, 105–
        108
    primitive, 104–105
    *Thermotogae*, 102–104
*Thermoplasma acidophilum*, 255–256
*Thermoproteus tenax*, 243
*Thermosipho africanus*, 104
*Thermotogae*
    example species, 103–104
    features of, 102–103
    taxonomy, 102
*Thermotoga maritima*, 103, 357–360